Computer Models of Speech Using Fuzzy Algorithms

ADVANCED APPLICATIONS IN PATTERN RECOGNITION
General editor: Morton Nadler

A STRUCTURAL ANALYSIS OF COMPLEX AERIAL
PHOTOGRAPHS   Makoto Nagao and Takashi Matsuyama

PATTERN RECOGNITION WITH FUZZY OBJECTIVE FUNCTION
ALGORITHMS   James C. Bezdek

COMPUTER MODELS OF SPEECH USING FUZZY ALGORITHMS
                                        Renato de Mori

A Continuation Order Plan is available for this series. A continuation order will bring delivery of each new volume immediately upon publication. Volumes are billed only upon actual shipment. For further information please contact the publisher.

Computer Models of Speech Using Fuzzy Algorithms

Renato De Mori
Concordia University
Montreal, Quebec, Canada

Plenum Press • New York and London

Library of Congress Cataloging in Publication Data

De Mori, Renato.

Computer models of speech using fuzzy algorithms.

(Advanced applications in pattern recognition)
Bibliography: p.
Includes index.
1. Speech processing systems. 2. Fuzzy algorithms. I. Title. II. Series.
TK7882.S65D45 1983   001.64′42   83-11082
ISBN-13: 978-1-4613-3744-7   e-ISBN-13: 978-1-4613-3742-3
DOI: 10.1007/978-1-4613-3742-3

©1983 Plenum Press, New York
A Division of Plenum Publishing Corporation
233 Spring Street, New York, N.Y. 10013
Softcover reprint of the hardcover 1st edition

To My Mother

General Editor's Preface

It is with great pleasure that I present this third volume of the series "Advanced Applications in Pattern Recognition." It represents the summary of many man- (and woman-) years of effort in the field of speech recognition by the author's former team at the University of Turin. It combines the best results in fuzzy-set theory and artificial intelligence to point the way to definitive solutions to the speech-recognition problem. It is my hope that it will become a classic work in this field.

I take this opportunity to extend my thanks and appreciation to Sy Marchand, Plenum's Senior Editor responsible for overseeing this series, and to Susan Lee and Jo Winton, who had the monumental task of preparing the camera-ready master sheets for publication.

Morton Nadler

General Editor

PREFACE

Si parva licet componere magnis

Virgil, Georgics, 4,176 (37-30 B.C.)

The work reported in this book results from years of research
oriented toward the goal of making an experimental model capable of
understanding spoken sentences of a natural language. This is, of
course, a modest attempt compared to the complexity of the functions
performed by the human brain.

A method is introduced for conceiving modules performing
perceptual tasks and for combining them in a speech understanding
system.

Although some of the results reported here may be of practical
interest, the main goal of this work has been the conception of
data-structures and algorithms for representing and using knowledge
of speech analysis and perception in a speech understanding system.

The model proposed is mainly a flexible instrument for research,
rather than a system prototype. It is composed of modules which
are updated whenever new convincing results appear in the scientific
literature or are obtained by the system itself.

Many valuable papers related to this field contain interesting
results in areas such as pattern recognition, artificial intelligence,
signal processing, speech communication, logic, and information
processing systems. This book represents an attempt to unite in
one model some of the best achievements in the above mentioned
areas.

The contents of the book will show how other approaches can be
embedded in the framework proposed here, leading to the conclusion
that, although they may appear to be greatly different, they are,
in essence, very similar.

Another important point supported in this book is that most of the processes involved in speech understanding do not require precise numerical computations. It is shown how a flexible experimental model can be built using a fuzzy algebra for composing evidences.

I hope that the book will offer a ground for a critical view of the conception of speech understanding models as knowledge-based systems without forcing the reader to consider all the proposed solutions as the best possible ones.

ACKNOWLEDGEMENTS

This research on speech understanding models was started in 1971 at the Centro per l'Elaborazione Numerale dei Segnali (CENS) of the National Research Council of Italy (CNR). R. Sartori and A.R. Meo have contributed with many discussions at the beginning and during the development of the research.

King Sun Fu (Purdue University), Lotfi Zadeh (University of California at Berkeley), and Jonathon Allen (MIT) have given me important suggestions and advice in the fields of pattern recognition, appropriate reasoning, and speech perception, respectively.

The Groupement des Acousticiens de Langue Francaise (GALF) has offered me the opportunity of examining my ideas during the annual meetings. I would like to express my gratitude especially to Guy Mercier (Centre National d'Etudes des Telecommunications) and Jean-Paul Haton (University of Nancy-I).

I have learned a lot working with V.A. Makhonine, A.V. Knipper (IPIT Moscow), R. Gubrynowicz (Polish Academy of Sciences, Warsaw), and M. O'Kane (University of Canberra) during their long visits to Turin.

I am also indebted to T. Sakai (Kyoto University) and H. Fujisaki (Tokyo University) for their valuable help when I was a guest of the Japanese Society for the Promotion of Sciences.

Many thanks are due to Morton Nadler for his careful review of this book.

Finally, as it is impossible to mention all the colleagues and students who have collaborated with me in the past, I would at least like to thank warmly my colleagues who are still offering the contribution of their creativity to this research on speech understanding: Pietro Laface, Lorenzo Saitta, Pietro Demichelis, Giovanna Giordano, Attilio Giordana.

CONTENTS

LIST OF FIGURES   xix

1. COMPUTER MODELS FOR SPEECH UNDERSTANDING   1

    1.1 Motivations for speech understanding researches (1)

    1.2 Tasks, difficulties and types of models (3)

    1.3 A passive model for automatic speech recognition (8)

    1.4 Active models for speech understanding (10)

         1.4.1. Elementary psychoacoustic considerations (11)

         1.4.2. Interpreting the acoustic signal is problem
                 solving (12)

         1.4.3. Structures for Speech Understanding System
                 models (14)

         1.4.4. Functional description of the system (18)

         1.4.5. On the tools for representing and using
                 knowledge (19)

    1.5 On the use of fuzzy set theory (20)

    1.6 The structure of the book (23)

2. GENERATION AND RECOGNITION OF ACOUSTIC PATTERNS 25

    2.1 Speech generation (25)

    2.2 Techniques for generating acoustic patterns (28)

         2.2.1. The filter bank (29)

         2.2.2. The Fast Fourier Transform (31)

         2.2.3. Identification of the vocal tract parameters (33)

         2.2.4. Extraction of articulatory parameters (36)

         2.2.5. On the use of spectral representation of
                 speech (37)

2.3   Background on syntactic pattern recognition   (42)

2.4   Acoustic Cue Extraction for Speech Patterns.   (45)

    2.4.1   Silence interval between two sounds in a word (48)

    2.4.2   Quasi-stationary portions of the acoustic
           pattern   (49)

    2.4.3   Lines (50)

2.5   Classification of speech patterns (51)

    2.5.1   A brief history of automatic recognition of
           isolated words   (51)

    2.5.2   The dynamic programming approach   (52)

2.6   Automatic recognition of continuous speech.   (57)

2.7   References   (64)

3.   ON THE USE OF SYNTACTIC PATTERN RECOGNITION AND
    FUZZY SET THEORY   67

3.1   Introduction and motivations   (67)

3.2   The syntactic (structural) approach to the
    interpretation of speech patterns   (78)

3.3   The syntax for the recognition of the phonetic
    feature "vocalic"   (87)

3.4   Background on fuzzy set theory   (92)

    3.4.1   Definition of fuzzy sets   (92)

    3.4.2   Operations on fuzzy sets   (95)

    3.4.3   Fuzzy restrictions   (97)

    3.4.4   Possibility distributions   (98)

    3.4.5   A simple example   (99)

3.5   Fuzzy relations and languages   (101)

    3.5.1   Fuzzy relations   (101)

    3.5.2   The extension of principle   (103)

    3.5.3   Fuzzy languages   (104)

3.6   Use of fuzzy algorithms for feature hypothesization   (107)

    3.6.1   Fuzzy algorithms   (107)

    3.6.2   An example of application   (109)

3.7   References   (121)

4. DESIGN PRINCIPLES FOR CONTROLLING THE USE OF
    STRUCTURAL RULES FOR SEGMENTATION

    4.1 The meaning of the meaning (122)

    4.2 The control problem in the segmentation process (125)

    4.3 Computation with linguistic probabilities (130)

    4.4 Segmentation of continuous speech into pseudo-syllabic
         nuclei (137)

         4.4.2 Introduction (137)

         4.4.2 The segmentation grammar   (142)

         4.4.3 The segmentation algorithm (151)

         4.4.4 Examples (158)

    4.5 A parallel processing model for generating phoneme
         hypotheses (160)

    4.6 A review of previous work on phoneme recognition (167)

    4.7 References (171)

5. RULES FOR CHARACTERIZING SONORANT SOUNDS   172

    5.1 A fragmant of the structural knowledge source for
         pseudo-syllables (172)

         5.1.1 Generalities (172)

         5.1.2 Generation of hypotheses about sonorant
                sounds (174)

    5.2 Extraction of detailed spectral features for sonorant
         sounds (182)

         5.2.1 Extraction of a multilinked data structure from
                a spectrogram (183)

         5.2.2 Deletion of unsuitable links (187)

         5.2.3 Assignment of weights to the arcs (188)

    5.3 Generation of hypotheses about vowels (190)

         5.3.1 Algorithm SZDET (190)

         5.3.2 Recognition of the place of articulation of
                vowels (194)

         5.3.3 Hypothesis generation and problem solving (196)

    5.4 Use of formants for the recognition of liquids and
         nasals (201)

         5.4.1 Liquid-nasal classification (202)

         5.4.2 Applications to the classification of liquids (207)

5.5 Detailed recognition of nasal sounds (209)

    5.5.1 Introductory acoustical and perceptual
           considerations (209)

    5.5.2 Inference of the recognition rules (210)

        5.5.2.1 Speech material (210)

        5.5.2.2 Parameters of the atomic questions (211)

        5.5.2.3 The recognition rules (211)

    5.5.3 Experimental results (219)

    5.5.4 On the extension of the rules to other
           contexts (221)

    5.5.5 On the evaluation of binary features (222)

5.6 Structure of the procedural knowledge (222)

5.7 References (227)

6. RULES FOR CHARACTERIZING THE NONSONORANT SOUNDS 228

6.1 Introduction (228)

6.2 Recognition of the phonetic features of nonsonorant
    sounds (237)

6.3 Bottom-up generation of phonemic hypotheses of
    plosive sounds (247)

    6.3.1 Review of research concerning the plosive
           consonants (247)

    6.3.2 Recognition of plosive sounds (251)

6.4 Rules for the recognition of plosive sounds (262)

    6.4.1 Rules for formant loci, formant slopes and
           burst spectra (262)

    6.4.2 Rules for spectral characteristics of
           plosives (265)

    6.4.3 Rules for formant features (266)

    6.4.4 Rules for phonemic hypotheses (266)

    6.4.5 Composition, of evidences (267)

6.5 Experimental results (269)

6.6 References (270)

7. THE LEXICAL KNOWLEDGE SOURCE   271

7.1 Word recognition in continuous speech (271)

7.2 Dynamic programming for matching word patterns of
     quasi-continuous feature vectors (275)

7.3 Matching speech states (280)

     7.3.1 Minimum-distance models (280)

     7.3.2 Stochastic models (283)

7.4 Word detection by the hypothesize-and-test paradigm (285)

7.5 The lexical component as a problem solver (292)

7.6 The structure of the lexical knowledge (296)

7.7 Strategies for lexical access (307)

     7.7.1 Top-down constraints (307)

     7.7.2 Preconditions based on the first syllable (308)

     7.7.3 Precondition degradations (309)

     7.7.4 The lexicon as a content-addressable-memory (310)

     7.7.5 The syll-type tree (312)

     7.7.6 Precondition evidences (314)

     7.7.7 The algorithm for lexical access (315)

7.8 Selection of candidates and hypothesis evaluation (316)

     7.8.1 Evaluation of precondition evidences (316)

     7.8.2 Candidate selection (318)

     7.8.3 Other possible methods for hypothesis
            evaluation (321)

7.9 Strategies for the generation of lexical hypotheses (324)

7.10 References (327)

8. ON THE STRUCTURE AND USE OF TASK-DEPENDENT
    KNOWLEDGE   329

8.1 Introduction (329)

8.2 Finite-state language models (332)

8.3 Measuring evidences (341)

8.4 Search strategies (347)

     8.4.1 Branch-and-bound algorithms (347)

     8.4.2 Non-admissible search algorithms (351)

8.5 On the use of production systems for problem solving (356)

8.6 Scheduling of interpretation processes based on
     approximate reasoning (359)

8.6.1 Background (359)

8.6.2 On the use of truth functions and fuzzy logic (361)

8.6.3 Priority assignment and approximate reasoning (363)

8.7 Outline of a semantically-guided use of task-dependent knowledge (366)

8.7.1 System organization (366)

8.7.2 The semantic knowledge (371)

8.7.3 The syntactic knowledge (379)

8.7.4 Pragmatics (381)

8.8 Evaluating language complexity (382)

8.9 Review of recent work on task-dependent knowledge (386)

8.9.1 Representation (386)

8.9.2 Control of strategies and scoring philosophies (391)

8.10 References (399)

9.   AUTOMATIC LEARNING OF FUZZY RELATIONS   400

9.1 Introduction (400)

9.2 Formal definition of the problem and an example of application (403)

9.2.1 Generalities (403)

9.2.2 An example of application (407)

9.3 A simple preliminary learning case (408)

10. TOWARDS A PARALLEL SYSTEM   416

10.1 A new model for lexical access (416)

10.2 Description of acoustic cues (422)

10.3 The knowledge of the descriptor of the global spectral features (426)

10.4 Conclusions (434)

BIBLIOGRAPHY   441

INDEX   477

LIST OF FIGURES

Fig. 1.1   Transformation of information along the   5
            speech chain

Fig. 1.2   A passive model for speech recognition   8

Fig. 1.3   Levels involved in the speech understanding   15
            process

Fig. 1.4   Example of a heterarchical structure   15

Fig. 1.5   Example of a hierarchical structure   16

Fig. 1.6   Speech understanding system organization   17

Fig. 1.7   Types of imprecisions involved in hypothesis   21
            generation

Fig. 2.1   a) Speech production scheme   27

            b) Speech production model

Fig. 2.2   Spectrogram of the sentence "tell me"   38

Fig. 2.3   a) waveform of the vowel /e/ in "tell me"   39
               short-term spectrum of the vowel /e/
               obtained by:

            b) analog techniques

            c) synchronous FFT

            d) LPC

            e) cepstrum

Fig. 2.4   Pitch-synchronous FFT (solid line) and LPC
            (dotted line) spectra of the vowel /e/   40

Fig. 2.5   a) waveform of the vowel /i/ of "tell me"   40

Fig. 2.5   b) FFT   40
   (contd.)   c) LPC

                d) cepstrum

Fig. 2.6   Place of articulation   41

Fig. 2.7   Example of a parse tree   47

Fig. 2.8   a) Formants of the syllabic nucleus /can/   49

                b) Graphical representation of the second
                   formant F2 versus the first formant F1
                   of the syllabic nucleus /can/. A quasi-
                   stationary portion of the pattern
                   corresponds to points inside a circle.

Fig. 2.9   Slope code for line description   50

Fig. 2.10   Types of errors involved in a possible   54
                generation of the pattern P from the
                archetype A

Fig. 3.1   Spectrogram of the sentence: "vorrei   69
                cancellare una prenotazione" (I would like
                to cancel a reservation)

Fig. 3.2   a) Knowledge source involved in speech   72
                   preprocessing

                b) Knowledge sources involved in the   73
                   generation of phonetic hypotheses

Fig. 3.3   A simplified scheme for hypothesis   77
                information

Fig. 3.4   Relational graph for the word "prenotazione"   79
                (reservation)

Fig. 3.5   Relational graph for the syllable /ta/   82

Fig. 3.6   Example of a Problem Reduction Representation   85

Fig. 3.7   Primitives of the languages L_S (a) and L_A (b)   87

Fig. 3.8   Time evolutions of S,A and R for the pseudo-   90
                syllable segment /ama/

Fig. 3.9   Automata for the recognition of PS (a) and   93
                PA(b)

Fig. 3.10   Fuzzy sets defining the meanings of "high"   100
                for the three acoustic measurements
                characterizing a Vocalic Pattern Feature

Fig. 3.11   Compatibility function of the fuzzy variable   113
                x: = high consonant duration

Fig. 3.12   Time evolutions of the parameters S and R   117
                for the pseudo-syllable-segment /oni/

Fig. 3.13   Time evolutions of the parameters S and R   119
                for the pseudo-syllable-segment /edi/

Fig. 4.1   Scheme for the extraction of context-   127
                independent features

Fig. 4.2   A fragment of the procedural knowledge used   128
                for segmentation

Fig. 4.3   Probability densities for the hpotheses H_S 130
                and H_N as functions of $x = \mu_S - \mu_N$

Fig. 4.4   Definitions of some linguistic probabilities   131

Fig. 4.5   Example of calculation of probability/   133
                possibility distributions

Fig. 4.6   Linguistic evidences represented by fuzzy   134
                subsets

Fig. 4.7   Scheme for segmentation   141

Fig 4.8   Graphical representation of the segmentation   147
                grammar   148
                                                                    149

Fig. 4.9   Problem reduction representation of the   153
                segmentation grammar

Fig. 4.10   Data structures used by the segmentation   154
                algorithm

Fig. 4.11   Problem reduction representation of a   156
                consonantal interval

Fig. 4.12   Example of segmentation lattice   159

Fig. 4.13   Lattice of the phonetic features added   159
               because of the uncertain evidence of a
               VOCALIC hypothesis

Fig. 4.14   Lattice of hypothesized syllables   160

Fig. 4.15   Expert structure   162

Fig. 4.16   Auditory and syllabic expert societies   163

Fig. 4.17   Scheme for actor creation   166

Fig. 5.1   Example of data structure at the syllabic   180
               level

Fig. 5.2   Types of graph nodes that may appear in the   184
               extraction of formant patterns

Fig. 5.3   Establishment of a long link between two   187
               formant arcs

Fig. 5.4   Sample of transformational rules used for   188
               pruning the formant graph

Fig. 5.5   Example of extraction of a fuzzy graph   192/3
               representing a formant pattern from the
               spectrogram of the syllable /ma/

Fig. 5.6   A step of the algorithm for the SZ search   194

Fig. 5.7   Membership functions for the fuzzy   195
               restriction LF2, MF2, HF2

Fig. 5.8   AND-graphs of the problem HPV   197

Fig. 5.9   Vowel loci as fuzzy sets in the F1-F2   199
               plane

Fig. 5.10   Formant pattern of the syllable /ilu/   200

Fig. 5.11   Example of formant tracking on the syllable   202
               /emi/

Fig. 5.12   Formant frequencies and amplitudes of the   205/6
               syllables /eno/, /one/, /elo/, /ole/
               extracted from continuous speech:

Fig. 5.13   Formant pattern of the pseudo-syllabic   217
               segment

Fig. 5.14   Formant frequencies (a) and amplitude   218
               (b) of the pseudo-syllabic-segment /ine/

Fig. 5.15   ROC for phonetic features: 223
               1: sonorant
               2: nonsonorant
               3: liquid
               4: nasal

Fig. 5.16   Problem space representation for interpreting 224
               a consonant interval

Fig. 6.1   PRR representation for the segments labelled   236
               UN during segmentation

Fig. 6.2   Definition of the acoustic features HFE, DDTE, 238
               and DRFDS

Fig. 6.3   Linguistic variables defined over the acoustic 239
               parameters of Table 6.2

Fig. 6.4   Time evolutions of some gross spectral   245
               features of the syllable /agu/

Fig. 6.5   Field of existence of the pseudo-loci of   256/7
               the plosive sounds

Fig. 6.6   Fuzzy restrictions defined over the pseudo-   258
               loci values of the third formant frequency F3P
               before the plosive sound

Fig. 6.7   Fields of existence of the parameters of the   260/1
               burst spectra

Fig. 6.8   Time evolutions of the formants of the   267
               syllable /agu/

Fig. 7.1   Tree representation of the lexicon   286

Fig. 7.2   Pointers used by an algorithm for lexical   287
               access

Fig. 7.3   Example of phone network   289

Fig. 7.4   Graph of a piece of a word obtained after   291
               application phonological rules

Fig. 7.5   First level of the hierarchical network for   295
            generating the lexical knowledge

Fig. 7.6   Detailed subplans for the generation of lexical   297
            items

Fig. 7.7   Subplan for the generation of a word item   298

Fig. 7.8   Components of lexical item   299

Fig. 7.9   Details of the PRR plan for W(w)   300

Fig. 7.10   PRR for the Italian word /tono/ ("tone"   301
            in English)

Fig. 7.11   Plan for storing W(w)   304

Fig. 7.12   Plan for generating lexical hypotheses   306

Fig. 7.13   SSR for a set of vistas   308

Fig. 7.14   Relation between evidence, level, and   315
            possibility of solution

Fig. 7.15   Subplan for generating hypotheses on a vista   317

Fig. 7.16   Definition of linguistic evidences   321

Fig. 7.17   Subplan for setting descriptions of   322
            maximal and minimal evidences

Fig. 7.18   Expansion of the state V(ji) in the SSR   325

Fig. 8.1   Finite-state language model   333

Fig. 8.2   a) Finite-state language model   334

            b) Word subnetworks

            c) Integrated network   335

            d) Graph representing the integrated
               network

Fig. 8.3   a) Simple graph respresenting a language   337
               model

            b) Lattice of phone hypotheses and its
               corresponding graph

Fig. 8.4   A matching path between a finite-state   338
            automaton and a lattice of input hypotheses

Fig. 8.5   Model of a Markov source   343

Fig. 8.6   Phonological model of the word "band"   344

Fig. 8.7   Example of the application of the algorithm   346
                SAL1

Fig. 8.8   Example of the application of the algorithm   348
                SAL2

Fig. 8.9   Examples of best-first, breadth-first, and   352
                depth-first algorithms

Fig. 8.10   Example of beam-search algorithm   354

Fig. 8.11   Performance of the beam-search algorithm   355

Fig. 8.12   Decomposition of the plan: "interpret   368
                sentence"

Fig. 8.13   Scheme of the "interpret-sentence" subsystem   372

Fig. 9.1   Scheme of the learning system   405

Fig 9.2   Domains Q_j^t and Q_j^f for all possible assignments   411
                (x,y) in the x-y plane

Fig. 9.3   Error distribution for all possible   413
                assignments (x,y) in the x-y plane,
                for $f \epsilon Q_x'$, Q_x'', Q_y', and Q_y'' respectively

Fig. 9.4   Error domains $Q_k (Q_N)$ for the entire sample   414
                Q_N

Fig. 10.1   Iterative network of computational   417
                activities for lexical access in continuous
                speech

Fig. 10.2   Tree structure for lexical access   419

Fig. 10.3   Messages exchanged between computational   421
                activities

Fig. 10.4   Example of a network of plans   427

Fig. 10.5   Example of time evolution of gross   433
                spectral features

Fig. 10.6   Example of frame instantiation   435

Fig. 10.7   Example of output queue configuration   436/7

1. COMPUTER MODELS FOR SPEECH UNDERSTANDING

1.1 Motivations for Speech Understanding Research

Research on Automatic Speech Recognition (ASR) began in the early fifties by the attempt to extract significant features from acoustic data and to classify them, using methodologies developed in the area of Pattern Recognition.

Later, in the seventies, more complex techniques available from research in Artificial Intelligence (AI) were applied to the design of Speech Understanding Systems (SUS).

These techniques will be introduced later in this book; they mostly concern the representation and use of various kinds of knowledge related to phonetics, phonology, syntax, semantics, and pragmatics.

Humans use all these types of knowledge very effectively in everyday life in the process of understanding the meaning of a spoken sentence and trying to profit from the redundancies of information implicit in the speech signal. This is essential for settling the ambiguities with which the sentence components are detected and recognized.

Roughly speaking, no reliable methods are available at the moment, for separately detecting and recognizing all the words of a sentence by the extraction of purely acoustic features from the signal and mapping them into elementary symbols (phonemes), similar to the characters of a written text. Rather, a number of ambiguous hypotheses can be generated from a set of data.

1

These hypotheses have to be selected and combined in order to form a sentence that is meaningful and coherent with the discourse which is being carried on.

A question arises at the very beginning: what are the motivations for such research?

There are many answers to this question with different priorities given by the research people in this area where a lot of activity is going on.

A relevant motivation is the attractive perspective of gathering a deeper understanding about the complex brain mechanism underlying human perception of the spoken language and the characterization of speech sounds in terms of physically detectable features. The results of research oriented towards this goal are of great utility in psychology, the field of humanities, development of aids for the handicapped.

Klatt (1979) has pointed out how recent results achieved in the conception of SUS have contributed to progress in speech perception research. Particularly stimulating are the methods for structuring the knowledge acquired heuristically by experiments, for representing each piece of it, for dealing with ambiguities of the features extracted from the data, for using knowledge to correct partial interpretations based on erroneous perceptions.

Another motivation is that SUS research offers a good opportunity for investigating complex parallel processing systems capable of modeling human perception. This concept of parallel processing is relevant in Computer Science, Perception, Cognitive Sciences, Pattern Recognition, and Artificial Intelligence.

Further motivation for research in automatic speech recognition lies in industrial and government applications, where it would be a big step forward to simplify communication between man and machines, avoiding intermediate keying or handwritten steps.

To achieve this would mean the elimination of errors due to the coding of the human messages and would give an operator "multiple-task capability", that is to say he would be able to communicate with the machine while his hands or eyes were left free to carry out other functions.

Among the current applications of voice-input terminals, the following are worth mentioning: factory source data collection, quality control and inspection, parts programming for numerically controlled machine tools, inventory control, materials handling, production and process control, industrial robots, computer-aided design, keyboard replacement, financial reporting, air traffic control, aids for the handicapped, cartographic and hydrographic data entry, computer-aided instruction, office automation. Martin (1976) discusses the potential benefits for some of these applications.

For example, in the case of cartographic and hydrographic data entry, the cartographer's voice can enter the data directly into the computer, while his hands and eyes remain on the source of data for verification.

More generally, examples of applications where spoken input capabilities are important cover situations where the user's arms are occupied by other tasks (in this case an ASR system can also be useful to prevent accidents), situations requiring remote access (perhaps by ordinary telephone) to computerized information systems, instructional situations (like air traffic control), situations where the computer may act as an expert consultant, situations in which human creative efforts are intermixed with communication to machines (speech input is inherently faster than other methods).

1.2   Tasks, Difficulties, and Types of Models

Figure 1.1 shows a very simple scheme of the process of spoken language generation and understanding. This scheme shows the essential transformations of information involved in the process. A sentence generator produces a representation S of the sentence. This representation is an abstract one and we may assume, for the sake of simplicity, that it is an orthographic form (i.e., related to spelling). The representation S is converted by the speaker into a sequence of sets of discrete articulatory commands which drive the vocal tract actuators, producing a continuously time varying pressure signal $x_1(t)$.

The signal $x_1(t)$ is transmitted through a noisy acoustic channel, resulting in a different signal $x(t)$ which is perceived by the listener.

The listener transforms the signal $x(t)$ into an acoustic pattern Ω .

The purpose of this transformation is to have a representation of the spoken message which better exhibits the linguistic features than the signal itself. Furthermore, an attractive hypothesis about the usefulness of this transformation is that it could allow the application to speech patterns of the same interpretation strategies that are used for vision.

The acoustic pattern is then interpreted by an understanding system which first transforms the acoustic information in Ω into an abstract, linguistic representation Λ . We may assume this representation Λ to be in an orthographic form.

Unfortunately Λ is not S. Rather, it may be a continuous string or a lattice of characters or of word hypotheses which have to be interpreted in order to produce $\hat{S}$, the best interpretation of $x(t)$. Hopefully, the system would produce an $\hat{S}$ which is S most of the time.

Bahl and Jelinek (1975) have represented the scheme in Fig. 1.1 as a source transmitting S into a noisy channel which presents Λ at the receiver. The receiver has to decode Λ in order to produce $\hat{S}$, using a source model. Crucial problems are how to model the source and which strategy allows the best use of the model. The aim of this book is to discuss such problems and to present an original solution.

The solution consists in modeling the source and the sentence interpreter as rule-based systems, whose rules represent the apriori knowledge we have about human speech generation and understanding.

There are many dimensions affecting the feasibility and performance of a speech recognition system.

One of the main features affecting the complexity of a speech recognition task is whether the speech is connected or is spoken one word at a time. In the latter case, the system complexity mainly depends on the range of vocabulary. Even the recognition of isolated words is a difficult task because some acoustic ambiguity may be present in the spoken word to be recognized. This ambiguity may depend on the speaker, his personality, his dialect, his speaking rate.

Some of the difficulties may be removed by asking the speaker to be cooperative, pronouncing the words carefully, with sufficiently long pauses between them. But even when speakers are cooperative, there may be ambiguities in the recognition system due to the limitations of the acoustic analyzer and to our imperfect knowledge of the features to be used and the way they have to be extracted. Fortunately, contextual (inter- and intra-syllabic) information can be used to resolve such ambiguities.

The beneficial effects of contextual information are particularly evident if the redundancy of the lexicon is high. For this reason efficient systems for the recognition of isolated words have been developed and are actually operating in many real-life situations. They have a lexicon of hundreds of suitably chosen words (see, for example, Chiba et al., 1978, Interstate, 1978, Martin, 1976, Rivoira and Torasso, 1978, Velichiko and Zagoruiko, 1970, White and Neely, 1976).

Such systems accept isolated words with an average speaking rate of about 50 words per minute, require the speaker to be cooperative,

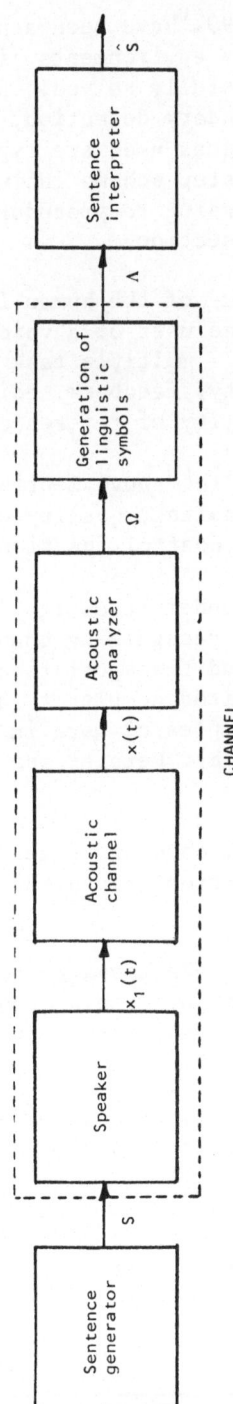

S: sentence representation

$x_1(t)$: generated signal

$x(t)$: received signal

Ω : acoustic pattern

Λ : lattice of hypotheses composed of linguistic symbols

$\hat{S}$: interpretation

Fig. 1.1 Transformations of information along the speech chain

and to adapt the system to his voice in a preliminary training phase.

Recognition rates higher than 99% have been achieved in operational situations typical of factory environments, for which the
following problems have been successfully solved: background and
breath noise cancellation, word boundary detection, operator-origi-
nated bubble[1] detection. The techniques used are typical of pattern
recognition and are based on a two-step scheme including feature
extraction and classification, according to procedures that will be
described and discussed in a later section.

It is interesting to report some of the human factors, mentioned
by Martin (1976), which influence the user of a voice-input system.
The most interesting among them are: multiple task capability, operator mobility, vocabulary flexibility, feedback, editing and interaction, recognition accuracy, stability of reference data.

Applications of systems having the above mentioned performances
may be in the areas of materials handling, quality control, numerical
control, aid to the handicapped for control functions.

Far more difficult, and still under investigation through laboratory prototypes, is the problem of recognizing connected speech.
Tasks such as data base retrieval and transportation schedule inquiry
can be achieved by protocols of limited complexity and vocabulary
sizes of less than 1000 words. Examples of work in this area are
described by Kohda and Nakatsu (1978), Mercier and Quinton (1980),
Levinson and Shipley (1980).

More complex tasks, requiring vocabulary sizes of thousands of
words, are computer assisted instruction, dictating machines, and
visual display for the deaf.

There are several reasons why these tasks are difficult ones.
One of them is that speech processing by machine requires several
orders of magnitude greater computational resources than interfaces
designed for man-machine communication.

In fact, these interfaces normally operate with a sequential
decoding and encoding of information that is context-independent
and is based on time-invariant symbols.

On the contrary, different people speak in different ways, a
noisy acoustic environment may interfere with reliable interpretation of the acoustic data, and even the same single talker will vary
from time to time in his enunciation.

---

[1]"Bubbles" are non-speech sounds due to globules of gas passing
through the lips (ed. note).

Furthermore, in naturally flowing connected speech, there are no pauses between words and the articulation of a speech sound is strongly affected by its context.

Thus, the recognition of complex protocols cannot be based on the comparison between the uninterpreted data and a collection of prototype patterns. This is possible for recognizing a limited set of isolated words, requiring every user to speak all words into the system in a learning phase so as to generate a personal set of word patterns. For complex tasks, the set of prototypes would require a prohibitive amount of memory and their learning and comparison with the input data would require too much time.

Depending on the task complexity, two different types of models can be used for speech decoding: a passive model and an active model.

In the *passive* model, human reception of speech is viewed as consisting of sensation followed by perception, followed by cognition. Sensation deals with the raw signal, perception classifies the sensations into words or objects and cognition establishes relationships among the words or objects. This model has been used for designing systems for the recognition of isolated words and consists of acoustic preprocessing, feature extraction, and pattern matching.

It is clear that human listeners use expectation for understanding what is being said. In a model of such behavior, all facets of the listener's knowledge, such as syntax, semantics, pragmatics, phonology, prosody, are used to aid the decoding. Human perception is thus likely to be an *active* process in which cognition may even guide the lower levels of decoding.

None of the components using different types of knowledge can currently be made to perform reliably. A strategy must be found to combine information from all the components in order to profit from the redundancy to resolve the ambiguities.

An *active* model for speech understanding involves the representation of knowledge at various levels (Knowledge Sources: KS): a procedural knowledge containing rules on how to use the KSs effectively in order to solve the problem of interpreting a signal and a set of data structures where the interpretation hypotheses are written.

An essential feature of the active model is that cognition and expectation may drive decoding.

Machines based on the passive model for the recognition of limited-size vocabularies have been available since the early seventies. Laboratory prototypes appeared in the fifties (see Davis et al., 1952).

Moreover, interesting laboratory prototypes for Speech Under-standing Systems have been developed (Woods, et al., 1976; Lowerre, 1976; Erman, 1977; Nakagawa, 1976).

Machines for the recognition of limited-vocabulary continuous speech are appearing on the market now. Machines for the recogni-tion of unconnected natural language are expected around 1985 and machines for the recognition of continuous natural language have still far to come for real-life application.

1.3   A Passive Model for Automatic Speech Recognition

A passive model for automatic speech recognition follows the scheme of Fig. 1.2.

The microphone roughly corresponds to the ear, the cochlea may be considered as a sort of preprocessor performing sensation, the auditory nerve acts as a feature extractor performing perception, the left cerebral cortex may contain the classifier, whose knowledge is a memory of templates.

Dashed lines show the flow of information when the system is learning its templates for each class of patterns to be recognized.

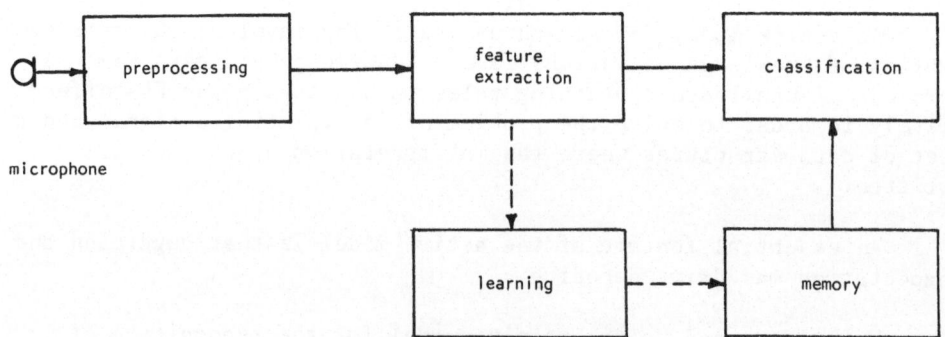

Fig. 1.2 A passive model for speech recognition

Two main approaches have been proposed for conceiving a pattern recognition scheme like the one shown in Fig. 1.2. The first one follows the so called "parametric" approach and consists in extracting features, which are measurements taken on the acoustic pattern. Each unknown pattern is then represented by a point in an N-dimensional space and its classification depends on the position of the point.

The space may be partitioned by linear or non-linear functions based on statistical distributions of the points for each class or a distance measure following a suitable metric defined for that space.

The pattern recognition literature is rich in algorithms for learning templates and performing classification in such a framework (see Fu, 1968, for example). Trainable classifiers (Damman, 1965) have been also proposed for automatic speech recognition.

A second approach, the "syntactic" one, consists in describing a pattern by a *language*. Such a language is generated by a generative *grammar* whose syntactic categories contain the classes of the pattern to be recognized. *Parsing* a description sentence consists in obtaining its syntactic structure, which contains the pattern class as a syntactic category. A generative grammar is a syntactic representation of the pattern classes because it contains a finite set of rules capable of generating an infinite variety of patterns. The book by Fu (1974) is an excellent introduction to these concepts. Learning can be performed in the syntactic approach by using grammatical inference techniques (see Fu and Booth, 1975).

A grammar can be used either for recognizing pattern descriptions or for predicting features which should be present in the pattern. This aspect makes the syntactic approach suitable for use in an active model for speech understanding as well.

The development of this concept for application in designing speech recognition and speech understanding systems is one of the characteristic features of this book.

Both the parametric and the syntactic approaches may be used together by adding numerical attributes to the symbols introduced for describing pattern primitive forms. Two main problems have to be solved, in performing pattern matching: one is that of time normalization, due to the fact that the same word may have different durations when pronounced at different times; another problem is that of discarding non-essential information that may cause confusion in recognition.

Early approaches to speech recognition have used a passive model with a parametric representation of a speech pattern. Hyde (1972) has very well reviewed the work done before the seventies. Some of the points he made are briefly summarized in the following.

One problem in comparing complete word patterns is the natural variability which can occur in the speed at which a word can be spoken.

When a word is spoken at different speeds the different sound elements in the word do not undergo the same degree of change. For instance a vowel sound can easily be prolonged while the sound of air escaping at the release of a stop consonant cannot.

When the recognizer has to be used by several different people, particularly when both sexes are involved, a further degree of normalization may be necessary in the frequency dimension. This is referred to as talker normalization. Alternatively, it may be necessary to store a different set of reference patterns for each type (e.g., sex or age range) of speaker. Talker normalization is required because of consistent spectral differences between speakers of different size, etc. This is due to systematic differences in the geometry of the vocal tract.

Another problem in pattern matching recognition is in the initial alignment of the two patterns being compared. A particular difficulty is found in detecting the start of the pattern when the first speech sound is weak.

While the problem of detecting the end points of an utterance has been satisfactorily solved (see Rabiner and Sambur, 1975, for example), the problem of talker normalization is far from solved. This is the reason why most of the available machines have stored speaker-dependent templates and require a training phase for each user.

Recently, attempts have been made to cluster the templates of different speakers to obtain speaker-independent performances. Good results have been obtained so far, at the price of a strong reduction in the vocabulary size (a few tens of words). Rabiner (1978) is a good reference for this subject.

The system proposed in this book does not try to solve the problem in terms of clustering templates. Rather, the research is focused on the characteristic cues of the sound patterns of a language and on the detection of these cues.

1.4 Active Models for Speech Understanding

When continuous speech has to be recognized, more complex models

than the passive ones have to be conceived. Following an approach
of Artificial Intelligence (A.I.), a great many of the acoustic ambiguities can be overcome by using various levels of contextual information.

1.4.1 Elementary psychoacoustic considerations

Psychoacoustic experiments in which speech sounds are selec-
tively masked with noise show that listeners use semantic, syntactic,
prosodic, pragmatic, as well as acoustic knowledge to understand
acoustically corrupted speech (see White, 1976, for details).

The non-acoustic sources of knowledge referred to by White
(1976) as "semiotic" information can be incorporated in a speech
understanding system. A fundamental problem, that stimulated a
fruitful merging of ideas from speech research and Artificial Intelligence, is how to represent knowledge in such a way as to facilitate
understanding with computational efficiencies.

Liberman et al. (1971), assert that language is composed of segments arranged in hierarchically ordered layers. The structure
formed by the language segments and layers is described by two types
of rules. One set of rules characterizes phonology and applies to
segments that are, in themselves, devoid of meaning; the other set
of rules covers many levels varying from morphology to pragmatics.
Among the phonological segments there are, basically, from 30 to 50
units, called *phonemes*, themselves composed of a smaller number of
phonetic features; from these units thousands of syllables can be
obtained.

A large vocabulary of words and a vast repertory of phrases can
be generated from syllables, whose composition gives an infinity of
sentences.

The acoustic correlates of a given phoneme change considerably
with different phonetic contexts but two requirements are imposed on
its perception by the nature of linguistic structure. The first is
that the phonemes must not lose their identity as they enter into
combinations with other phonemes. The second requirement is one of
speed. The capacity to produce an infinite number of utterances
using a small number of phonemes implies that phonemes are rapidly
identified. Furthermore, there is a complex relation between perceived language and the acoustic signal which conveys it and the cues
for the phonemes are encoded into units of approximately syllabic
size.

1.4.2 Interpreting the acoustic
signal is problem solving

Interpreting the acoustic signal x(t) is a complex problem to
be solved. The A.I. approach consists in decomposing such a problem into a hierarchy of subproblems   leading to a Problem-Reduction-
Representation (PRR) of the original problem. Such a representation
may be an AND/OR or a relational graph. A formal definition of PRR
will be given later in this book. What is interesting to notice here
is that problem decomposition can be performed with the help of the
available knowledge about spoken language generation and understanding. The same set of levels into which this knowledge can be subdivided can be assumed as a basis for the decomposition of the general understanding problem.

Subproblems at a given level can be solved by using some structural knowledge characterizing that level, a procedural knowledge
describing how to use the structural, apriori knowledge, and a set
of hypotheses about previously solved subproblems.

Acoustic ambiguities and the complexity of the process of speech
perception suggest it is worthwhile designing a speech understanding
system, applying the so-called "hypothesize and test" paradigm by
dynamically allocating computational resources to the appropriate
interpretation tasks (Reddy, 1976).

System organization, hypothesis evaluation, priority assignment
to competing interpretation processes, decision-making in the presence of a corrupted message, error correction are among the most important problems that need to be solved in designing speech understanding systems. Again, the complexity of the system and the number
of difficulties to be overcome depend on how ambitious the task is.

The first attempt to make something more complex than isolated
word recognition is what Reddy (1976) calls "*restricted connected
speech recognition systems*". In such systems the vocabulary and the
syntax are very restricted and the speaker is requested to speak
clearly.

One such system has been implemented at Bell Labs by Sambur and
Rabiner (1976) for the automatic recognition of connected digits over
a telephone line, with good performances.

More complex are the restricted speech understanding systems,
for which it is assumed that the speech signal does not have all the
information necessary to uniquely decode the message and, to be successful, all the available sources of knowledge must be used to infer
the meaning of the message.

Without any constraints, given a vocabulary of W different words
and admitting sentences having a maximum of S words each, the number
of possible sentences would be of the order of W^S.

The exponential explosion of alternatives can be constrained by
realizing that the sentence to be interpreted respects a grammatical
structure defined by a *syntax*, that relations among objects and
events are in accordance with *semantic rules*, and that the meaning
of the sentence is in accordance with the conversation-dependent contextual information referred to as pragmatics. All these constraints
are represented by *task-dependent* knowledge sources (KS) of the systems. The best path through the constrained network of alternative
phrase hypotheses depends on how well the hypothesized words match
the acoustic data.

It is well known that even a perfect phonetic transcription of
a spoken sentence can create difficulties and ambiguities in the
detection of word boundaries because of the mutual phonological influence across adjacent words and the enunciation, which is often
imperfect. The negative effect of these ambiguities is increased by
the errors and ambiguities made in assigning phonetic features and
phonemic interpretations to speech waveform.

Because of the errors and ambiguities, a process of performing
word hypothesis and verification has to be incorporated in the Speech
Understanding Systems (SUS). This approach is known in Artificial
Intelligence (AI) as the *hypothesize-and-test paradigm*.

Hypothesis evaluation is made by numerical relations associated
with the structural rules relating words, syllables and phonemes to
their possible physical realizations through speech waveforms.

From the above general considerations it appears that the main
problems involved in the design of speech understanding systems cover
the following items:

- knowledge representation at various levels
- methods for evaluating hypotheses,
- policies for scheduling interpretation processes on the basis of
  the values of the hypotheses previously generated and on the plausibility of possible predictions.

By analogy to computer systems, the knowledge sources and the
hypotheses they permit can be viewed as components and information
of a hierarchical machine. Fig. 1.3 shows a possible model for such
a machine. The interactions between levels of the hierarchical
machine mostly concern exchange of information instead of commands
for evaluating functions. The latter situation is typical of operating systems where a level uses lower levels as virtual machines.
Furthermore, in SUS, processes at various levels may operate con-

currently, provided that some conditions regarding the data they are
to process are verified.

Each level may correspond to a knowledge source. A knowledge
source (KS) operates on hypotheses generated at lower levels and
generates hypotheses at its own level. A KS activation is a process
that uses lower level hypotheses as data, Each time a KS is activated for processing some data, a process called "KS instantiation"
is created. This creation can be stimulated by the data (*data-driven
KS instantiation*) or by a request of verification from a higher level
(*model-driven KS instantiation*).

1.4.3 Structures for speech understanding system models

The execution of a KS instantiation can be decided by a control
component which supervises the entire understanding process or can
be stimulated asynchronously by the data.

The former control structure is called hierarchical, the latter
is called heterarchical.

Central to the design of heterarchical structures is the competition of KS instantiations for access to a three-dimensional
data-base (blackboard), where all kinds of hypotheses about the interpretations of the input sentence are written. The three dimensions of the data base are time, levels of interpretation, and hypotheses. Usually, and particularly in HEARSAY II (Erman, 1974a)
linkages of hypotheses at various levels can be established.

Fig. 1.4 shows the scheme of a heterarchical structure (Erman,
1974a). Here data on the blackboard act as stimuli to the various
KSs, causing the creation of interpretation processes whose principal product is the writing of a hypothesis onto the blackboard.
Execution and synchronization of the various processes as well as
monitoring of the access to the blackboard are among the main problems involved in the implementation of such a structure.

Fig. 1.5 shows the scheme of a hierarchical structure. The
controller decides the execution of the various components. Each
component produces data only for the component immediately above it.
The controller itself keeps track of the partial interpretations
which have to be completed (theories).

Theories (see Woods et al., 1976a) are queued in order of priority.

Each configuration of a theory represents a state of solution

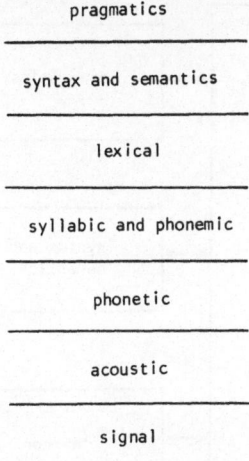

Fig. 1.3 Levels involved in the Speech Understanding Process

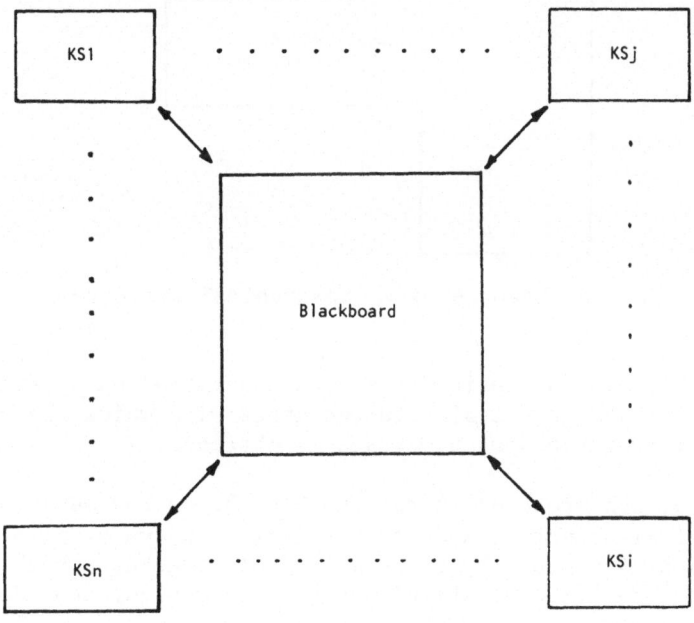

Fig. 1.4 Example of a heterarchical structure

of the interpretation problem.

All the possible states of this type form the State Space Representation (SRR) of the problem solver.

The evaluation of priorities for the existing theory is another

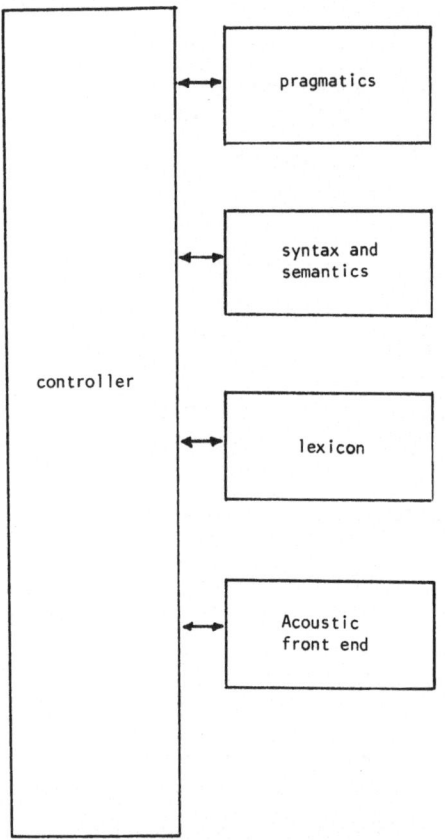

Fig. 1.5 Example of a hierarchical structure

crucial problem in the design of Speech understanding Systems, because priority assignment affects the speed with which the right
interpretation of the spoken message is obtained.

The approach proposed in this book combines the advantages of
a heterarchical structure and the simplicity in synchronization of
a hierarchical structure by using a control structure which con-
strains the generation of stimuli but allows asynchronous data-driven
stimulations of Knowledge Sources as well.

A scheme of the structure is shown in Fig. 1.6.

The blackboard is sub-divided into levels. Each level can be
accessed by higher level KS instantiations for model-driven verification of predictions or can activate higher level KSs by the generation of data-driven stimuli.

The former use of knowledge is *top-down*, the latter is *bottom-up*.

Each level of knowledge has associated a local controller which

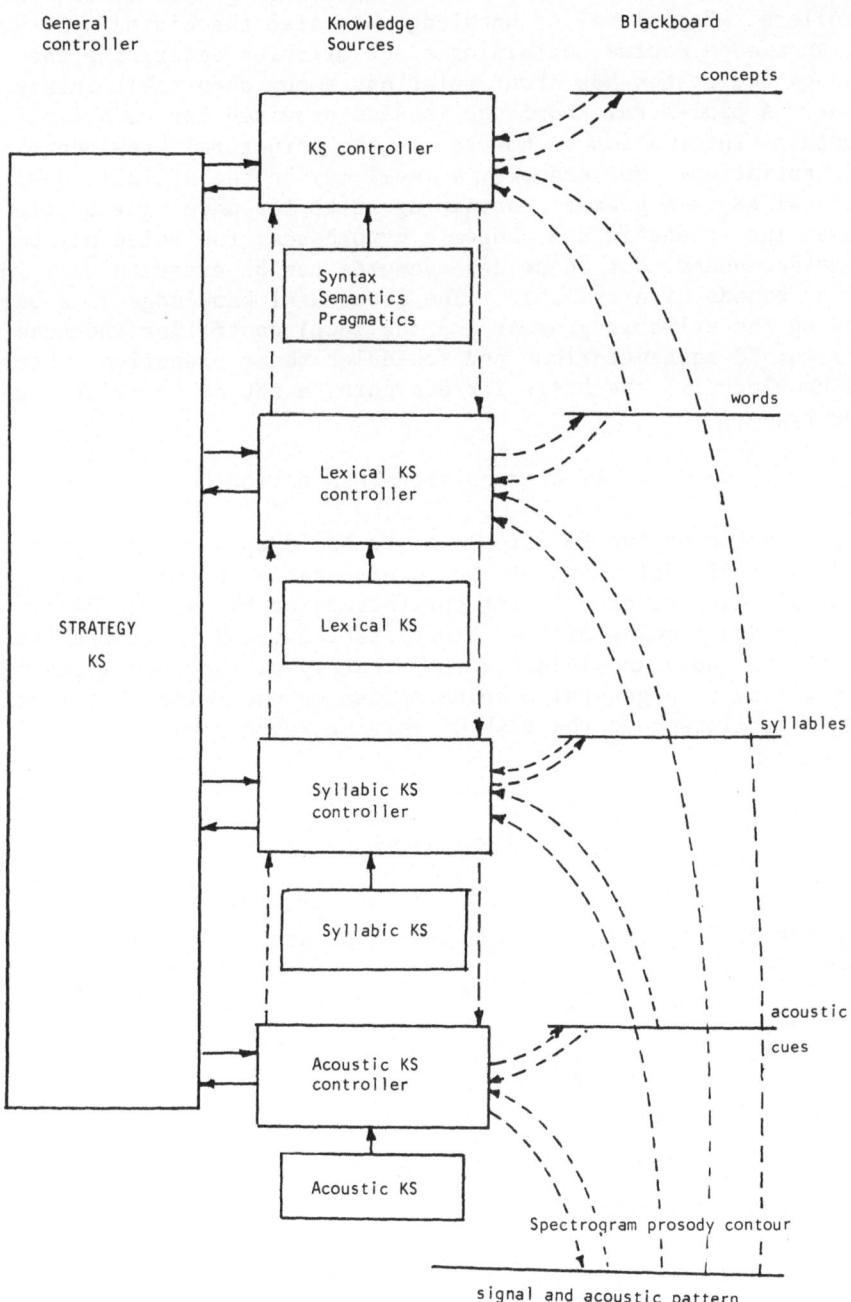

Fig. 1.6 Speech Understanding system organization

receives stimuli and requests, either from the general controller
or from another local controller, creates KS instantiations, schedules their execution, and may send request messages to other local
controllers. Each level of knowledge has also associated a *structural Knowledge Source* containing a set of rules describing the
knowledge the system has about relations among acceptable interpretations. A *procedural Knowledge* is also provided for each level.
It contains information on how to use the structural knowledge in a
KS instantiation. For example, a level may be the syllable one; the
structural KS is a grammar containing rules for mapping acoustic
features into phonetic and phonemic hypotheses; the rules may be
context-dependent, but these dependencies can be extended only inside the bounds of a syllable. The procedural knowledge is a parser
for using the syllabic grammar and the local controller enqueues requests for KS instantiations and schedules their execution. Structural knowledge is the basis for designing a set of procedural rules
of the type:

IF precondition THEN action

The application of the KS allows one to build up a graph representing the possible solutions of the interpretation problem. This graph
may contain many acceptable interpretations of the speech data. The
generation and growing of the interpretation graph is controlled by
a supervisor whose knowledge is the strategy of the understanding
system acting as a general *problem solver* or controller, while the
local controllers have the task of solving subproblems.

1.4.4 Functional description of the system

Functionally, the following relations are established by the
systems:

concept = f_1 (sentences)

sentence = f_2 (words, prosody)

word = f_3 (syllables, stress)

syllable = f_4 (acoustic cues)

acoustic cue = f_5 (acoustic pattern)

The acoustic patterns containing spectra, stress and prosody
information are obtained from the signal by algorithms which are
mostly based on numerical transformations. The functions $f_i(\alpha)$,

just introduced, concisely represent the types of hypotheses which
may be generated under the control of the structural and procedural

KS at the i-th level based on the hypotheses $\alpha(i = 1,2,\ldots, 5)$.

For limited tasks, some of the functions $f_i(\alpha)$ may be composed and used for generating all the acceptable sequences of arguments of the function with the highest index. For example, by the following composition ($\circ$ is an operator indicating composition):

$$F_{14} = f_1 \circ f_2 \circ f_3 \circ f_4$$

the following relation may be established:

$$\text{concept} = F_{14} \text{ (acoustic cues)}$$

Given a task (the set of concepts allowed in the language) and F_{14} (the way the concepts can be expressed in terms of acoustic cues) the search space can be built. The search space can be a network where every arc is labelled by a symbol of acoustic cues.

Each path from the initial state to one of the final states of the network represents a sequence of acoustic interpreted by the system. Precompiling the network allows one to speed up the recognition process. If the task is very limited all the functions can be composed in a single one that can be used for generating reference patterns to be stored into the template memory in the passive model.

Knowledge representation at various levels and strategies for its use are the main points to be investigated for designing Speech Understanding Systems. These aspects will be discussed in the successive Chapters of this book.

Integration of various KSs, depending on the system task, may improve the system efficiency (as in the HARPY system designed by Lowerre, 1976).

1.4.5 On the tools for representing and using knowledge

Most of the structural knowledge will be represented by grammars. A brief background on formal language theory will be given in Chapter 2. Methods for representing complex problems as compositions of simpler subproblems will be introduced in Chapter 3.

A theoretical result will be recalled, based on which some types of grammars can be represented in the same way as problem decomposition, that is by AND/OR graphs. This representation will be used to describe the ordering of a non-left-to-right application of syntactic rules to descriptions of the acoustic patterns. Applications of such rules for the interpretation of ambiguous descriptions will be seen as a hierarchical execution of complex plans. It will be shown in Chapter 8 how these plans can be described as sets of production rules of the type:

IF some conditions hold THEN some actions have to be performed.

These rules are derived in a straightforward way from the grammars and their parsers.

1.5 On the Use of Fuzzy Set Theory

Most of the structural knowledge is represented in an algebraic form, i.e., by rules expressing the possibility of generating a hypothesis as a set of concatenations of hypotheses which have to be previously generated by some lower level KS. Each rule may be assigned a degree of worthiness representing the imprecision of the apriori knowledge.

Furthermore, each element of a concatenation of hypotheses may have been hypothesized with uncertainty. A degree of worthiness is assigned to it for representing such an uncertainty. The evidence of a hypothesis has to be evaluated by taking into account the imprecision of the rule which has been used and the vagueness by which the elements involved in the rule have been detected. This evaluation can be performed within the framework of a fuzzy algebra which deals with possibility values. The situation is sketched in Fig. 1.7; here it is shown how an interpretation process, created by a control KS, uses an imprecise knowledge for interpreting vague data and generating a hypothesis which is represented by a triplet:

$$H = < T, (a,b), q >$$

where T is the hypothesis description, (a,b) is the time interval of the signal of which the hypothesis is an interpretation, and q is its possibility.

The use of an algebraic approach based on possibility theory for hypothesis evaluation is an original aspect of the research described in this book.

The possibilities of the hypotheses are used for scheduling the activitation of the KSs by the system that controls the process of understanding.

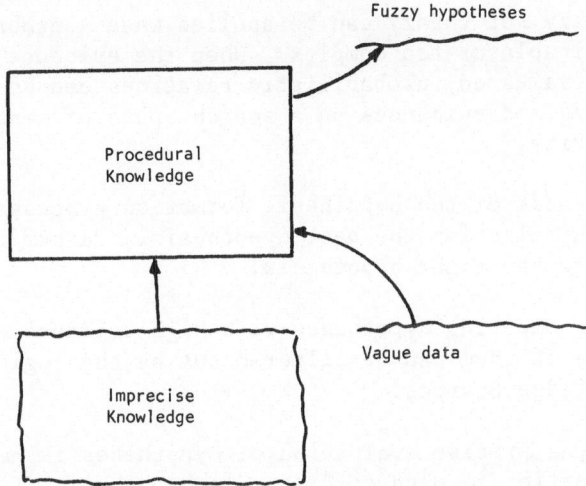

Fig. 1.7 Types of imprecisions involved in hypothesis generation

The reasons for using possibilities instead of probabilities or purely heuristic methods for scoring hypotheses can be summarized as follows.

Firstly, the theory of possibility, in contrast to heuristic approaches, offers algorithms for composing hypothesis evaluations which are consistent with axioms in a well-developed theory; the theory of possibility, rather than the theory of probability, relates to the perception of degrees of evidence instead of degrees of likelihood or frequency. The aim of the approach proposed in this book is to express the evidence of a hypothesis concerning a speech pattern by the possibility of having high evidence measurements for clearly interpretable patterns, even if the patterns and the features considered are scarcely probable.

Secondly, the theory is very flexible and allows easy and simple computation of evidences in complex relations.

Thirdly, for applications like speech understanding, a probabilistic approach requires more complex computation and more distributions to be known. Many distributions have to be subjectively decided and the estimations are affected by a high degree of imprecision.

Fourthly, fuzzy decision making is not poorer than statistical decision making in some cases.

Fifthly, the flexibility of fuzzy set theory allows good learning algorithms to be designed.

Sixthly, fuzzy set theory can be applied when a probabilistic approach is unsuitable or too complex. When the evidence of hypotheses has been evaluated, probabilistic relations can be established between hypotheses and evidences on a search space of reduced dimensions and complexity.

Finally, in most of the hypothesis formation processes we are not interested in selecting the best hypothesis. Rather the attempt is to avoid losing the right hypothesis.

Having many competing hypotheses with high evidence can be tolerated because some of them can be filtered out by the logical structure of the Knowledge Sources.

Furthermore, a precise evaluation of hypotheses is not required because the evaluation is used to discard a hypothesis if its evidence is very low or to schedule successive actions. Scheduling can be based on a sort of Approximate Reasoning (see De Mori and Saitta, 1979) for which only qualitative evidence judgments are required.

This kind of Approximate Reasoning uses inference rules of the type:

IF a hypothesis is *evident* THEN the priority of a consequent action is *high*.

Notice that Approximate Reasoning (AR) based on qualitative judgments models the behavior of humans when they apply uncertain knowledge to interpret something they do not know precisely.

A different type of motivation for using fuzzy rules is that as Oden (1978) has recently shown, fuzzy logic models very well the relations between acoustic cues of plosive consonants.

A final remark is that a serious model of speech understanding has to contain ill-defined relations and restrictions defined over ill-detected cues. Whether this imprecision has to be described in a probabilistic or an algebraic framework is a minor matter because the evaluation of vagueness cannot and need not be precise. Fuzzy algebra and fuzzy logic offer a tool which is simple and flexible enough for computing evidences.

Evidences are measurements of hypotheses. Statistics of evidences will be collected and used for making decisions on the elimination of hypotheses which are unworthy of belief. Decision-making is based on evidence probabilities because it has to minimize a probabilistic quantity which is the expected number of errors.

1.6   The Structure of the Book

The aim of the book is to describe research on the characterization of the knowledge sources involved in the speech understanding process. Most of the structural knowledge is a set of rules
representing the designer's experience, acquired from disciplines
like phonetics, psychoacoustics and psycholinguistics, and from a
limited set of experiments. The degree of worthiness of each rule
component is obtained by a learning algorithm that attempts to optimize some system performances.

In this way the structural knowledge contains most of the essential elements of the spoken language which are known to be
speaker independent. Only the importance of the combinations of
these elements is learned by attempting to reach a speaker indepen-
dency.

The details of the algorithm for learning imprecisions will be
given in Chapter 9.

Chapter 2 reviews the essential information about speech generation, signal processing, feature extraction, the possible implemen-
tations of a passive model, and the research activity on Speech
Understanding. Chapter 3 describes the fundamentals of fuzzy set
theory and introduces its application to speech understanding.

Chapters 4, 5, and 6 describe in great detail the characterization of the syllabic knowledge which is a fundamental item in
Speech Understanding.

Chapter 7 introduces the lexical knowledge as a bridge between
acoustic and orthographic information.

Chapter 8 deals with the representation and use of task dependent knowledge, such as syntax and semantics.

1.7   References

Fundamental books dealing with speech problems are those by
Fant (1960, 1973) and Flanagan (1972).

An excellent review of the work done in automatic speech recognition before the seventies has been done by Hyde (1972) in an interesting book by David and Denes (1972).

Design concepts for Speech Understanding Systems are given in

the book by Newell et al. (1973).   A good collection of tutorial
papers for Speech Understanding can be found in the book by Reddy
(1975).   Recent results of important projects are described in books
by Lea (1980), Walker (1978), Dixon and Martin (1979) and in reports
by Lea and Soup (1979), Woods et al. (1976), Reddy (1976).

Interesting works for relating Phonetics, Phonology, and Perception models with Automatic Speech Recognition are due to O'Malley
(1976), White (1976), Oshika et al. (1975), Cohen and Mercer (1975);
fundamental in this area are also the books by Chomsky and Halle
(1968) and Fant (1973).

Recent reviews of the state of the art in the field have been
provided by Reddy (1976), Martin (1976), Jelinek (1976), Klatt
(1977), Medress et al. (1977), Zagoruiko (1977), De Mori (1979),
Erman et al. (1980).

An excellent review of the work done in Japan is contained in
the Ph.D. thesis by Nakagawa (1976), available in English.   Interesting monographs describe the work done in the Soviet Union; among
them are worth mentioning those by Tsiemiel (1971), Trunin-Donskoi
(1975), Fain (1977), Zagoruiko (1976), Chitavichius (1977).

Progress in France is reviewed in the thesis by P. Quinton
(1980).

A recent paper by Kaplan (1980) describes available industrial
products and the most promising laboratory prototypes and research
items.

Of particular interest are also the progress reports of the
Massachussetts Institute of Technology (RLE) in Cambridge, Mass.,
the Haskins Laboratories in New Haven, Conn., the Electrotechnical
Laboratory in Tokyo, the Royal Institute of Technology in Stockholm.

There are many journals in which the research results in Speech
Understanding can be found.   Among them, the IEEE Transactions on
Acoustic, Speech and Signal Processing (ASSP), on Pattern Analysis
and Machine Intelligence (PAMI), the Journal of Problems for Information Transmission, Signal Processing, Pattern Recognition, Artificial Intelligence, the International Journal on Man-Machine Studies and the forthcoming journal entitled Speech Communication.

Among the International Conferences, those on Pattern Recogni-
tion and those on Artificial Intelligence usually contain papers on
Speech Understanding.   Other important conferences are ICASSP (USA),
ARSO (USSR), EUSIPCO (Europe), GALF (France).

2. Generation and Recognition of Acoustic Patterns

2.1 Speech Generation

Before a sentence is pronounced, concepts, represented in our mind in an abstract way, are first coded into sentences respecting the syntactic rules of the Language we intend to speak. The structure of the sentence determines the so called suprasegmental features of the corresponding spoken message. These features affect durations of phonemes (the units of speech) and the time evolution of the frequency of the signal exciting the vocal tract for producing the speech signal.

Furthermore, a discrete sequence of phonemes is derived from a sentence and each phoneme corresponds to a limited set of commands to the articulators which control the speech production mechanism.

An ample presentation of speech production and its mathematical models can be found in the books by Fant (1960) and Flanagan (1972). Only a brief discussion of certain basic facts will be attempted here.

A speech sound is the result of a signal generated by the vibration of the vocal chords or by some noise sources, and modified by its passage through the vocal tract and, for some sounds, the nostrils. The main components of the vocal tract are the lips, the tongue, the lower jaw and the velum; the velum is the valve which closes off the nasal tract. All these components are movable, making the vocal tract assume a variety of configurations, some of them corresponding to speech sounds.

Fig. 2.1a shows a scheme for speech production; Fig. 2.1b shows a model of the speech production system based on resonance cavities.

For vowels and the so called "voiced" consonants, the source signal is generated by the oscillations of the vocal folds in the larynx due to pressure built up in the lungs; when the muscles of the larynx, which keep the vocal folds closed, are at the proper tension, the pressure forces the vocal folds apart and a puff of air emerges. This flow of air is called the glottal waveform; this waveform is quasi-periodic, and a period is called "pitch period".

For many consonants, an important component of the sound is due to turbulent air flow through some constriction in the vocal tract. This source consists of random noise injected at some position along the vocal tract. The signal is again modified by its passage through the vocal tract.

A third type of noise is due to sudden release of built-up pressure in stop consonants.

During speech, the articulators are continuously moving, generating an acoustic waveform that is difficult to interpret.

Each speech sound (a phoneme) is produced by a discrete set of stimuli.

From an information theoretical view, speech is generated by a source S which converts concepts into a sequence of stimuli for the articulators which move the speech production system shown in Fig. 2.1a.

The effect of the stimuli is a continuous movement of the vocal tract, whose configuration at a given instant depends on the phoneme to be uttered and on its context. These context dependencies, due to the inertia of the vocal tract organs, generate complex speech patterns exhibiting the so-called "coarticulation effects".

Phonemes, words, or any other arbitrary item can be seen as units of information. The average amount of information per source symbol is called entropy, $H(\sigma)$, and depends on the probability $p(S_i)$ of occurrence of each symbol S_i $(1 \leq i \leq n)$.

For an alphabet V_σ of symbols:

$$V_\sigma = (S_1, \ldots, S_i, \ldots, S_n)$$

the source entropy is defined as follows:

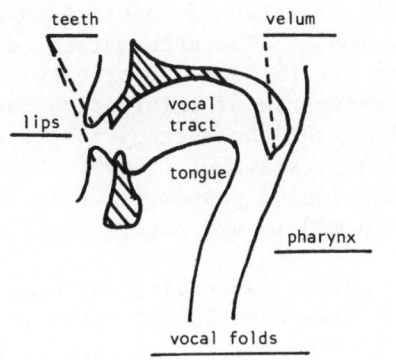

a)

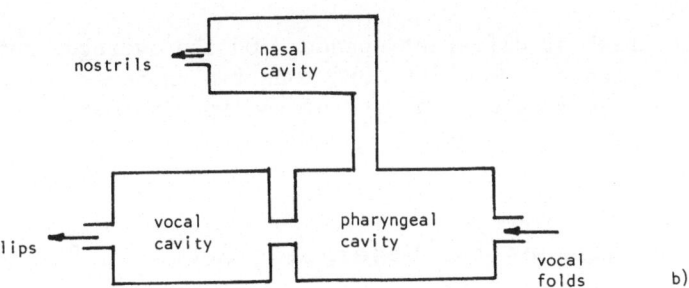

b)

Fig. 2.1 (a) Speech production scheme
(b) Speech production model

$$H(\sigma) = \sum_{i=1}^{n} -p(\sigma_i) \log_2 p(S_i) \qquad\qquad (2.1)$$

and is measured in bits. For example, American English has 27 letters (including space) with an entropy of 4.1 bits/symbol. If the
symbols were equiprobable, the entropy would achieve the maximum
value of 4.8 bits/symbol.

If the occurrence of a symbol S may depend upon a finite number
m of preceding symbols, the source is called an m-th order Markov
source and the entropy is the average of information carried by each
symbol, given its past m symbols. The mathematical details of this
definition are omitted here for the sake of brevity. The book by
Abramson (1963) is a good reference of Information Theory. For a
written text in American English, the entropy is 3.6 bits/symbol if
m=1; the entropy becomes 3.3 bits/symbol if m=2. Considering now
the spoken language and assuming 42 phonemes for American English,
the entropy of a source with m=0 is 4.9 bits/symbol.

Let us consider now a source generating words as information
units. For a vocabulary of more than 20,000 words in American
English, entropies of 7.5 bits/symbol for the spoken language and
of 9.4 bits/symbol for a written text have been calculated. Accounting for the context in a written English text, an entropy of 4.5 bits
bits/symbol has been calculated for a word generating source.

We can speak at different speeds. On the average, normally
speaking, we produce something more than ten phonemes per second.
Thus, the average information rate of the speech source is about
50 bits/second.

2.2   Techniques for Generating Acoustic Patterns

It is commonly believed that non-linear operations are performed after the uninterpreted speech signal has been transformed
to obtain a representation in the frequency domain. There is experimental evidence (see, for example, Flanagan, 1972) that even
the ear performs such a type of transformation; techniques for preprocessing speech are briefly reviewed in this section.

An attempt to extract acoustic cues from the speech waveform was made back in the fifties, using short-term statistics of
the axis-crossings of waveforms obtained by filtering the speech
signal (Davis et al., 1952).

More recently, Baudry and Dupeyrat (1978), following syntactic
approaches to signal processing described by De Mori (1973), have

proposed a grammar for the interpretation of waveform segments generated by plosive sounds.

2.2.1 The filter bank

Because of the scarce possibility of extracting linguistic
cues directly from the speech waveform, time-domain techniques have
rarely been used alone.

Another approach, that has been followed in many practical realizations of speech recognition systems, uses a signal processor
made of a bank of 1/3 octave band-pass filters in a frequency range
of 200-10,000 Hz. The outputs of such filters are rectified and integrated with a low pass filter, having a cut-off frequency of about
50 Hz. The reason for this solution is that the human ear acts as a
power spectrum analyzer and that all acoustic information relevant
to speech recognition is represented by the time evolution of
the power spectrum, while the phase component is relatively unimpor-
tant.

Taking the power spectrum and discarding the phase component
is an attempt to transform the signal in such a way that unessential
information for recognition is neglected and what remains gives more
evidence to the linguistic features.

The frequency bands of the filters increase with frequency approximately following a logarithmic law. It has been found that the
ear also has a sensitivity in frequency which follows a logarithmic
law.

Table 2.1 contains, as an example, the frequency ranges of the
filter bank used in the French speech understanding system called
KEAL (Mercier, 1978).

The output of a filter bank is usually sampled every 10 ms
and each sample is coded with 8 bits. This corresponds to an information rate in bits per second of 800 times the number of filters.

The use of the filter bank models the auditory analysis performed by the basilar membrane of the human ear. The frequency components of the signal to be analyzed produce maxima of membrane displacements in some places. Intensities of membrane displacements
are integrated within small segments of the basilar membrane before
being further processed by the auditory system. This integration reduces the frequency resolution of the analysis performed by the ear.
These limitations in frequency resolutions can be modeled by a system
having a set of contiguous critical bands within which frequency

TABLE 2.1

channel	band
1	250 - 450 Hz
2	450 - 650 Hz
3	650 - 850 Hz
4	850 - 1050 Hz
5	1050 - 1300 Hz
6	1300 - 1600 Hz
7	1600 - 1900 Hz
8	1900 - 2200 Hz
9	2200 - 2500 Hz
10	2500 - 2800 Hz
11	2800 - 3100 Hz
12	3100 - 3400 Hz
13	3400 - 3800 Hz
14	3800 - 4200 Hz

variations are not perceived. One such system can be a filter bank
whose filter bands correspond to the critical bands of the ear.

More details on the use of filter banks for speech processing
can be found in the book by Flanagan (1972) and in a recent book
chapter by Ruske (1982).

The limitations of such systems lie in the difficulty of extracting articulatory cues such as formants, which are the time
evolutions of the major peaks of the power spectrum of speech. Other
limitations of such systems are due to their meager time resolution
that makes difficult the extraction of some acoustic cues of the
plosive sounds. Furthermore, the system accuracy is affected by the
time variation of the characteristic parameters of the analog components.

In spite of these limitations, filter banks have been successfully used in automatic recognizers of hundreds of words; such systems are actually available on the market (see Tsuruta, 1978, as an
example).

White and Neely (1976) have performed an experiment on isolated
word recognition, comparing several analysis methods. The experiment
showed that recognition based on the output of a filter bank was not
poorer than recognition based on the coefficient of an articulatory
model. Davis and Mermelstein (1980) have a different opinion. They
found that a representation of speech based on a series expansion of
the so called mel-scale *cepstrum* (see Subsection 2.2.5 for definition)
gives better recognition results on isolated words than the output of
a filter bank.

The problem of designing the best speech analysis system is
still open. Certainly, at the actual level of research, acoustic
cues related to articulatory parameters are very useful.

2.2.2 The fast Fourier transform

A digital computation of speech spectra can be obtained by
Discrete Fourier Transform.

For this purpose, the input signal is sampled at discrete points
in time and the sample values are represented digitally.

The sampling rate is usually constant and must be at least twice
the maximum frequency in the analog input. If frequency components
are present above half the sampling rate such components are folded
back onto lower frequencies by the sampling process. If $f(s)$ is the
sampling frequency and there is a component at frequency $f(s)/2+a$,
this component will give contributions, after sampling, at frequency
$f(s)/2-a$. This "aliasing" effect can be avoided using a low-pass
analog filter before sampling. Usually, the sampling frequency is
chosen according to the following conditions:

$$10 \text{ kHz} \leq f(s)' \leq 20 \text{ kHz}$$

The number of bits used to represent a signal value is related
to its dynamic range and affects the signal-to-noise ratio (quanti-
zation introduces noise). With linear representation, 11 bits are
sufficient for reconstructing speech with a quality comparable to
that of the telephone; this corresponds to a dynamic range of 60 dB;
16 bits correspond to a dynamic range of 90 dB. Sampling at 10 kHz

and quantizing with 11 bits corresponds to an information rate of
110,000 bit/sec which is much higher than the source rate; the
latter is of the order of 50 bits/sec. Elimination of such a redundancy is the task of a Recognition System.

The Fourier transform cannot be performed on a long interval of
the speech waveform, otherwise the spectrum obtained will exhibit an
average of the characteristics of the sounds contained in the interval. To analyze portions of the speech waveform corresponding to a
single sound, a running window must be introduced to select only
small intervals of the waveform for the analysis. There are different types of windows, introducing different types of noise into the
computed spectra; furthermore, the duration and the position of the
window may affect the features extracted from the speech spectra to
a remarkable extent.

For the sounds generated by the vibration of the vocal chords
exciting the vocal tract, which is shaped in such a way as to exhibit
a few noticeable resonances, a pitch-synchronous analysis is very
efficient. It consists in making the window the same length as the
period of the excitation signal and in positioning it on the portion
of signal following the glottal closing instants in such a way that
the effects of the excitation signal on the computed spectra are
minimized. This brings out the spectral features related to vocal
shapes characterizing the speech sounds.

The time evolution of the pitch during the enunciation of a sentence is called *prosody* and contains useful information about the
syntactic structure of the sentence. A detailed discussion on the
effect of windowing is provided by Harris (1978).

If $x(t)$ is the speech signal, $w(t)$ the window, the windowed
signal is given by:

$$s(t) = x(t) \, w(t) \qquad\qquad (2.2)$$

Let T be a sampling period and $\hat{s}(t)$ the sampled signal:

$$\hat{s}(t) = s(t) \sum_{n=-\infty}^{+\infty} \delta(t-nT) = \sum_{n=-\infty}^{+\infty} s(nT)\delta(t-nT) \qquad (2.3)$$

where $\delta(t)$ is the ideal sampling function.

The spectrum of $\hat{s}(t)$ can' be computed as follows:

$$S(k\Omega) = \sum_{n=0}^{N-1} s(nT)e^{-jnk\Omega T} \qquad\qquad (2.4)$$

where $\Omega = 2\pi/NT$ is the spectral resolution, NT the duration of the window, and k≤N is a non-negative integer. $S(k\Omega)$ can be computed by a fast algorithm known as the *Fast Fourier Transform* (FFT).

More details on Digital Fourier Transformation can be found in textbooks on digital signal processing (Gold and Rader, 1969, Markel and Gray, 1976, Oppenheim and Shafer, 1975). A collection of programs is available in a book edited by the Digital Signal Processing Committee IEEE-ASSP (1979).

The ripple introduced into the spectra by the window can make erroneous or impossible the detection of some acoustic cues. For this reason, appropriate methods have been introduced for performing a spectral smoothing on each spectral frame. One such method, named "*cepstral analysis*", consists in computing the logarithm of the magnitude of the spectrum and performing an FFT on this logarithm. The spectrum obtained is called "*cepstrum*" and its frequencies are called "*quefrencies*".

The effects of the excitation give contributions mainly at high quefrencies. These contributions can be eliminated by low-pass filtering of the cepstrum. The FFT of the filtered cepstrum is a smoothed version of the logarithmic spectral magnitude of the original signal. The method, described in detail in many books (see Oppenheim and Shafer, 1975, for example), is very time consuming even allowing that pitch, too, can be obtained by high-pass filtering the cepstrum itself. The peaks of the ripple in the original spectra have, in fact, a distance depending on the fundamental frequency.

2.2.3 Identification of the vocal tract parameters

A faster method for obtaining smoothed spectra consists in modeling the vocal tract by an all-pole transversal filter and computing short-time spectra from this model.

The parameters of the model are obtained by an algorithm performing a *Linear Predictive Coding* (LPC). This algorithm tries to predict a sample $x(nT)$ of the speech signal, given p previous samples:

$x(nT-T), x(nT-2T), \ldots, x(nT-pT)$

The predicted sample $x'(nT)$ is given by:

$$x'(nT) = \sum_{k=1}^{p} a_k \, x(nT-kT) \qquad (2.5)$$

For each sample, a prediction error is defined as follows:

$$e(nT) = x(nT) - x'(nT)$$

The prediction coefficients a_k are computed by minimizing the mean-square prediction error. If the signal is windowed, the minimization of the error can be extended only for the duration of the window. Three basic methods have been proposed for computing the coefficients of the model; they are named the direct, the autocorrelation, and the covariance method and are described and discussed by Makhoul (1975).

Given a series of samples $x(nT)$, $-\infty \leq n \leq +\infty$, its z-transform $X(z)$ (see Gold and Rader, 1969, for details) is defined as follows:

$$X(z) = \sum_{k=-\infty}^{+\infty} x(kT)z^{-k} \tag{2.6}$$

Taking the z-transform of the signal $x(nT)$ and the error $e(nT)$ one gets:

$$E(z) = X(z) \ (1-\sum_{k=1}^{p} a_k z^{-k}) = X(z)H(z) \tag{2.7}$$

with

$$H(z) = 1 - \sum_{k=1}^{p} a_k z^{-k} \tag{2.8}$$

$H(z)$ is the transfer function of a digital filter that receives the speech samples at its input and gives the samples of the prediction error at the output.

Equation (2.7) can be rewritten as follows:

$$X(z) = \frac{E(z)}{H(z)} \tag{2.9}$$

The prediction error can now be seen as the excitation of a filter having transfer function $1/H(z)$ that gives the speech signal at the output; the filter is a transversal one and is considered a possible model of the vocal tract. This model is rather good only for vowels and semivowels (see Flanagan, 1972, for a precise definition of these sounds) but it may also be useful for analyzing other sounds.

Many experiments have shown that the prediction error is, in many cases, very similar to the excitation signal of the vocal tract and can be approximated, for the voiced sounds, by a Kronecker delta.

This suggests that one consider an approximation $\hat{X}(z)$ of $X(z)$ given by:

$$\hat{X}(z) = \frac{A}{1 - \sum_{k=1}^{p} a_k z^{-k}} \qquad (2.10)$$

where A is the z-transform of a single input pulse.

An estimate of the smoothed spectrum can be computed by putting $z = e^{j\Omega T}$ as follows:

$$X(k\Omega) = \frac{A^2}{\left| 1 - \sum_{k=1}^{p} a_k e^{-jk\Omega T} \right|^2} \qquad (2.11)$$

Such a spectrum depends on the vocal tract configuration and on the poles of a pulse-shaping unit which transforms a pulse into the signal corresponding to a period of vocal chord vibration. For nasal and nasalized sounds, models with poles and zeros have proposed; the z-transform of their transfer function can be expressed as follows:

$$H(z) = \frac{1 + \sum_{k=1}^{p} b_k z^{-i}}{1 + \sum_{k=1}^{p} a_k z^{-1}} \qquad (2.12)$$

The coefficients b_i and a_i can be computed using a recursive algorithm based on Kalman filtering.

Using such models of the vocal tract, an information rate of thousands of bits per second can be obtained by preserving the possibility of reconstructing a speech message which is still understandable.

Although the reduction of information achieved by using linear prediction analysis is high with respect to the rate of the sampled waveform, a further reduction of about two orders of magnitude has to be performed to obtain the information rate of the speech source.

2.2.4 Extraction of articulatory parameters

Atal (1975) and others have shown how geometrical parameters of the vocal tract shape can be computed from the linear prediction coefficients.

Relevant advances on the estimation of the vocal tract characteristics have been achieved in the last two years. Interesting work has recently been done at the Electrotechnical Laboratory (ETL) of Tokyo by Nakajima (1976). It mostly concerns the identification of dynamic articulatory models by acoustic analysis. The model may contain both poles and zeros. The work by Shirai and Honda (1978) is also noteworthy.

An approach to pole-zero modeling of speech which avoids the problem of pitch synchronization is proposed by Kopec, Oppenheim and Tribolet (1977). The proposed analysis, resulting from combining homomorphic filtering with linear prediction, has been termed *homomorphic prediction*.

The idea is to first estimate nonparametrically a minimum-phase signal which has the same magnitude transform as the vocal tract impulse response, and then to apply a two-step algorithm (one step for estimating the poles and another for the zeros), to obtain a pole-zero model.

Steiglitz (1977) has proposed an interactive prefiltering method which simultaneously estimates poles and zeros; computer simulation has shown that this prefiltering method gives better spectral estimates than estimating the numerator of the transform by a single least-square fit.

Another approach for the simultaneous estimation of poles and zeros based on linear prediction analysis is proposed by Atal and Schroeder (1978). Their method is not recursive but requires three LPC analyses and other computations to get the coefficients of the numerator and the denominator of an estimation of the vocal tract transfer function.

A non-recursive method is also proposed by Yegnanarayama (1981) for the simultaneous estimation of the poles and zeros of a vocal tract model.

The properties of the normalized prediction error signal have been investigated by Rabiner, Atal and Sambur (1977) showing that the prediction error can vary significantly with the position of the analysis frame.

Interesting progress has also been achieved in the detection of

the glottal closing instants (De Mori et al., 1977) and in the identification of articulatory parameters such as tongue position and the
extent to which the lips are opened (Shirai and Honda, 1978).

An excellent review of pitch extraction algorithms and their
performances has been recently provided by Hess (1980).

2.2.5 On the use of spectral representation of speech

Using the methods presented above, a spoken sentence can be
visualized by a *spectrogram*.

A spectrogram is a three-dimensional representation that has
time, frequency and energy-density as dimensions. In practice, spectrograms are plotted on two dimensions with the energies represented
by gray levels.

Figure 2.2 shows the spectrogram of the sentence "tell me" obtained by an analog device. Figure 2.3b shows the short-time spectrum of the vowel /e/ of "tell" obtained by the same device. Figure
2.3a shows the time waveform of the same vowel. Figure 2.3c shows
the FFT computed with a Hamming window over two periods (18 ms) of
the signal (see Harris, 1978, for the definition of Hamming window).
Figure 2.3d shows the spectrum obtained after LPC analysis. Figure
2.3e shows the cepstrum (below) and the smoothed spectrum (above)
obtained after filtering the cepstrum.

Figure 2.4 shows the pitch-synchronous FFT spectrum (solid line)
and the LPC spectrum (dotted line) of the vowel /e/ in the word
Venezia (Venice). Both spectra have been computed after pre-emphasis. The smoothing effects of the LPC are evident. The peaks of
the LPC spectrum are the formants. Usually more spurious spectral
peaks appear when FFT is computed asynchronously.

The examples of Fig. 2.3 show that, at least for the vowels,
the spectrum computed with LPC exhibits easily detectable peaks at
the formant frequencies, corresponding to the three main resonances
of the vocal tract. Figure 2.5 shows the same type of spectra but
for the last vowel of "tell me" and seems to confirm the above considerations.

Unfortunately, LPC analysis does not perform so well on all the
consonants. For this reason, a combination of FFT and LPC has been
used for generating the speech patterns of the system described in
this book.

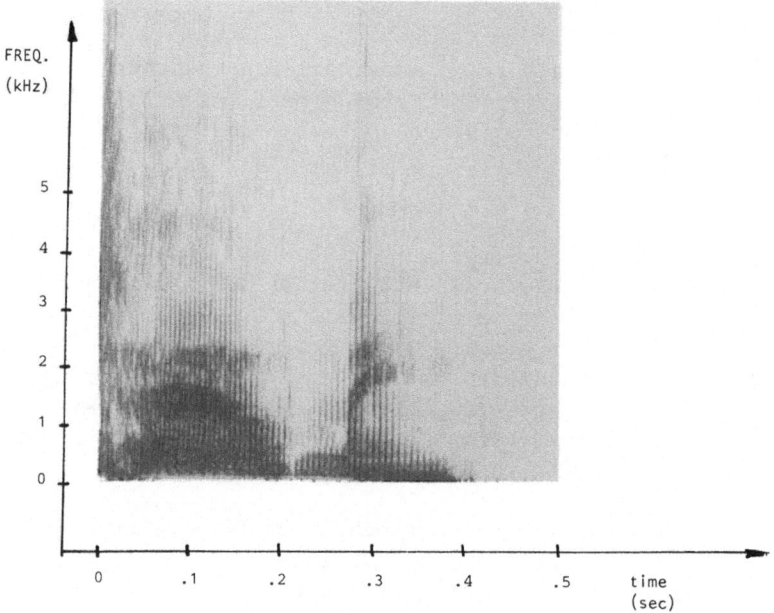

Fig. 2.2 Spectrogram of the sentence "tell me"

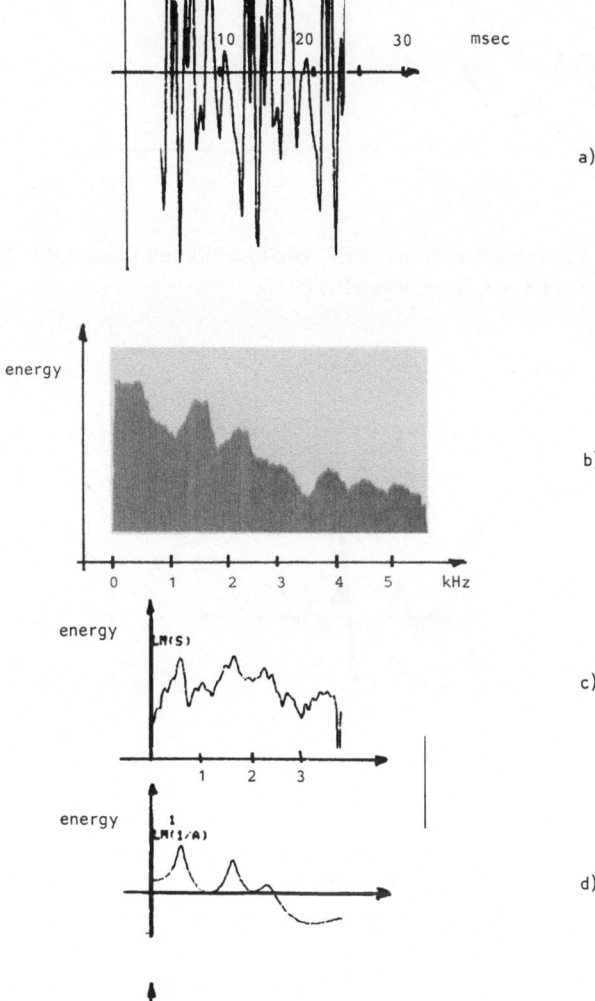

Fig. 2.3. a) waveform of the vowel /e/ in "tell me"
Short-term spectrum of the vowel /e/ obtained by:
b) analog techniques
c) synchronous FFT
d) LPC
e) cepstrum

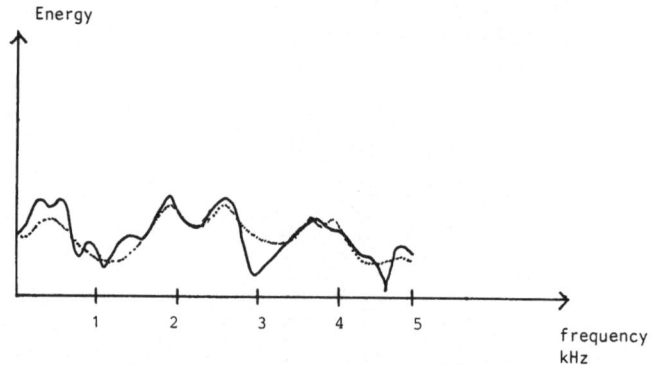

Fig. 2.4   Pitch-synchronous FFT (solid line) and LPC (dotted line)
           spectra of the vowel /e/

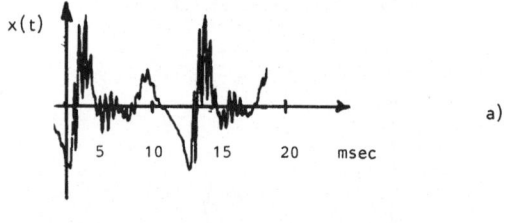

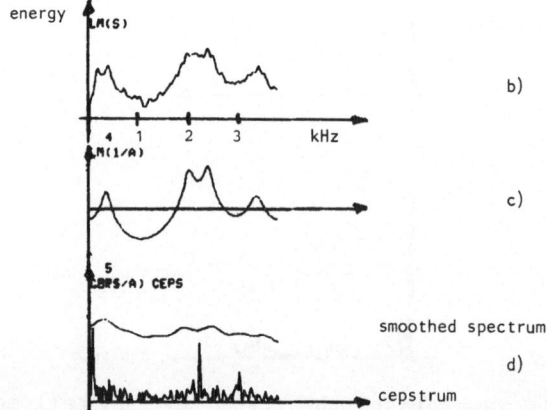

Fig. 2.5   a)   waveform of the vowel /i/ of "tell me"
           b)   FFT
           c)   LPC
           d)   cepstrum

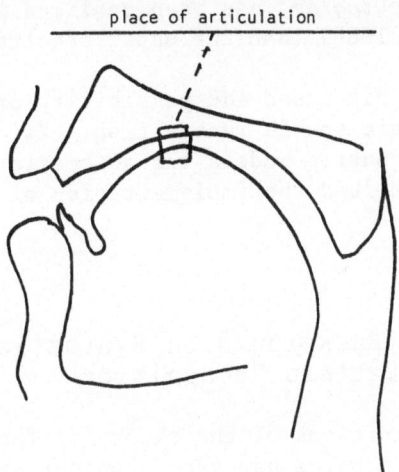

Fig. 2.6 Place of articulation

The frequencies of the peaks of the smoothed spectra are a good estimation of the formant frequencies. For the vowels and certain consonants, the second formant frequency is related to the place of maximum constriction of the vocal tract (*place of articulation*) and the first formant frequency depends on the tongue position (manner of articulation). An example of the place of articulation is shown in Fig. 2.6.

Time evolutions of the formant frequencies can be coded at a rate of hundreds of bits/sec. which approaches the order of magnitude of the information rate of the speech source.

An interesting result has recently been found by Davis and Mermelstein (1980). They have compared several methods of speech analysis, including LPC. They found that a mel-scaled cepstrum gives somewhat better results than an LPC representation. This representation is derived by first modifying the frequency-scale of the FFT spectra in accordance with the frequency selectivity of the human auditory system and then, after computing the cepstrum, smoothing in frequency by preserving the low-quefrency or inverse frequency components of the log-power spectrum.

It is of some interest to mention the results recently achieved in modeling acoustic perception, based on physiological experiments made on cats.

Modeling the discharge mechanism of neural fibers, a sort of spectrogram called "*neurogram*" has been realized with finer time-resolution (Delgutte, 1980) than the usual spectrogram.

Klatt (1979) has discussed the possibility of computing the neurograms for automatic speech recognition. Zwicker et al. (1979) have proposed psychoacoustic models for automatic speech recognition. Caelen (1979) has described the implementation of a numerical model of the ear.

2.3   Background on Syntactic Pattern Recognition

Because some definitions of the theory of formal languages will be used in this book, a brief background on phrase-structure grammars will be given in this section.

Definition 2.3.1

A phrase-structure grammar G is a four-tuple:

$$G(V_T, V_N, P, \sigma)$$

where:

V_T is a finite set of terminal symbols representing primitive forms; this set is called *terminal alphabet*;

V_N is a finite set of nonterminal symbols, representing syntactic categories or families of structures.

$\sigma \epsilon V_N$ is the start symbol of the class of patterns described by the language generated by a certain number of applications of the rules of G;

P is a finite set of rewriting rules or productions denoted by:

$$\gamma \rightarrow \delta$$

where γ and δ are strings composed of symbols of

$$V_T \cup V_N$$

and γ contains at least one symbol in V_N.

The following notation is often used:

1) if y is a string of symbols y^n is y written n times.

2) $|y|$ is the number of symbols (the length) in y.

3) V_T^* is the set of all strings of symbols in V_T, including ε, the string containing no symbols;

$$V_T^+ = V_T^* - \{\varepsilon\}$$

4) $\alpha \overset{G}{\Rightarrow} \beta$ means that a string α generates the string β in the following way:

$$\alpha = x\gamma y, \quad \beta = x\delta y$$

and

$$\gamma \rightarrow \delta$$

is a rule in P of G. Usually the label G is omitted.

5) $\alpha \overset{*}{\underset{G}{\Rightarrow}} \beta$ means that there exists a sequence of strings

$$x_1, x_2, \ldots, x_m$$

such that

$$\alpha \overset{G}{\Rightarrow} x_1, x_i \overset{G}{\Rightarrow} x_{i+1} \quad \text{for} \quad i = 1, 2, \ldots, m-1, \ x_m \Rightarrow \beta.$$

The sequence of strings $\alpha, x_1, \ldots, x_m, \beta$ is called a derivation of β from α.

Definition 2.3.2

The language L(G) generated by the grammar G is:

$$L(G) = \{y / y \in V_T^* \text{ and } \alpha \overset{*}{\underset{G}{\Rightarrow}} y\} \tag{2.13}$$

Definition 2.3.3

If the rules of the grammar having the form: $\gamma \rightarrow \delta$ have the property:

$$|\gamma| \leq |\delta|$$

the grammar is called "context-sensitive" or type-1 grammar.

If γ is a single nonterminal symbol for all the rules in G, the
grammar is called "context-free" or type-2 grammar. If for all the
rules, γ is a single nonterminal and δ = bA or δ = b, b$\in V_T$, A$\in V_N$, then
the grammar is a type-3 or "regular grammar".

The language generated by a grammar takes the same name (context-free, etc.) as the grammar.

Example 2.3.1

Consider a context-free grammar G with:

$$V_N = \{\sigma, \alpha, \beta\}, V_T = \{a, b\} \quad \text{and}$$

P :

1) $\sigma \rightarrow b\alpha$

2) $\sigma \rightarrow a\beta$

3) $\alpha \rightarrow a$

4) $\alpha \rightarrow b\alpha\alpha$

5) $\alpha \rightarrow a\sigma$

6) $\beta \rightarrow b\sigma$

7) $\beta \rightarrow a\beta\beta$

8) $\beta \rightarrow b$

For convenience, the rules can also be written as follows:

$\sigma \rightarrow b\alpha$ / $a\beta$

$\alpha \rightarrow a/b\alpha\alpha$ /aσ

$\beta \rightarrow b\sigma$ /a$\beta\beta$ /b

Generation of sentences includes (the number of a derivation indicates the production used):

(1)   (5)   (2)   (8)
$\sigma \Rightarrow b\alpha \Rightarrow ba\sigma \Rightarrow baa\beta \Rightarrow baab$.

An alternative method for describing any derivation in a context-free grammar is the use of a "*parse*" tree which can be constructed according to the following procedure:

1) Every node in the tree has a label which is a symbol in $V_N \cup V_T$.

2) The root has label σ.

3) If a node has at least one descendant other than itself and has the label α then $\alpha \epsilon V_N$.

4) If nodes with labels $a_1, a_2, \ldots, a_n$ are the direct descendants of a node with label α, then:

$$\alpha \rightarrow a_1 a_2 \ldots a_n$$

must be a production in P.

Parsing is the operation by which a parse-tree of a given string of terminal symbols is derived. *Parser* is an algorithm for parsing.

Parsing the string $s = baah$ under the control of the grammar of the example 2.3.1 gives the parse tree shown in Fig. 2.7.

2.4   Acoustic Cue Extraction
from Speech Patterns

A lot of inessential information is still present in the speech spectrogram even after the phase component of each short-term spectrum is discarded.

An early attempt to reduce pattern dimensionality with little loss in essential information consisted in transforming each frame of the speech spectrogram (for example any interval of a duration of 10 ms) into a few "principal components".

An interesting application of this principle has been proposed by Pols (1971). In such an approach, a word is described by a sequence of points in a multidimensional space. The 17 coordinate values of the points represent the signals obtained with a system of 17 band-pass filters. The output signals are sampled every 15 ms, after logarithmic amplification and envelope peak detection.

Using principal component analysis, a three-dimensional subspace is derived which explains most of the variance. The unknown word is thus represented as a sequence of points in three dimensions and is compared with reference traces.

In order to account for the different durations of the same word pronounced at different times, a sort of time normalization has to be performed. In the above-mentioned approach, linear time warping is performed to make the durations of the unknown pattern and the reference templates equal. Still keeping to the parametric approach, a non-linear time normalization can be performed concurrently with pattern matching by the use of dynamic programming. Such a method will be discussed later in this chapter.

Another way of performing time normalization consists in representing an utterance by a parametric graph with time as parameter. An application of this approach to pseudo-syllable segments has been proposed by De Mori et al. (1976).

Instead of extracting the principal components, formants, which are the time evolutions of the resonant frequencies of the vocal tract, can be extracted for the sonorant portions of the acoustic pattern. Reduced acoustic patterns are thus obtained. They are composed of the formants together with the time evolution of the fundamental frequency; these patterns contain enough information for reconstructing understandable speech with an information rate of a few hundred bits/sec.

Although a detailed description of an algorithm for formant tracking will be given in Chapter 5, some perceptual figures on human discrimination ability which may help in designing algorithms for obtaining reduced acoustic patterns are given in the following. Zagoruiko (1976) has used such data for extracting cues from digital spectrograms.

Differential intensity discrimination for wideband noise may be as low as 0.5 dB.

Discrimination of fundamental frequency changes is around 0.5% of the fundamental frequency. The minimum increment that is discernible in a formant frequency is around 5% of the formant frequency.

Intensity discrimination for a formant is about 3 dB. Formant bandwidth discrimination is rather poor (about 40% at -3 dB).

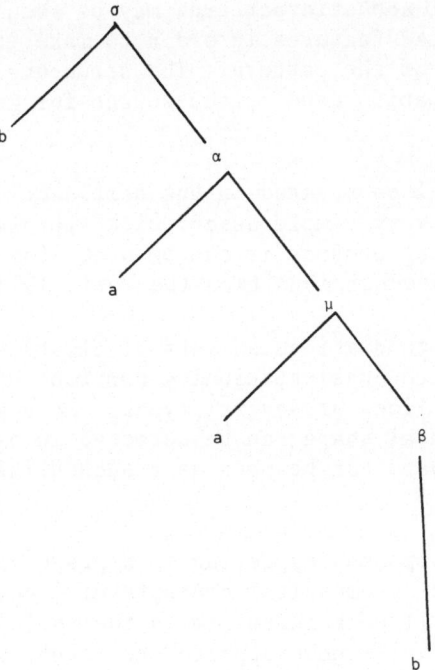

Fig. 2.7 Example of a parse tree

Formant frequencies are related to the place and manner of articulation of the vowels and the sonorant consonant.

    The sonorant sounds are those whose articulation can be de-
scribed in terms of resonances of the vocal tract. For these sounds
the time evolutions of the formant frequencies contain low rate information which is sufficient for reconstructing understandable
utterances. The information of these evolutions can be further reduced by coding in different ways the stationary and the transient
portions of a formant pattern.

    Figure 2.8a shows the formant pattern of the syllabic segment
/can/ extracted from continuous speech. Figure 2.8b shows the parametric graph obtained by plotting the second formant F2 versus the
first formant F1. Each point of the graph corresponds to a 10 ms
interval. Different pronunciation speeds mainly alter the density
of the points in the clusters of the graph, with minor influence on
the shape of the graph. Planar graphs, generated by the time evolution of acoustic parameters, can be described by a language. The
description of these graphs can be combined with descriptions of
other significant acoustic cues, giving a description of the acoustic
pattern.

Descriptions of acoustic patterns may be seen as an attempt to organize the extracted features in order to make it possible to infer the structure of the pattern. The structure may be seen as the essential information used by the source for generating the pattern.

This aspect will be treated in the next section and in Chapter 3. Nevertheless, a very simple descriptive approach will be introduced in the following because it can be used, for example, in a passive model for speech recognition (De Mori, 1973).

It is possible to distinguish sets of clustered points in the graph corresponding to quasi-stationary portions of the acoustic waveform, joined by lines of several types. More generally, some components of elemental shape can be detected in each graph, and the pronunciation of a word can be seen as a sequential generation of these components.

An elemental component appearing in a graph can be associated with one of the usual geometrical concepts of line, arc, circle, etc. These forms will be referred to in the following as "primitives", recalling a well-known approach to structural picture description.

In addition, the elemental components singled out in a graph will be called "atoms". An atom is a materialization of a primitive (for example a line) with certain numerical attributes (e.g. the length, the number of points, the slope, the starting point, etc.).

An algorithm operating on the coordinates of the points in the graphs looks for the atoms belonging to each primitive and gives a description for each atom. Such a description, which will be called "local aspect description", consists of a suitable symbol, specifying the nature of the atom, followed by a set of numerical attributes.

For limited tasks, the following minimal set of primitives can be used.

2.4.1 Silence interval between two sounds in a word

Atoms of this primitive, which will be referred to in the following as SL, are looked for by checking to see if there are some sequences of time frames of 10 ms whose energy is below a threshold. These points are labeled with a description of the type:

$$N; \; P_1; \; P_2 \qquad\qquad\qquad\qquad\qquad (2.14)$$

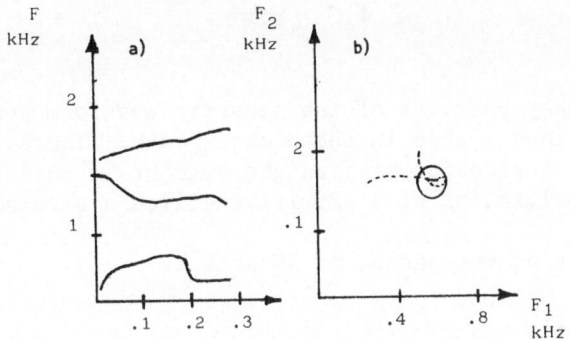

Fig. 2.8 a) Formants of the syllabic segment /can/
Fig. 2.8 b) Graphical representation of the second formant F 2
versus the first formant F1 of the syllabic segment
/can/. A quasi-stationary portion of the pattern corresponds to points inside a circle.

N is the symbol used for denoting SL, p_1 is a positive number indicating when the SL starts, and P_2 is also a positive number indicating
the SL duration.

2.4.2 Quasi-stationary portions
of the acoustic pattern

A relatively dense set of points in the planar graph is generated by a quasi-stationary configuration of the vocal tract.

When points representing successive segments of the speech
signal lie within a surface of relatively small and fixed dimensions
and their number is higher than an established threshold, a primitive, called "stable zone" (SZ), is assumed to be present in the
graph. Each detected SZ is assigned the following description:

$$Z; s_1; s_2; s_3; s_4 \qquad (2.15)$$

Z is the symbol used for denoting SZ, and s_1, s_2 have the same meaning as p_1 and p_2 for SL, and s_3 and s_4 are the coordinates of the
center of gravity of the SZ.

2.4.3. Lines

Nonstationary portions of the acoustic waveform generally lead to lines of various shapes in the planar graph. These lines are approximated by a succession of straight segments. Each segment is considered a realization of a primitive called a straight line LN.

The describing message of an LN atom is

$$L;l_1;l_2;l_3;l_4 \qquad\qquad (2.16)$$

L is for a straight line, l_1 and l_2 are numbers with the same meaning as p_1 and p_2 for SL, l_3 is the length of the line, and l_4 is the line slope according to the code of Fig. 2.9, which was introduced by Freeman (1961).

Using the description method just introduced, the graph of Fig. 2.8 b is described as follows:

L;0;70;200;g

Z;80 180;600;1800

L;190;240;300; e

This means that the syllable /can/ is represented by an acoustic pattern described by a line with slope-code g, length 200 and 70

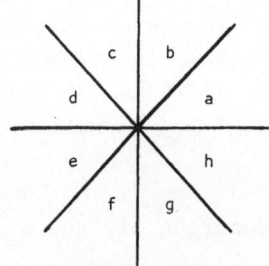

Fig. 2.9 Slope-code for line description

ms duration, followed by a stationary pattern whose center of gravity coordinates, called formant loci, are F1=600 Hz and F2=1800 Hz, followed by another line of slope e. Time measurements are in milliseconds, frequency measurements are in Hertz.

Distances between a pattern description and a prototype can be defined and the description can be classified in the class of the minimum distance prototype (see Levenshtein, 1966, for a detailed discussion of this item).

Other types of features can be extracted and used for recognition. Better feature extraction can be achieved if performed under the control of a Knowledge Source which makes specific requests only when some hypotheses have to be verified. This aspect of model-driven feature extraction will be largely discussed in other chapters of this book.

2.5 Classification of Speech Patterns

2.5.1 A brief history of automatic recognition of isolated words

Attempts at classifying speech patterns were made just after the development of the first sound spectrograph in 1947, and their history is almost as long as the history of computers.

In 1952, Davis et al. developed a laboratory model for the recognition of the ten digits in English. The system was capable of recognizing isolated words spoken by a single talker. The recognition method was based on a cross-correlation, performed by analog devices, between an unknown pattern and a set of templates. The patterns were based on the short-time energies at the output of a low-pass and a high-pass filter with a cutoff frequency of 900 Hz. Recognition rates of about 98% were obtained for a single talker and of about 60% for new talkers.

Dudley and Balashek (1958) improved the recognition performances by segmenting the speech signal into phonetic units, such as vowels and consonants.

Later, Forgie and Forgie (1959) attained good results in the recognition of vowels.

Wiren and Stubbs (1956) made the first attempt to extract distinctive features from segments of the speech patterns. Along this line, Hughes and Hemdal (1965) proposed a model for extracting phonetic features from continuous speech and Itahashi et al. (1973) developed a speech recognizer based on this concept.

Damman (1965) conceived a trainable classifier, that is a system with learning capabilities.

In 1960 Denes and Mathews made a speech recognizer on a digital computer by introducing a time-normalization algorithm for improving the matching between patterns and introduced a primitive form of linguistic constraints on expected pronunciations, based on digram probabilities.

In the seventies, systems based on dynamic programming techniques were developed for handling, with a single algorithm, the problems of the time normalization and matching (Velichiko and Zagoruiko, 1970). At the same time, a big research program on continuous speech (Newell et al., 1973) was started.

Systems based on dynamic programming were produced in the last ten years with very high recognition rates using speaker-dependent templates for vocabularies of up to 1000 words (Kohonen et al., 1980).

The concept of dynamic programming and its application to automatic speech recognition will be introduced now in a rather general form. Most of the applications differ only in the way this concept is applied.

2.5.2 The dynamic programming approach

The acoustic pattern   to be classified, represented by a set of features that are parameters or descriptions, is indicated as

$P_x \epsilon P$, where P is the universe of possible patterns. P_x is a collection of subpatterns, adjacent in time:

$$P_x = P_x(t_0,t_1) \; P_x (t_1,t_2) \; \ldots \; P_x (t_{N_x-1},t_{N_x})$$

where $P_x (t_{j-1},t_j)$ is the frame of the acoustic pattern corresponding to the time interval (t_{j-1},t_j).

Let us denote $P_x (t_{j-1},t_j)$ by $p(x,j)$. In the general case

$p(x,j)$ is a triplet:

$$p(x,j) = <D,(t_{j-1},t_j),q> \qquad (2.18)$$

where D is a pattern description which can be a vector of numbers
(for example, the energies of signals at the output of a set of filters during the interval (t_{j-1},t_j), a single symbol describing the
pattern in a time frame (for example a stable-zone), a vector of
binary variables, each one of which expresses the presence or the
absence of a feature, a set of symbols representing qualitative
interpretations of the time frame ordered by degrees of evidence,
etc. The "sampling period" of subpatterns can be constant or variable. In the former case, i.e., synchronous sampling, $t=N_x b$ is the
pattern duration; b is the duration of each frame; q is a measure of
evidence or a vector of evidence measures.

The classifier has to assign P_x to a class C_i $(1 \le i \le N_c)$. For
each C_i there are one or more prototype patterns (archetypes)
A_{ik} $(1 \le k \le N_i)$ where N_i is the number of the archetypes of the class
C_i; N_C is the number of classes.

A metric is introduced for computing the distance:

$$D (P_x, A_{ik}) \quad (1 \le i \le N_C, 1 \le k \le N_i).$$

Let $D(x,i,k)$ be such a distance. In order to compute $D(x,i,k)$, the
distance d_{jl} $(p(x,j),a(i,k,l))$ between an element $p(x,j)$ of P_x and
an element $a(i,k,l)$ of A_{ik} has to be defined. In this way, P_x is
seen as a possible distortion of A_{ik} due to the transmission of P_x
through the information channel shown in Fig. 1.1.

This distance depends on the similarity between the two subpatterns expressing how much $p(x,j)$ may be seen as a distortion of
$a(i,k,l)$. Eventually the distance may be obtained by weighting the
similarity measure by a term depending on the duration of $p(x,j)$ and
its evidence.

Similarities between subpatterns can be measured in different
ways, depending on the nature of the representations. For example,
if $p(x,j)$ is a vector of numbers, P_x is a matrix and the distance
$D(P_x,A_{ik})$ can be a weighted sum of the partial distances:

$$d (p(x,j), a(i,k,l))$$

where $a(i,k,l)$ is the l-th vector of A_{ik}. The measure of a partial distance depends on the metric of the space in which the vectors are represented. If $p(x,j)$ and $a(i,k,l)$ are a symbol of strings of symbols, the definitions of the partial distances and the global distance are more complex.

In order to compute the total distance between an unknown pattern P_x and a prototype A_{ik} it may be assumed that P_x is a distorted version of A_{ik} for which some elements $p(x,j)$ of P_x are distorted elements of A_{ik}, some other elements of P_x do not have any corresponding element in A_{ik} and are due to errors of insertion and some elements of A_{ik} do not have any corresponding element in P_x because they have been deleted during the transmission through the information channel.

Fig. 2.10 shows the types of errors involved in a possible generation of the pattern $P = P_x$ from the archetype $A = A_{ik}$. The path

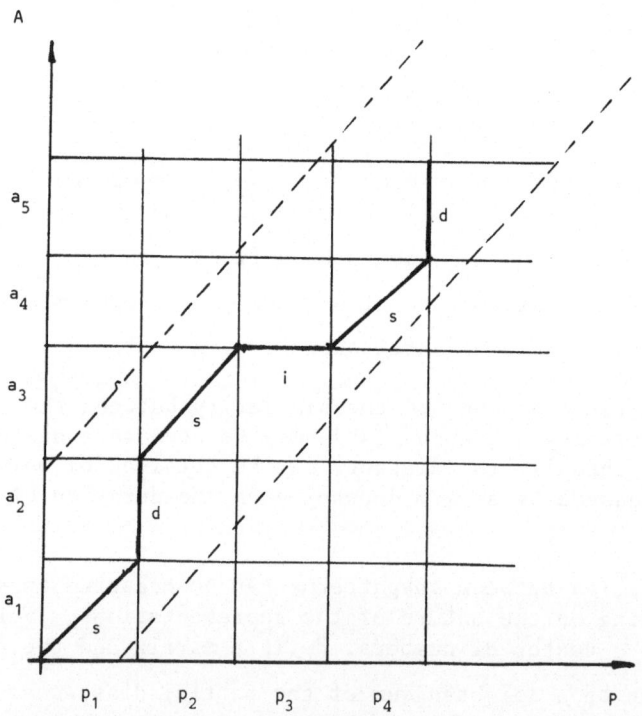

Fig. 2.10   Types of errors involved in a possible generation of the
            pattern P from the archetype A.

in Fig. 2.10 represents a sequence in which the operation of insertion (i), deletion (d) and substitution (s) are represented by their
first character. In the example in Fig. 2.10, p_1 is a distortion of
a_1,a_2 has been deleted, p_2 is a distorted version of a_3, p_4 has been
deleted. Following each path π_m (A,P) of the type shown in Fig.
2.10, a distance D_m (A,P) between A and P can be computed; the distance D(A,P) between the unknown pattern P and the prototype A is
defined as follows:

$$D\ (A,P)\ =\ \min_{m\ \in\ M}\ D_m\ (A,P) \tag{2.19}$$

where M is the set of the indices of all possible paths π_m.

Given a point (i,j) corresponding to the comparison of a(i,k,l)
and p(x,j), the global distance from the first comparison up to
(i,j) can be computed by evaluating a resulting distance g(i,j) as
follows:

$$g\ (i,j)\ =\ \min\ \begin{bmatrix} g(i-1,j)\ +\ d_{ij} \\ g(i-1,j-1)\ +\ d_{ij} \\ g(i,j-1)\ +\ d_{ij} \end{bmatrix} \tag{2.20}$$

where $d_{ij}\ =\ d(p(x,j),a(i,k,l))$ and $g(1,1)\ =\ d(1,1)$. Levenshtein
(1966) has proposed as distance measure the minimum total weight of
symbol insertions, deletions and/or substitutions required to convert
one string into another. By assuming the same weight d for insertions and deletions, the weighted Levenshtein distance (WLD) between
two patterns can be computed by using the equation:

$$g(i,j)\ =\ \min\ \begin{bmatrix} g(i-1,j)+\ d \\ g(i-1,j-1)\ +\ d_{ij} \\ g(i,j-1)\ +\ d \end{bmatrix} \tag{2.21}$$

Ackroyd (1980) notices that the WLD is highly sensitive to the
durations of different utterances of the same word. This happens
when strings of the type: A=abc and P=aabbcc are compared. He suggests eliminating the penalty of words of differing lengths by the
subtraction of $|N(x)\ -\ N(i,k)|d$ from the WLD to give a "modified
WLD".

$N(x)$ is the length of the unknown pattern description; $N(i,k)$ is the length of the (i,k)-th prototype.

Kohonen et al. (1980), have proposed another type of distance, called the "*feature distance*", defined as the difference between the maximum number of symbols in the strings to be compared minus the number of matching symbols. The authors propose a fast lexical access method based on hash functions. Haltsonen et al. (1978), describe a method for simultaneous segmentation and phoneme labeling. This method is used for generating the symbols based on which the feature distance is computed.

The calculation of $D(A,P)$ requires the knowledge of all $D_m(A,P)$, $m \in M$. This calculation can be simplified by considering a subset $WD(A,B)$ of all the possible paths $\pi_m(A,P)$; this subset can be characterized by a window in the plane with the constraint that all the acceptable paths must be inside the window (see Fig. 2.10).

Each pair of elements to be compared $(p(x,j),a(i,k,l))$ can be represented by a point $x(f) = (j(f), (f))$. The sequence of points $x(f)$, $f \in (1,F(m))$ corresponding to a path $\pi_m(A,P)$ defines a function called the *warping function*. $F(m)$ is the number of points of the m-th path. This function approximately realizes a mapping from the time axis of P_x onto A_{ik}.

Several constraints can be imposed on the warping function, particularly on its slope.

A good discussion of these problems can be found in a paper by Sakoe and Chiba (1978).

An interesting application of dynamic programming to word recognition has been proposed by Itakura (1975). He uses the linear prediction coefficients as basic features and the linear prediction error as a measure of distance.

Christiansen and Rushforth (1977) have extended this technique to the detection and location of key words in continuous speech.

Significant contributions to the application of dynamic programming to the recognition of speech patterns have also been provided by Velichiko and Zagoruiko (1970), White and Neely (1976).

Modern designs for pattern recognition systems use a linguistic approach in which P_x is a string and the class C_i is represented by a language $L(G_i)$ generated by a grammar $G_i = G(C_i)$. In this case, P_x is recognized as belonging to the class C_i if it can be generated

by $G(C_i)$. A more flexible approach is one in which P_x is not recognized by any grammar because of noise and distortions. Classification can be performed by computing the distance between P_x and $L(G_i)$
for every class and assigning P_x to the class for which the distance
is minimum. Examples of such approaches are given by Fu and Fung
(1975), Kashyap (1979), Fu and Lu (1977). A grammar for describing
the significant acoustic cues in a pattern is proposed in Appendix C.

    Significant efforts have been made recently (see, for example,
Furui, 1980) for obtaining speaker-independent templates.

    The system, developed at the Musashino Electrical Communication
Laboratory in Japan (Furui, 1980) for the recognition of isolated
words, segments the input speech into phoneme segments represented
by parameters which are function of the Linear Prediction coefficients. For each phoneme, a vector of *Log Area Ratios* (LAR) is estimated from the samples of the training set.

    The Log Area Ratios are coefficients related to the shape of
the model of the vocal tract and are obtained by "arctanh" transformation of the PARtial CORrelation coefficients (PARCOR) obtained
during the calculation of the LPC parameters. Using a 67 word dictionary of Japanese airports and a training set of 12 utterances of
a subset of 25 words for each speaker in a set of 30 male talkers,
Furui has obtained a recognition rate of 98.2 percent.

2.6 Automatic Recognition
of Continuous Speech

    An attempt to recognize continuous speech may follow the
so-called two-level matching principle. It consists in matching
sequences of word hypotheses by accumulating the distances obtained
for each word match. Such a method, used by Nippon Electric Company
(Sakoe, 1979) is very time expensive because it requires matching
the input signal with the chain of templates of every possible sequence of acceptable words. For this reason it may be applied only
to limited tasks with small-size vocabularies and short word sequences.

    For more complex tasks, task-dependent knowledge has to be used
for prefiltering the sequences of words to which algorithms for
searching the best interpretation have to be applied.

    In the early systems developed under the ARPA project, especi-
ally SPEECHLIS of BBN (Woods and Makhoul, 1974) and HEARSAY I of
CMU (Erman, 1974a) there was a tendency to represent separately

the knowledge about the grammatical structure of acceptable sentences
(syntax) and the meaning and interrelationships of words within a
sentence (semantics).

For a detailed presentation of the objectives of the ARPA project, see Newell et al. (1973).

Interesting papers describing knowledge representations are
those by Bates (1975), Woods (1975), Nash-Webber (1975) and Hendrix
(1975).

Recognition under the control of a grammar is more difficult in
SUS than in formal language processing for the following reasons:

- the input to be parsed is not a string but a matrix of
  hypotheses;

- the input matrix generates a set of strings and it is possible
  that none or several of these strings are recognized by the
  grammar and the recognition process must perform error corrections and selection of alternatives;

- it is difficult to represent naturally spoken languages in
  a formal way.

A new tendency has appeared recently with the attempt to inte-
grate syntax and semantics into a single knowledge source. Generally
knowledge is represented by different types of regular or context-
free grammars augmented in various ways to account for the context
sensitivities due to the structure of natural language and to the
constraints imposed by semantic restrictions.

Parsing under the control of the grammars differs from one project to another. It depends on the form of the grammar (regular,
context-free, dependency, pragmatic, etc.), on the way the hypotheses
are evaluated based on the units considered as terminal symbols
(words, phonemes, etc.'), and on the purpose of parsing. Among the
different parsing strategies and purposes, Levinson (1977) proposes
a syntax-directed recognition by transforming an input string of word
hypotheses into a grammatically correct sentence for which the total
Itakura distance (1975) between the hypotheses and the sentence is
minimum. More precisely, let G be a regular grammar generating a
finite language $L(G)$; let $W \epsilon L(G)$ be a sentence encoded in a waveform,
$x(t)$; let $w_1 w_2 \ldots w_i \ldots w_n$ be the words composing W. Each word w_i
$(i=1,2,\ldots,n)$ belongs to the terminal alphabet V_T of G.

The acoustic recognizer processes $x(t)$, giving a sequence $\hat{W}$ of
word hypotheses:

$$\tilde{W} = \tilde{w}_1, \tilde{w}_2, \ldots, \tilde{w}_i \ldots, \tilde{w}_n;$$

where $\tilde{w} \epsilon V_T$ $(i=1,2,\ldots,n)$ but generally $\tilde{W} \notin L(G)$.

The syntax analyzer produces a sentence $\hat{W} = \hat{w}_1 \hat{w}_2 \ldots \hat{w}_i \ldots \hat{w}_n$, $\hat{W} \epsilon L(G)$, satisfying the following equation:

$$D(\hat{W}) = \min_i \left\{ \sum_{i=1}^{n} d_{ij} \right\} \tag{2.22}$$

where the d_{ij} is the Itakura distance between $\tilde{w}_i$ and all the words w_j that may appear in the i-th position of sentences in the $L(G)$. The Itakura distance $d(a,\hat{a})$ between the sets of linear prediction coefficients (LPC)a and $\hat{a}$ is defined as follows:

$$d(a,\hat{a}) = \log \left[\frac{a \, R \, a^t}{\hat{a} \, R \, \hat{a}^t} \right] \tag{2.23}$$

where R is the autocorrelation matrix of the speech frame with LPC set $\hat{a}$.

A more diffuse use of syntax is for word prediction. Word predictions are performed under the control of a regular grammar in
Backus Normal Form (BNF) in the HEARSAY I system. The knowledge in
such a system is represented in a procedural form and is shared between three modules, the recognition controller, the syntax, and the
acoustic module. The recognition proceeds in a left-to-right fashion
at the word level. A partially completed path is built to contain
all the words recognized at a certain moment and an indication on
where the unrecognized portion starts.

A best-first search is used. The best path is defined as the
partial path with the highest provisional rating.

The provisional rating for a partial path is obtained by aver-
aging the ratings of all words currently recognized in the path.

The unrecognized portion of the sentence is called <FILLER>.
The syntax module is called to decide what words may be predicted,
given the two words immediately left of <FILLER>. Scoring is done
heuristically.

One of the main advantages of the best-first strategy is that

it allows the system to follow one syntactic path to completion very
quickly. Disadvantages are due to backtracking that leads to large
standard deviation of the time needed for recognition. This is due
to the fact that the presence of many heuristics can cause complete
rejection of a correct path and that the system does not remember
what it has learned during the search of one path in order to apply
it to another path. Improvements on the HEARSAY I have been achieved
by two other systems developed at Carnegie-Mellon University (namely
the HARPY System, based on a no-backtrack search algorithm, and
HEARSAY II). Details on the HARPY system are described by Lowerre
and Reddy in Lea (1980).

Word prediction under the control of a grammar is performed
by the speech recognition system developed by Saito and Kohda (1976)at
Musashino Laboratory of N.T.T. near Tokyo. Here phrase prediction
is made heuristically by searching a pragmatic and a syntax list of
structure alternatives. A hypothesis is evaluated by transforming
a phoneme lattice produced by an acoustic front-end into a word hypothesis. As a result of the transformation, an orthographic representation of words is obtained by using phonological rules. When a
phonological rule is applied, a negative penalty corresponding to
the rule is imposed. The score of the hypothesis is the sum of the
applied penalties. Hypotheses are built up sequentially and a hypothesis is rejected when the score falls below a given threshold.

Word prediction is also performed by the speech regonition system of FORTRAN programs designed by Sekiguchi et al. (1977) at
Yamanashi University.

Word prediction consists in the selection of groups such as
digits, letters, functions, etc., based on the previously recognized
word and the characteristic features of the statement extracted from
the previously recognized word string. Predictions are based on
deterministic transition networks; recognition errors are recovered
by simple error correcting routines. Parsing follows a best-first
strategy with backtracking when a hypothesis score falls below a
threshold.

Takeya and Kawaguchi (1976) of Kyushu University use a dependency grammar to describe the syntactic structures of input utterances.
Such grammars generate context-free languages and the authors consider them particularly convenient for a top-down parsing of sequences of word hypotheses. The reasons for this are the following:

- dependency grammars directly represent dependency relations
  between words considered as nonterminal symbols;

- the nonterminal symbols correspond to word classes;

- every expansion of a nonterminal symbol produces a word

corresponding to it.

Sakai and Nakagawa (1977) of Kyoto University designed and implemented the LITHAN (LIsten THink ANswer) System in which the application of rules of a context-free grammar is restricted by additional information regarding pragmatics, phonological rules, and context dependencies.

In the prediction process, a word class is regarded as a terminal symbol. Among the nonterminal symbols, there are special symbols that are rewritten in a context-sensitive manner. The context sensitivities are represented by predicates that represent sentence structures.

Another interesting approach is due to J.K. Baker (1975) and Jelinek (1976). They view grammar as a probabilistic function of a Markov process and represent it as a finite-state network with transition probabilities. A similar approach has been proposed by De Mori et al. (1975a). They present a methodology for describing spectral patterns by an artificial language and for learning probabalistic or fuzzy relations between such abstract descriptions and syllabic, lexical, and syntactic units.

The most important component of the recognition system proposed by Jelinek (1976) is a linguistic decoder whose task is to determine what sentence W has most probably caused the observed sequence U of phones. Phones are the recognition units hypothesized by an acoustic front-end.

The linguistic decoders' task is to find the path through the language model for which

$$P(W)P(U/W) \qquad\qquad (2.24)$$

is maximized. This is accomplished by a search process utilizing a stack and an evaluation function L that assigns values to partial paths. A language model provides estimates of $P(W)$ for all word strings W. In order to compute the probabilities $P(U/W)$, it is necessary to have a model of the performance of the speaker and of a memoryless acoustic processor that relates, in a probabilistic way, spoken and recognized phones.

The most recent and interesting application of the stochastic model is to a minicorpus of a task named Laser Patents. Such minicorpus has a vocabulary of more than 1000 words.

The language is described by a model called WPP3-gram. WPP is for word-part-of-speech pairs. In decoding, language-model predictions are based on the conditional probability of a WPP, given the two WPP's immediately preceding it, as found by the decoder.

A short-phrase recognition system for carrying on a dialogue between an operator and a computer has been developed at the Institute for Problems of Information Transmission of the Soviet Academy of Sciences (Knipper and Makhonine, in Fain, 1977).

A sequential decoding algorithm due to Zigangirov has been applied to the recognition of continuous speech (1974).

Interesting projects based on context-free grammars have been developed in France. The most important of them are the KEAL project (Mercier, 1978, Quinton, 1980), the MYRTILLE project developed by Haton (1976) at the University of Nancy, the ESOPE (Mariani, 1980) of LIMSI and the projects of the Universities of Toulouse (Perrenou and Tep, 1979) and Grenoble (Groc and Tuffelli, 1980).

At the University of Turin, Coppo and Saitta (1976) have developed a system based on a semantic graph whose nodes represent concepts.

To each node is associated a set of relations with other nodes, as well as syntactic and lexical constraints for concept realizations.

An algorithm is given for generating a graph of possible interpretations of the input utterance. Phrase hypotheses are scored by fuzzy relations and the graph is pruned on the basis of the degree of worthiness reached by the competing hypotheses.

Among the most complex systems, the HEARSAY II replaced HEARSAY I and was designed on the basis of the hypothesize-and-test paradigm using cooperating independent knowledge sources, communicating through a global data structure where all types of hypotheses about the input sentence are written. The knowledge sources are self-activating, asynchronous, parallel processes; the contextual and support connections are explicitly specified; the system structure is modular and suitable for execution on a parallel processing system.

Another interesting system developed at CMU is named HARPY. It is a combination of the best features of HEARSAY I and the DRAGON system, conceived on the basis of the Markov process. The knowledge source of the HARPY system is a finite state network whose input symbols are phonetic transcriptions of the input utterance. The most interesting feature of the HARPY system is the use of the so-called focus model of search. It is a graph-searching technique in which all, except a beam of near-miss alternatives around the best path, are pruned from the search tree at each segmental decision point, thus containing the exponential growth without requiring back-tracking.

The Stanford Speech Understanding System (Walker, 1978) has been

designed with two major objectives: integration, the process of
forming a unified system out of the collection of components, and
control of the dynamic direction of the overall activity of the
system during the processing of an input utterance. System integration gives a central role to the input-language definition which
is based on augmented phrase-structure (APS) rules. A rule consists
of a phrase-structure declaration, which specifies the possible constituents of a phrase, and an augmentation which is a procedure for
computing attributes and factors. Attribute statements determine
the properties of particular phrases constructed by the rule; they
may compute values for attributes that relate to syntax, semantics,
or discourse.

Factor statements make acceptability judgements on phrases and
are scored by integers referenced symbolically by linguistic variables. Factors are used for traditional syntactic tests, such as
agreement for person and number.

A large contribution to the representation of syntactic and
semantic knowledge for speech understanding has been provided by
Woods et al. at Bolt, Beranek and Newman (BBN) Inc., Cambridge,
Mass. (1976).

The evolution of their project has led to a complete integration of syntax and semantics by the design of a parser capable
of producing a syntactic tree of a sentence and its semantic interpretation.

The HWIM (Hear What I Mean) system follows SPEECHLIS, the first
system developed at BBN. In the old system, the parser was driven
by a modified Augmented Transition Network (ATN) grammar which permitted parsing to start anywhere, not necessarily left-to-right.

A definition and an application of ATN grammars will be given
in Chapter 4 because they are also used in the project described in
this book.

The semantic component uses case frame tokens to check for the
consistency of completed syntactic constituents and the current
semantic hypotheses. In HWIM, the Syntax component embodies the
syntactic, semantic, and pragmatic knowledge sources. The parser is
built around a pragmatic grammar which accepts only those utterances
that are grammatical, meaningful, and appropriate to the pragmatic
circumstance, given the previous conversation.

Understanding speech under the control of a grammar is not
purely a parsing action, because the input message to be interpreted
is often corrupted and ambiguous. If the grammar of admissible sentences is simple and can be represented by a finite state automaton,
a network can be conceived to represent all the possible interpret-

ations of the speech messages. The problem is that of pruning the
network in order to obtain, hopefully, a single path that best
matches the set of features extracted from the speech waveform.

Syntactic decoding and focus model of search solve the problem
for networks with low complexity. In these cases, the redundancy
of the speech message compensates for the imprecision of knowledge
and feature extraction. For more complex and less restricted protocols, other approaches in the area of Artificial Intelligence have
to be used. They require complex structures for knowledge representation, complex rules of inference and sophisticated strategies
for assigning priorities to the processes that compete for the interpretation of the verbal message.

These problems have been tackled only in recent years but their
full solution is far from being achieved. A contribution towards a
better understanding of them will be given in the following chapters.

2.7   References

Detailed descriptions of signal processing techniques are given
in the books by Gold and Rader (1969), Markel and Gray (1976),
Oppenheim and Shafer (1975), Rabiner and Shafer (1978), Shafer (1979),
Kunt (1980).

The basic source for syntactic pattern recognition is the book
by Fu (1974).

Other important books in the field are those by Tou and Gonzales
(1974), Gonzales and Thomason (1978), Pavlidis (1977), Duda and Hart
(1973). The book by Aho and Ullman (1972) contains an excellent
presentation of the mathematical theory of formal languages.

Other books by Fu (1975, 1977, 1982) describe applications of
syntactic pattern recognition.

The paper by Fu and Booth (1975) contains an extended review of
grammatical inference, i.e., the problem of learning the rules of a
grammar, given a finite set of data which have to be described by the
grammar and, possibly, a finite set of data that should not be described by the grammar.

Further to the general references given at the end of Chapter 1,
the following journals deal specifically with the problem of speech
production and perception:

Language, Language and Speech, Journal of Phonetics, Journal of
Experimental Psychology, Journal of Speech and Hearing Research,
Cognitive Psychology.

Table 2.2 gives a picture of the groups which are or have recently
been involved in speech recognition and understanding research.

TABLE 2.2

Groups actively involved in research in speech recognition

Country	Research group
Australia	- University of Canberra
Belgium	- University of Brussels
Canada	- University of Alberta - Calgary - Concordia University - Montreal - Bell Northern - Quebec
China	- Academy of Sciences - Peking
France	- CNET - Lannion - University of Nancy - ENST - Paris - LIMSI - Orsay - University of Toulouse - University of Grenoble - CEA - Saclay - University of Marseille
Finland	- University of Helsinki
Germany	- University of Munich, Erlangen
India	- Indian Statistical Institute - Calcutta
Italy	- University of Turin - CENS
Japan	- Universities of Tokyo, Kyoto, Sendai, Yamanashi, Kyushu - Electrotechnical Laboratory - Nippon El. Co. - Musashino Labs.
Poland	- Institutes of Cybernetics of the Academy of Sciences (Warsaw and Poznan)
Sweden	- Royal Institute of Technology
UK	- imperial College, University of London, Aston, Joint Speech Research Unit (JSRU)
USA	- BBN-Cambridge Mass. - Carnegie Mellon University-Pittsburgh - Stanford Research International - IBM Research Center-Yorktown - MIT - Lincoln Labs. - SDC - Sperry Univac - Xerox Corp.

- Bell Laboratories-Murray Hill
- University of Washington
- Purdue University
- Interstate Electronics Corp.-Anaheim, Ca.
- Threshold Technology Inc.-Delran, N.J.
- Dialog(Verbex) Syst. Newton, Mass.
- Centigram Corp.-Sunnyvale, Ca.
- Heuristics Inc.-Sunnyvale, Ca.

USSR
- IPIT ANSSSR - Moscow
- Computing Center ANSSSR - Moscow
- Institute of Mathematics, University of Novosibirsk
- University of Lvov
- Politechnic of Kiev
- Politecnic of Kaunas
- Institute of Cybernetics - Tbilisi

3. ON THE USE OF SYNTACTIC PATTERN RECOGNITION
AND FUZZY SET THEORY

3.1 Introduction and Motivations

The scheme introduced in Section 4 of Chapter 1 represents an active model for understanding speech.

Most of the problems involved in the design of the various components of a system based on such a model will be treated in detail in this and in the following chapters. The various components of the system contain structural and procedural knowledge and may be described in different mathematical frameworks.

Nevertheless, two types of mathematical background are of general interest for the model proposed in this book. They are the mathematical theory of languages, which is widely used here for representing the structural knowledge, and the algebra of ill-defined sets, which is believed to be the proper tool for representing the imprecision of the knowledge incorporated in the model and the vagueness with which the hypotheses are often generated. Furthermore, it allows one to introduce an effective type of Approximate Reasoning, within whose framework the procedural knowledge can be conceived.

Some of the engineering approaches to Automatic Speech Recognition prefer to use a mathematical background which is mainly based on the manipulation of numbers instead of symbols and descriptions.

The motivations for this are that such numerical techniques as statistical classification, dynamic programming, etc., have been

proven useful for limited-task speech recognition systems even if
they have very little relation with such disciplines as phonetics or
psycholinguistics.

These techniques consider the decoding system as a black box
for which input-output relations have to be learned in order to
optimize the system performances. Such an integrated approach has
the advantage of being well established from the mathematical point
of view and of allowing an automatic learning.

Disadvantages are that the "black box" does not distinguish
between speaker-dependent and speaker-independent features and
learning requires a great amount of experimental data to obtain
relations which nevertheless do not guarantee good performances on
protocols different from those used for their inference.

Furthermore, a purely numerical approach is a "brute-force"
one; it requires a large amount of unessential computation, which
becomes prohibitive when the tasks of the recognition system go beyond the bounds of limited vocabularies of isolated words or well
established protocols of connected speech. Furthermore, such a
"brute-force" approach offers very little help to understand the
reasons of a recognition error, allowing very often only one possibility for error recovery: the addition of new prototypes into the
system memory or an updating of the error statistics.

The purpose of this book is to describe and discuss a
rule-based model for speech understanding. The task is mainly basic
research, with the hope of helping to gain a deeper insight in the
investigation of such a complex and fascinating phenomenon as speech
understanding.

The main task of the model is the interpretation of speech
patterns, using all the available knowledge gained in many related
fields. These patterns are obtained by numerical transformations of
the speech signal. These transformations, performed by the acoustic
controller, using the knowledge of signal processing techniques,
model the operations performed by the ear.

The interpretation of speech patterns is mostly a complex set
of non-numerical operations consisting in the generation and verification of hypotheses about the linguistic message carried by the
speech signal. This message is non-numerical in nature. The speech
pattern is usually a three-dimensional representation of
energy-frequency-time obtained by short-term spectral analysis of a
spoken sentence.

Fig. 3.1 shows the spectogram of the Italian sentence: "vorrei
cancellare una prenotazione" (I would like to cancel a reservation).
Time is represented along the horizontal axis, frequency along the

Fig. 3.1 Spectrogram of the sentence "vorrei cancellare una prenotazione" (I would like to cancel a reservation)

vertical axis, and the energy is represented by gray levels. The
short-term spectra have been obtained by Fast Fourier Transformation.
A first glance at the picture reveals that, even if a speech sentence
is generated by a discrete sequence of stimuli, the acoustic pattern
is not a sequence of steady-state configurations. Rather, it dis-
plays continuously varying picture elements. Furthermore, there is
no immediate evidence of the beginning and the end of each word in
the sentence.

A visual inspection of the pattern made by an expert spectogram
reader would reveal complex interactions among adjacent sound units
(phonemes) and also across word boundaries.

It is impractical to store prototypes for every sentence; thus
some segmentation has to be performed on the sentence in order to
reduce the dimensions of the interpretation problem.

A simple approach to segmentation could be to split the entire
sentence into segments of equal duration.

Such an approach is not the best, because we speak at different
speeds. Very short segments have little significance if their context is ignored because their spectra may be strongly affected by
coarticulation. Large segments of fixed duration may have a large
and unpredictable variety of contents. Their interpretation may
become practically impossible if it is not known a priori whether
their content is a phoneme, a syllable or a word.

Because of these difficulties, an asynchronous segmentation is
more suitable. Segments may partially overlap each other and should
allow one to generate interpretation hypotheses based only on data
which are *inside* the segment. These segments have been called
Pseudo-Syllable-Segments (PSS). Before starting the interpretation
of a PSS, their bounds have to be established.

In the fourth chapter, the segmentation problem will be discussed in great detail. A key point underlying such an approach is
discussed now because it is useful for introducing the mathematical
framework of this design.

Segmentation of continuous speech into pseudo-syllabic segments
can be performed after some general phonetic features have been
detected in the acoustic pattern. The detection of these features
must be made by algorithms which do not require any knowledge about
the context. These general phonetic features will be indicated in
the following as "context-independent-features". Some of these
features can be introduced informally by an inspection of the
spectogram represented in Fig. 3.1.

Some portions of the spectogram exhibit, in the 0 to 3 kHz

band, lines of high gray level; these lines represent the time evolutions of the main resonances of the vocal tract and correspond to
a phonetic feature which will be called "sonorant". All the vowels,
the nasal consonants, the semivowels and the glides have the feature
"sonorant". Other portions of the spectogram exhibit silences or
"buzz-bar" traces at very low frequencies, frication or aspiration
noise traces, for which the areas of high gray-levels are not con-
centrated into lines but are diffused over large frequency intervals.
All these acoustic cues are correlates of the phonetic feature
"nonsonorant" which belongs to the fricative, affricate and plosive
consonants. The book by Flanagan (1972) contains the articulatory
description of each speech sound (phoneme) of American English.
Similar descriptions for the phonemes of other languages can be
found in books of phonetics. The distinction between sonorant and
nonsonorant phonemes is not specific of a particular language;
rather it is a general feature for characterizing the speech sounds
in every language. Using information on the time evolution of
broad-band energies which can be extracted from spectograms, some
other context-independent features useful for segmentation can be
extracted. One of these features is "vocalic"; it belongs only to
vowels and diphthongs. Another feature is "nonsonorant-interrupted-
lax" which belongs only to the plosive consonants /b/,/d/,/g/. When
these sounds are pronounced in vowel-consonant-vowel (VCV) utterances,
the consonant has coarticulation effects on both vowels. The same
extension of the coarticulation effects is often observed when sonorant consonants are pronounced in (VCV) utterances.

    The detection of the "context-independent features" is a sort
of "precategorical classification" of the elements of the speech signal. This detection will be indicated by this name in the sequel.

    A "categorical classification" will refine the precategorical
one by applying algorithms which may be context-dependent.

    The context-independent features may be used for setting con-
text constraints.

    The acoustic cues of a spectogram are mainly qualitative and,
even when they are obtained from precise quantitative measurements,
their extraction requires the application of non-numerical (syntactic) methods. This implies that a description of the speech pattern
has to precede its interpretation.

    Following the scheme of Fig. 1.5, the general controller
(strategy KS) schedules, at the beginning, the execution of a group
of "bottom-up" processes having the purpose of preparing stimuli for
the higher-level knowledge sources.

    Fig. 3.2a shows the components used by the system when it per-
forms a task that will be called "signal preprocessing". A process

which may be considered to be composed by a set of partially concurrent sub-processes (whose details are not relevant to this introduction) is started by a request from the general controller to the
acoustic KS controller.

Signal preprocessing contains:

1)   sampling and quantizing;

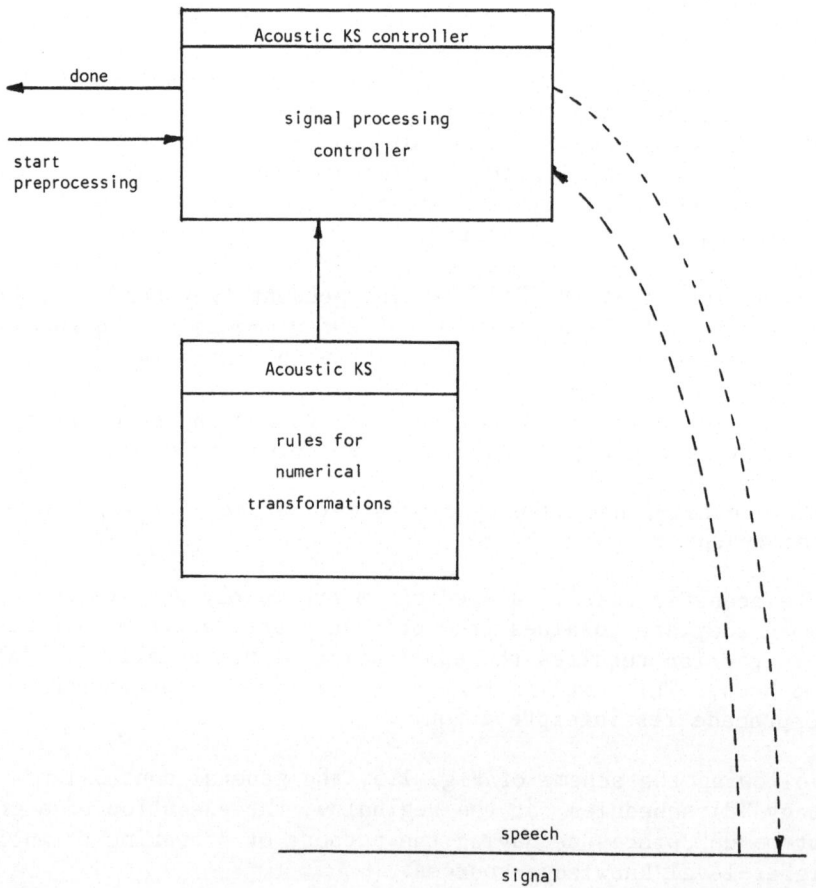

Fig. 3.2a   Knowledge source involved in speech preprocessing

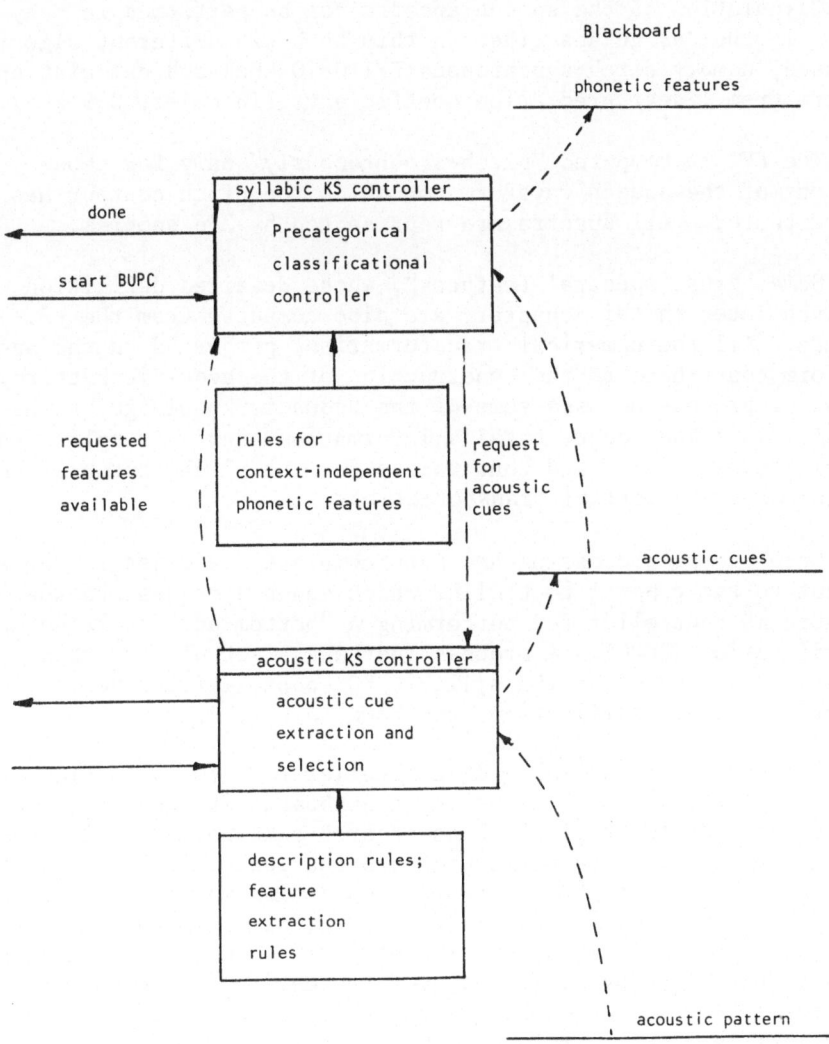

Fig. 3.2b   Knowledge sources involved in the generation of phonetic
            hypotheses

2)   detection of the start and end points of a sentence;

3)   evaluation of the pitch contour, when it exists;

4)   calculation of speech spectra.

Calculation of the speech spectra can be performed in many ways. In the design described in this book two different algorithms are used, namely pitch-synchronous FFT(0-10 kHz) and calculation of spectra from linear prediction coefficients (16 poles, 0-5 kHz).

The FFT is computed "pitch-synchronously" only for those portions of the speech waveform for which the pitch contour has been detected. All spectra are represented by 256 samples.

Some "gross spectral features", whose detailed definition will be given later in this chapter, are also computed from the FFT spectra. All the numerical transformations performed on the speech waveform contribute to the construction of the acoustic pattern. Signal preprocessing uses some of the acoustic knowledge of the system. This knowledge contains information such as sampling rate, signal windows, rules and constraints for the pitch contour, coefficients for the numerical transformations.

Once the preprocessing has been completed, a message: "done" is sent to the general controller which sends a request to the syllabic KS controller for performing a "bottom-up" precategorical classification (BUPC). A process (perhaps a set of concurrent processes) is scheduled by the syllabic KS controller for performing precategorical classification, as shown in Fig. 3.2b.

Every time the syllabic KS controller requires acoustic information that it does not find on the blackboard, it sends a request to the acoustic KS controller, which creates a KS instantiation process. When the process has been executed and the requested data have been written into the blackboard, a synchronization message is sent to the syllabic KS controller which, in turn, will send a synchronization message to the KS strategy controller when context-independent features have been hypothesized over the entire speech signal. The structural knowledge used by the syllabic KS controller for these operations is a set of rules establishing relations between phonetic features and acoustic cues, while the knowledge used by the acoustic KS controller is a set of rules for describing shapes of the speech pattern or for extracting acoustic cues.

Due to the context-independence of the features involved in precategorical classification, many KS instantiation processes may be executed concurrently. The problem of their synchronization is not treated here.

Pattern description requires a pattern grammar to be known and a syntactic procedure to be used to analyse a pattern and to make hypotheses about its structure. The proper background for this is the mathematical theory of languages.

Hypothesizing the structure of a pattern is a difficult task because any vagueness may affect the features extracted from the acoustic data and there is a degree of imprecision in the relations between acoustic cues and their phonetic or phonemic interpretations.

For example, a sonorant intervocalic consonant may be characterized by certain shapes in the gross spectral features, such as a marked dip in the time evolution of the signal energy, and in the energy in a frequency band from 3 to 5 kHz; furthermore, for such consonants, the low-frequency energy (roughly below 1 kHz) is much higher than the high-frequency energy (about 5 kHz). Given a specific acoustic pattern of an intervocalic consonant with its adjacent vowels, a speech understanding system trying to interpret the pattern may consider the possibility that the consonant is sonorant.

This possibility can be evaluated numerically, following the theory of possibility proposed by Zadeh (1978). To evaluate this possibility, the vagueness inherent in the terms "marked dip" and "much higher energy" has to be numerically represented by "degrees of compatibility" between these statements and the acoustic pattern. This can be done by applying Zadeh's theory of possibility.

To evaluate an interpretation, it has to be established how to combine the degrees of compatibility (or membership) of a marked dip in signal energy, in the 3-5 kHz energy, and the degree of compatibility with which the low-frequency energy is much higher than the high-frequency energy, in order to obtain the possibility of the interpretation: "the consonant is sonorant".

A method based on fuzzy restrictions is proposed for evaluating the evidence of acoustic cues (such as "marked dip") from the description of acoustic patterns.

A method based on fuzzy relations is also proposed for relating acoustic cues with phonetic and phonemic interpretations. Finally, the composition of restrictions and relations to compute the possibility of a hypothesis (a phonemic interpretation of a speech pattern) will be illustrated.

The concepts of fuzzy restrictions and fuzzy relations are due to Zadeh (1973,1975) and will be briefly recalled later on.

The algorithms used for evaluating the possibility of a hypothesis will be referred to as fuzzy algorithms.

The approach proposed in this book is an algebraic one, attempting to formalize the intuitive logic used by a phonetician.

The advantages of such an approach lie in its flexibility for modeling ill-defined non-numerical information.

Each relation and each restriction can be established to represent as closely as possible the knowledge attained in research on phonetics and to optimize the recognition performances; the emphasis is on evidence evaluation and composition rather than on classification.

Fuzzy algorithms are used whenever a KS is invoked to generate a hypothesis. The hypothesis is a syntactic category H related by rules to items previously hypothesized and belonging to lower levels of interpretation. Each rule used is associated with a fuzzy relation and each lower level item used by the rule has been previously associated with a possibility of being a correct interpretation of a speech pattern $\underline{p}$. To use H as an interpretation of $\underline{p}$ as well, it is necessary to evaluate the possibility that H represents $\underline{p}$ correctly. This possibility is computed by composing the fuzzy relation between H and its constituents with the possibility of the constituents themselves.

Details on the composition of fuzzy relations and restrictions will be given in a following section.

The possibilities of the hypotheses are used for scheduling the activation of the KSs by the system that controls the process of understanding.

Possibilities are evidence measurements related to similarities rather than to frequencies of occurrence. Decision making is based on probability distribution over evidence values because its task is that of keeping small the probability of losing a hypothesis which is true.

Before the details of the algorithms for generating hypotheses at various levels are presented, a simplified scheme for hypothesis formation is introduced. The scheme is shown in Fig. 3.3.

After the speech waveform has been preprocessed, using the methods described in Chapter 2, a precategorical classification of the signal into vocalic, sonorant and nonsonorant segments is performed, using algorithms whose details will be given later in this chapter.

After precategorical classification, pseudo-syllabic segments are hypothesized on the signal, with a parser described in Chapter 4.

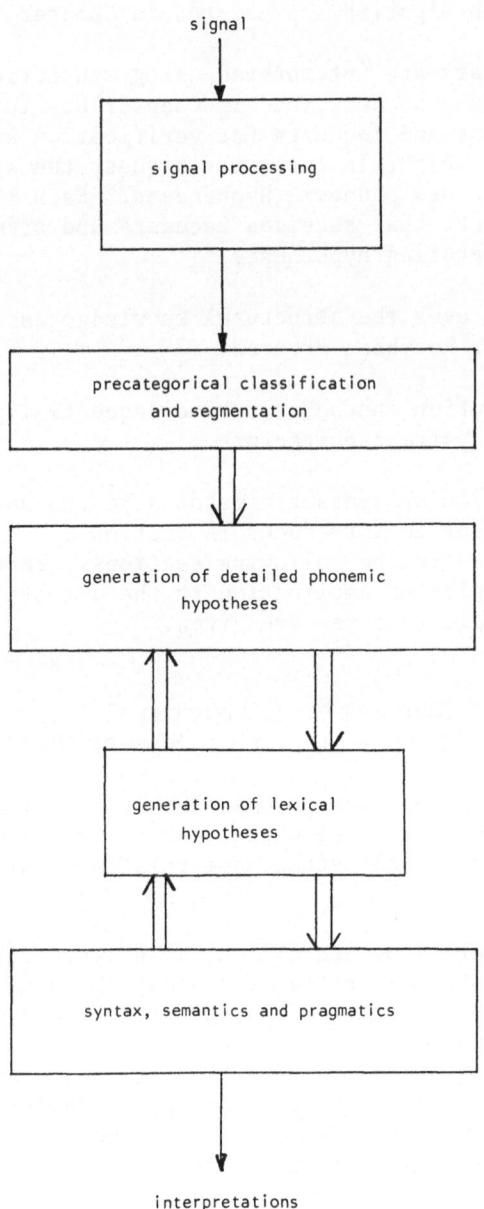

Fig 3.3 A simplified scheme for hypothesis information

Sequential or parallel algorithms are then used for generating detailed phonemic hypotheses, as described in Chapters 5 and 6.

Based on phonemic and phonetic hypotheses, lexical hypotheses are generated with algorithms presented in Chapter 7.

Word hypotheses are interpreted using syntactic, semantic and pragmatic knowledge, as described in Chapter 8. Interpretation can involve predictions and requests for verification advanced to the lexical component which, in turn, may request the syllabic knowledge source to generate new phonemic hypotheses. Each KS controller can be seen as an expert that receives requests and stimuli and creates processes for generating hypotheses.

Each process uses the structural knowledge associated with the expert by which it has been created.

Process execution can be performed sequentially or in parallel depending on the system architecture.

The application of syntactic methods to the analysis of continuous speech patterns is introduced in Section 3.3. Fuzzy set theory is briefly reviewed in the following sections. Furthermore, some introductory examples of application to the detection of context-independent phonetic features are given.

3.2 The Syntactic (Structural) Approach to the Interpretation of Speech Patterns

The structural approach proposed by De Mori (1973) to the automatic recognition of isolated words, can be extended to continuous speech by considering that structural relations exist between words and speech patterns.

Structural relations are also used in synthesis for generating the speech signal from a written text and can be used in recognition for controlling the generation of interpretation hypotheses.

Although an extended and valuable literature exists that describes the relations between the concepts of a message and its orthographic representation, very little has been achieved so far in the attempt of relating concepts to speech.

In contrast with the use of a structural approach is another approach, based on a direct matching between phoneme prototypes and spectral patterns.

Generation of phonemic hypotheses based on pattern matching is very unreliable because all the possible coarticulation effects cannot be taken into account. Furthermore, even when generation of hypotheses in spectogram analysis is based on a single spectrum (see for example Stevens and Blumstein, 1978), the detection of the significant spectrum and its interpretation are performed using non-numerical rules.

In order to describe and recognize coarticulation instances, an approach is now introduced which is more flexible and powerful than a parametric matching of patterns.

For the sake of simplicity, the structural relations between a word and its acoustic signal will be considered first.

Let us take, as an example, the word "prenotazione" (reservation) of the spectogram shown in the previous section.

Because this word starts with the tense plosive /p/ and is the last word of the sentence, its acoustic pattern, except for the pitch, is little affected by the previous words in the sentence. Following the classical syntactic pattern recognition approach (Fu, 1974), an acoustic realization of the word can be represented by a "relational graph" whose relations describe coarticulation effects and temporal precedences. The word components are partially overlapped syllabic segments and each segment is composed of phonemes, each of which, in turn, is composed of phonetic features.

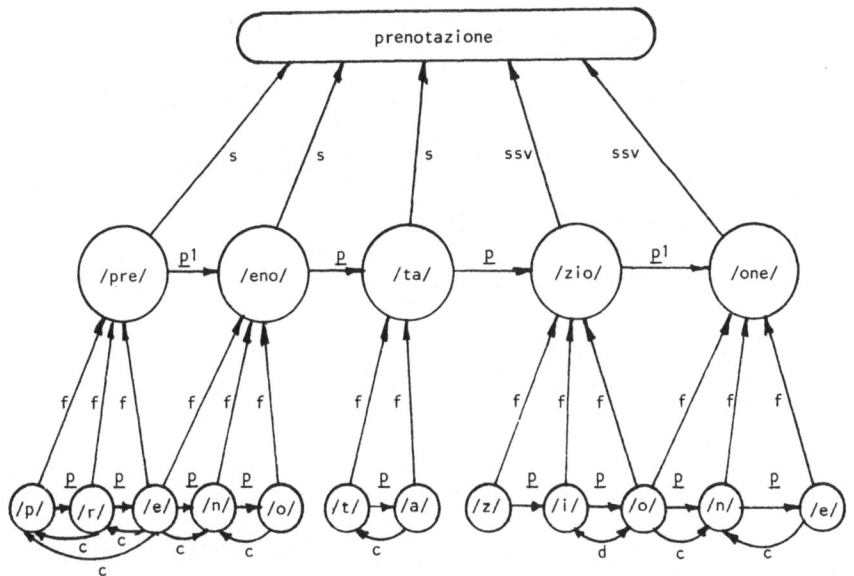

Fig. 3.4 Relational graph for the word "prenotazione" (reservation)

These relations are shown in Fig. 3.4. Notice that pseudo-syllabic segments may overlap but they are defined in such a way that coarticulation relations may appear only inside the segments themselves.

The segmentation of a phrase or a word, written in an ortho-graphic form, into pseudo-syllabic segments can be performed by a process (a computer program or a hand-made procedure) controlled by a segmentation grammar. This grammar has to be augmented to incor-porate relations between phonemes, phonetic and acoustic features if it has to control a segmentation process acting on an acoustic pattern. The rules of a segmentation grammar for the Italian lan-guage will be introduced in Chapter 4.

The relation labels appearing in Fig. 3.4 are defined by the following Table 3.1.

TABLE 3.1

relation symbol	meaning
s	is a syllabic nucleus of
ssv	is a syllabic segment containing the stressed vowel of
f	is a phoneme of
p	precedes
p1	precedes with the last vowel in common
d	makes a dipthong with
c	affects by coarticulation
sv	stressed vowel

Other relation symbols may appear in the graph; one of them may indicate, for example, the syllable containing a vowel with a secondary accent.

These details are omitted here for the sake of simplicity.

Now, for the sake of brevity, only the structure of the syllable /ta/ will be described in detail. The corresponding graph is shown in Fig. 3.5. Each phoneme of the syllable is represented in terms of phonetic and phonemic features; for example t(VC) represents the phonemic feature t in a prevocalic position followed by a central vowel, VC (/a/ is a central vowel because central is the *place of articulation*, i.e., the position of the main constriction along the vocal tract).

The phonetic and phonemic features are then expressed in terms of acoustic cues extracted from acoustic patterns, obtained by numerical transformations of the speech signal. In Fig. 3.5 each phoneme is related to subphonemic features. The label for these is "sp". For example, /t/ has the following subphonemic features: <nonsonorant-consonant>, <interrupted>, <tense>, <features of a dental consonant followed by a central vowel>, presented symbolically as t(VC).

The subphonemic features are related to acoustic cues AFij $(i = 1,2; 1 \le j \le 4)$.

The details of these relations are not given for the moment. Some of these relations are "context-independent" because they relate a phonetic feature of a phoneme with a set of acoustic cues which enter only into the relations of the same phoneme (e.g. <interrupted>). Other relations are "context-dependent". For example, t(VC) has a relation with a proper set of acoustic cues but also with the phonetic feature <central vowel>. This contextual relation is labelled "ctx" in Fig. 3.5.

A subset of the context-independent features can be used as "preconditions" to be met before applying rules which are "context-dependent".

The graph shown in Fig. 3.5 can be seen as the description of a complex problem which can be decomposed into sub-problems whose individual solutions are simpler. Each sub-problem can be described by a set of rules relating phonetic features and acoustic cues. These rules contain acoustic cues for phonetic features based on the knowledge attained by phoneticians.

For a better understanding of problem decomposition, a definition of a Problem Reduction Representation (PRR), a formalism used in Artificial Intelligence, is given in the following.

Definition 3.2.1

A *problem-reduction representation* (PRR) is a 5-tuple <P,r,t,u,B>, where:

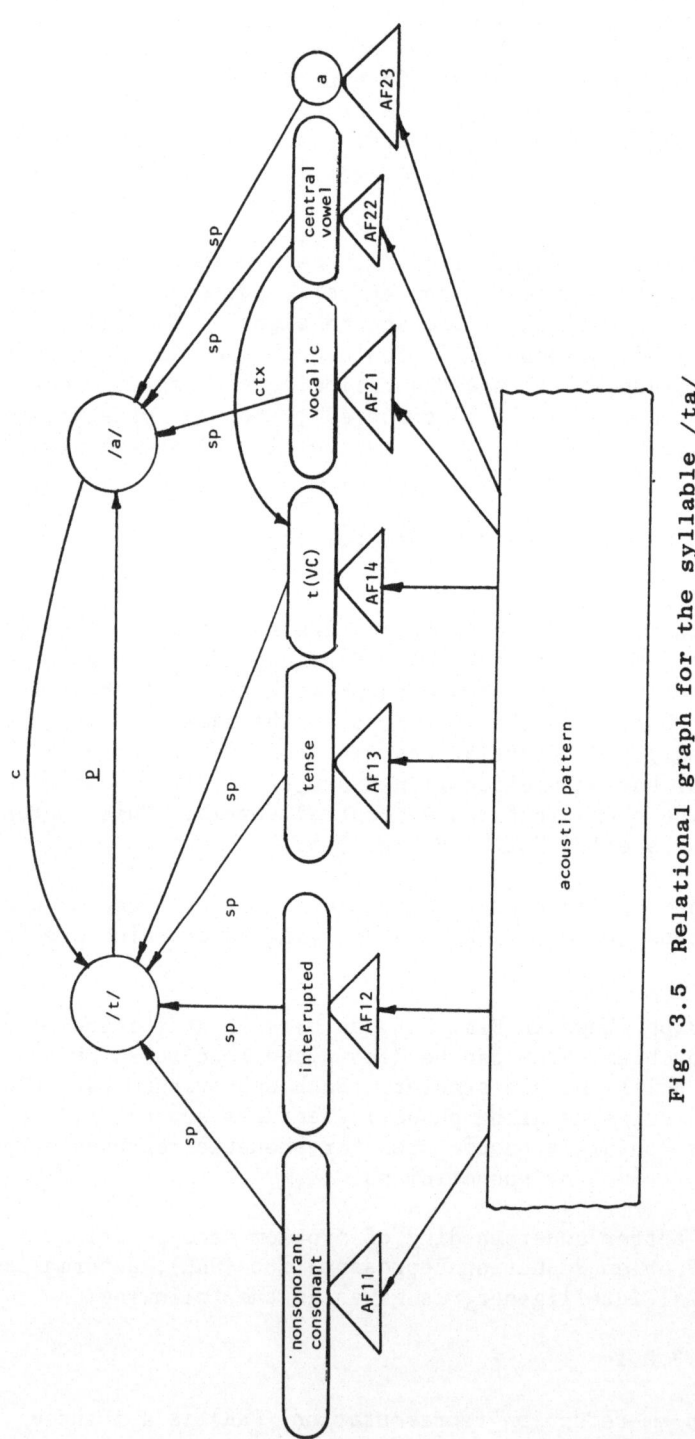

Fig. 3.5 Relational graph for the syllable /ta/

$P = \{P_i\}$ is an enumerable set of problem descriptions.

$B \subseteq P$ is a set of problems, only one of which needs to be solved.

$r: P \times N \rightarrow P$ is the ordered successor function with $r(p_i,j)$ = p_k interpreted as "p_k is the j-th successor of problem p_i". It is assumed that the set of all successors of problems p_i is $p_k : p_k = r(p_i,j)$, $j\epsilon \left(1, R\ (p_i)\right)$, by definition. N is the set of natural numbers. $R\ (p_i)$ is the number of successors of p_i, $R(i)$ is the set of the indices of successors of

$t: P, \rightarrow \{v,\wedge\}$ is the node type function with the usual interpretation that solvability of 'v' nodes implies solvability of predecessors and unsolvability of '∧' nodes implies unsolvability of predecessors.

$u: P, \rightarrow \{'L', 'S', 'D'\}$ is the node solution function interpreted as 'live', 'solved', and 'dead'. A solved problem is known to have a solution while a dead problem is known not to have one. A problem is live when it is not known to be solved or dead.

Definition 3.2.2

A problem i ϵ PRR is *solved* if and only if:

1) $u(i) = 'S'$

or   2) $u(i) = 'L'$ and there exists an index
$k \epsilon R(i) | \exists u(k) = 'S'$ and $t(k) = 'v'$

or   3) $u(i) = 'L'$ and for every successor
$k \epsilon R(i)$, $u(k) = 'S'$ and $t(k) = '\wedge'$

Definition 3.2.3

PRR has a solution if and only if some problem i ϵ B $\subseteq$ P is solved.

Definition 3.2.4

A state-space representation (SSR) is a 4-tuple $\langle S,I,F,q \rangle$ where:

$S = \{s_i\}$ is an enumerable set of state descriptions.

$I \subseteq S$ is the set of initial states.

$F \subseteq S$ is the set of final states.

$q: S \times N \to S$ is the ordered successor function, with $q(s_i, j) = s_k$, interpreted as "there exists a unit length path from state s_i to state s_k in the space". The set of all successors of s_i is denoted $Q(s_i) = \{s_k \mid s_k = q(s_i, j)$ for some $j\}$

Definition 3.2.5

A *path* exists between states s_i and s_w if and only if there is a finite sequence of states $s_i = s_1, s_2, \ldots, s_n = s_w$, such that $s_i + 1 \in Q(s_i)$.

Definition 3.2.6

SRR has a *solution* if and only if there exists a path between some state $s_i \in I$ and some states $s_f \in F$.

It has been shown (see Stockman, 1977, for example) that a problem reduction search can be converted to a state space search by letting a state encode a path in the problem reduction representation.

Hall (1973) has shown that every context free grammar (CFG) can be represented by a PRR. Given a PRR corresponding to a CFG, the conception of a strategy for problem solving will correspond to the design of a parser. In Pattern Recognition and, especially, in Speech Understanding, a grammar is often used for driving feature extraction before controlling recognition. For these cases a "non-directional" strategy, for which the analysis is not performed "left-to-right" along the time axis, can be more effective. Semantic information may suggest which subproblems have to be solved first because they require features which are expected to be easily detectable without any hypothesis about the context. Fig. 3.6 shows a PRR of the example of Fig. 3.5.

The solid lines below the nodes indicate AND nodes; the dashed lines indicate also AND nodes whose descending problems have to be solved in successive time intervals (they represent the p relation of Fig. 3.5).

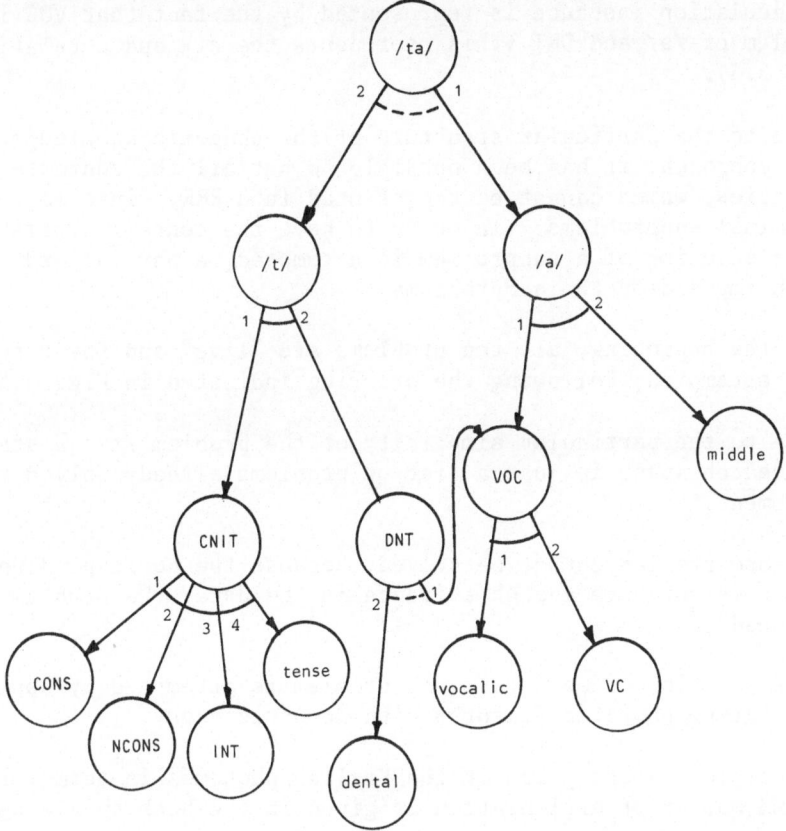

Fig. 3.6   Example of a Problem Reduction Representation

Arcs joining a problem with its subproblems are numbered according to the order in which the solution of the subproblems has to
be attempted.   Subproblems are labelled by phonetic features.

The subproblems of the time sequence /a/ have to be solved in
the order indicated in Fig. 3.6, that is, the problem of the vowel
has to be solved first, followed by CNIT, followed by DNT.   Notice
that "middle" describes the manner of articulation.   In this case
"middle" is related to the tongue position of the articulation of
/a/.

With a little effort, the complex relational graph of Fig. 3.5
has been transformed into an AND graph with two types of AND nodes.

A coarticulation instance is represented by the fact that VOC is a
subproblem of /a/ and DNT (this represents the ctx and c relations
in Fig. 3.5).

Due to the particular structure of the phonetic knowledge used
in this approach, it has been possible to put all the context-
dependencies, which cannot be represented in a PRR, into some of
the terminal subproblems. In order to have the context available
when the solution of a subproblem is attempted, a partial ordering
has been imposed on the subproblems.

At the beginning, all the problems are "live" and their solution is attempted, following the ordering indicated in Fig. 3.6.

Due to the particular simplicity of the problem /ta/ a state
in the search space is just a list of problems already solved at a
given moment.

If one problem cannot be solved, because the corresponding
feature does not have enough evidence in the data, the problem becomes "dead".

The solution of each terminal problem is attempted by applying
rules relating phonetic features with acoustic cues.

A complete description of the English phonemes in terms of
place and manner of articulation is given in the book by Flanagan
(1972).

Due to the correspondence between context-free grammars and
AND/OR graphs, non-directional parsers will be designed as problem
solvers.

Control (procedural) knowledge for using grammars in interpreting partially ambiguous and corrupted descriptions of the speech
patterns will be introduced in the following chapters.

In Chapter 8 it will be shown how the grammar rules, the non-
directional parsers, the rules for composing evidences, and the
decision rules can be represented by a "Production System".

A Production System is a widely used tool in Artificial
Intelligence for combining structural and procedural knowledge.

Production Systems for speech understanding will be discussed
in Chapter 8, the attention being focused in Chapters 4 to 7 on the
structure and use of the knowledge sources.

3.3   The Syntax for the Recognition
on the Phonetic Feature "Vocalic"

As an example of relations between phonetic and acoustic cues, a syntax which may control the hypothesization of the feature "vocalic" is presented.

The total energy S of the signal, the energy A in the band 3 - 5 kHz and the ratio R between low- and high-frequency energies are extracted from the signal every sampling period T = 10 ms.

The time series of the samples of these parameters, denoted respectively $S(nT)$, $A(nT)$, $R(nT)$, belong to the acoustic pattern obtained by a numerical transformation of the speech signal. Two describing languages, L_S for $S(nT)$ and L_A for $A(nT)$ are introduced. The alphabets of these languages are labels for the primitive forms shown in Figs. 3.7a and 3.7b.

These primitive forms are detected in a bottom-up way from the analysis of the time evolutions of the parameters. P1, P2, P3, and P4 are all defined for high values of R ($R(nT)$ must be above a threshold for all the points of all these primitives); the difference

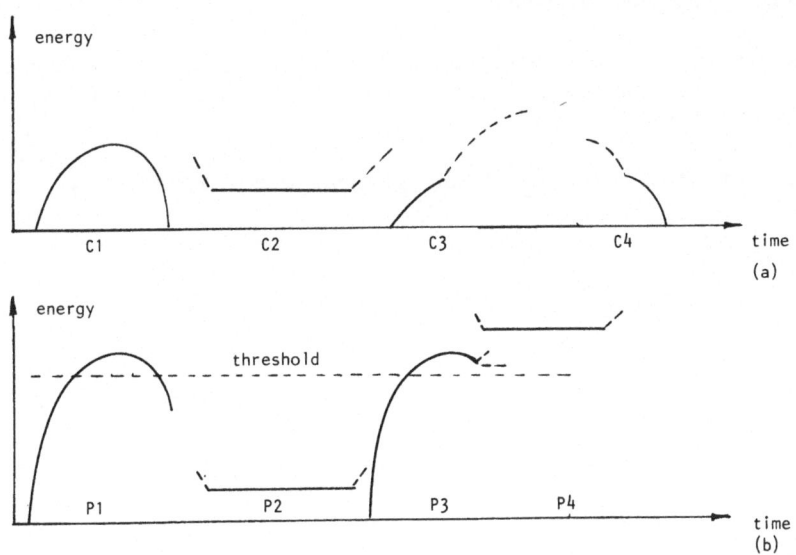

Fig. 3.7   Primitives of the Languages L_S (a) and L_A (b)

between P2 and P4 is that while the former is a quasi-horizontal tract below a given threshold, the latter is a similar segment but all its points must be above the threshold. P3 is a peak whose descending edge is interrupted and followed by at least a point with a non-negative first derivative. The ascendent tract of P2 and P4 may start at any level of energy.

A symbol P5 is also introduced to describe those segments of $A(nT)$ for which the ratio $R(nT)$ is below the threshold. Rather than the grammar for generating all the acceptable strings of L_S and L_A, we are interested in another grammar G_V which is capable of generating all the possible combined descriptions in L_S and L_A characterizing a "vocalic" tract.

Let VPF (Vocalic Pattern Feature) be the syntactic category of all descriptions of a vocalic pattern. The grammar G_V may be defined as follows:

$$G_V = (VPF, V_T, V_N, P_V)$$

$$V_T = \{C1, C3, C4, P1, P3, P4\}$$

$$V_N = \{VPF, PS, PA, \alpha_1, \beta_1, \gamma_1\}$$

The concatentation of symbols represents, in the language generated by G_V, the fact that the symbols describe subpatterns which are contiguous along the time axis.

The symbol s is a predicate regarding two items. The predicate is true when the two items describe subpatterns of S and A obtained in overlapped time intervals.

The rewriting rules P_V of the grammar are listed below:

P_V:

VPF $= s$ (PS,PA) $\alpha_1 \rightarrow$ C4

PS $\rightarrow$ C1 $\beta_1 \rightarrow \beta_1$ C4

PS $\rightarrow \alpha_1$ C1β_1 PA $\rightarrow$ P1

PS $\rightarrow \alpha_1$ C1   PA $\rightarrow \gamma_1$ P1

PS $\rightarrow$ C1 β_1 $\gamma_1 \rightarrow$ P3

$\alpha_1 \rightarrow$ C3 $\gamma_1 \rightarrow \gamma_1$ P3

$\alpha_1 \rightarrow \alpha_1$ C3 $\gamma_1 \rightarrow$ P3 P4

                                        $\gamma_1 \rightarrow \gamma_1$ P3 P4

In practice, a description process reads the samples $S(nT)$, $A(nT)$, $R(nT)$ of the entire sentence and generates two descriptions of them with the languages L_S and L_A respectively. To each detected primitive is associated a set of attributes containing the beginning and the end and the coordinates of the points allowing the primitive form to be reconstructed with a parabolic approximation.

A parsing algorithm is applied to the two descriptions. It looks first for PS and when it recognizes one of them, it looks for a possible, simultaneous PA; if it finds one, the hypothesis VPF is generated and the execution of the algorithm is restarted until the end of the descriptions is reached.

Example 3.3.1

Fig. 3.8 shows, as an example, the time evolutions of the parameters S (solid line), A (dotted line, R (dashed line) for the pseudo-syllable /ama/ extracted from a sentence. As R is always above its threshold, the descriptions in L_S and L_A, are the following:

L_S : C1   (70, 26, 40, 28)

C4   (110, 24)

C2   (170)

C3   (190, 27)

C1   (350, 20, 330, 28)

L_A : P1   (100, 15, 40, 30, 10)

P2   (160, 0)

P1   (350, 5, 260, 25, 7)

Attributes of the terminal symbol are defined as follows:

C1 (t_e, a_e, t_m, a_m)

C4 (t_e, a_e)

C2 (t_e)

C3 (t_e, a_e)

P1 $(t_e, a_e, t_m, a_m, R*)$

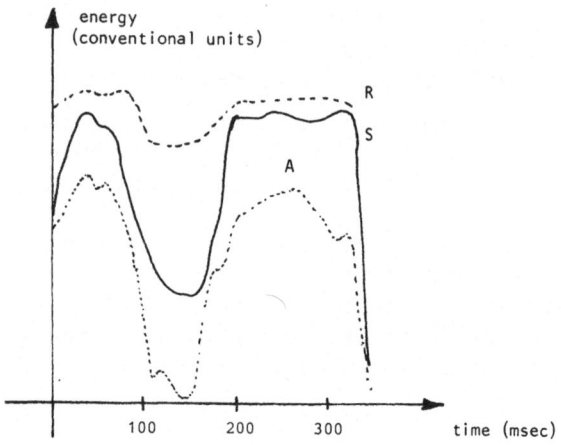

Fig. 3.8 Time evolutions of S,A and R for the pseudo-syllable
          segment /ama/

P2 $(t_e$, R*)

P3 : same as P1 ; P4: same as P_2 .

where: t_e is the time (in ms) of the end of the detected cue;
a_e is the amplitude of the cue at t_e ; t_m is the abscissa
where the amplitude reaches its maximum, a_m is the value of
such a maximum, R* is the minimum value of R in the interval
of the cue. Amplitude values are in conventional units.

Minor details in the description allow one to construct an
approximation of the signal patterns from their descriptions. This
description method can be used for a very low bit rate transmission
of speech as shown by De Mori, Laface and Piccolo (1976) with a
slightly more complicated coding system.

Applying grammar G_V and ignoring, for a moment, the attributes
of the descriptions, it is easy to verify that VPF is recognized
twice. In fact, the two descriptions C1 C4 and C3 C1, describing
respectively the time intervals (0,110) and (170,350), are recognized as PS by the use of the following derivations:

PS ⇒ C1 β1 ⇒ C1 C4

PS ⇒ α1 C1 ⇒ C3 C1

Now, because:

PA → P1

two PAs are recognized on the time intervals: (0,100) and (160,350). As slight time shifts are allowed when the truth of s(PS, PA) is evaluated, the feature VPF is recognized in two places with the time bounds defined by the following relations:

VPF (0,100) = s (PS(0,110), PA(0,100))

VPF (170,350) = s (PS(170,350), PA(160,350)).

A duration dur(VPF) can be defined as the difference between the end and the beginning of the time interval in which the feature has been detected; taking the minimum of R* for the primitives in VPF, the value mR of VPF is obtained, and taking the maximum of the maxima of P1 and the P3s, the maximum M of A in the vocalic tract is also obtained. Thus, a VPF can be represented with three attributes:

(VPF) = (dur, mR, M) .

The rules just introduced for computing the attributes of VPF are called *semantic rules* associated with G_V. For the example of Fig. 3.8 one obtains:

(VPF) = (100, 10, 30) for the interval (0, 100)

(VPF) = (180, 7, 25) for the interval (170, 350) .

The durations are in ms, the maxima and minima are in conventional units.

When a VPF hypothesis is generated, its attributes are evaluated and used for computing the evidence of the hypothesis <VOCALIC>, which is defined by introducing the linguistic label "high" appearing in three propositions connected by the logical conjunction <and> as follows:

<VOCALIC> = high (dur(VPF)) <and> high (mR(VPF)) <and> high (M (VPF)).

Thus, the evidence of the hypothesis <VOCALIC> is consistent only when a VPF is found and its attributes respect the conjunction of the three propositions appearing in the definition.

The linguistic value "high" has not yet been precisely defined. It will be shown in the next paragraph that its meaning can be represented by a fuzzy subset of the universe of the values of the acoustic parameter (duration, maximum peak energy etc.) it refers to.

The problem of recognizing PS and PA in the descriptions L_S and L_A is different from the classical recognition problem in the mathematical theory of formal languages.

The classical problem consists in deciding whether a given string belongs or not to a language generated by a grammar. Here a description containing an unknown number of substrings generated by a grammar has to be analyzed and whenever a feature is recognized its attributes have to be evaluated. These operations can be efficiently done for G_V using finite-state automata whose states are labeled by nonterminal symbols and whose transition arcs are labeled by primitive symbols. Semantic rules are applied by "augmenting" the automata with actions that have to be performed whenever an arc is consumed. Figs. 3.9a) and b) show the automata for PS and PA; I1 and I2 are the initial states of these automata.

The labels are represented above the arcs, while the actions are represented below the arcs. The variables T_b and T_e represent respectively the beginning and the end of the interval labeled by PS (Fig. 3.9a); T_B and T_E are the bounds of PA (Fig. 3.9b).

The symbol t_e represents the time of the end of the input primitive indicated above the arc. The time unit is 10 ms; thus 1 corresponds to 10 ms.

The automaton for PA recognizes a string starting at T_b (the beginning of a recognized PS and ending at T_e (the end of a recognized PS). The last symbol of the string read by the automaton PA is the one describing the cue ending at T_e followed by the "end of string" symbol λ.

3.4   Background on Fuzzy Set Theory

3.4.1 Definition of fuzzy sets

Trying to understand or model human perception, an important point has to be taken into account; much of human thinking and, particularly, the ability to recognize similarities, is approximate in nature. Mechanistic systems have usually units (of force, pressure, mass, etc.). Human systems have, in some instances, ill-defined scales without units.

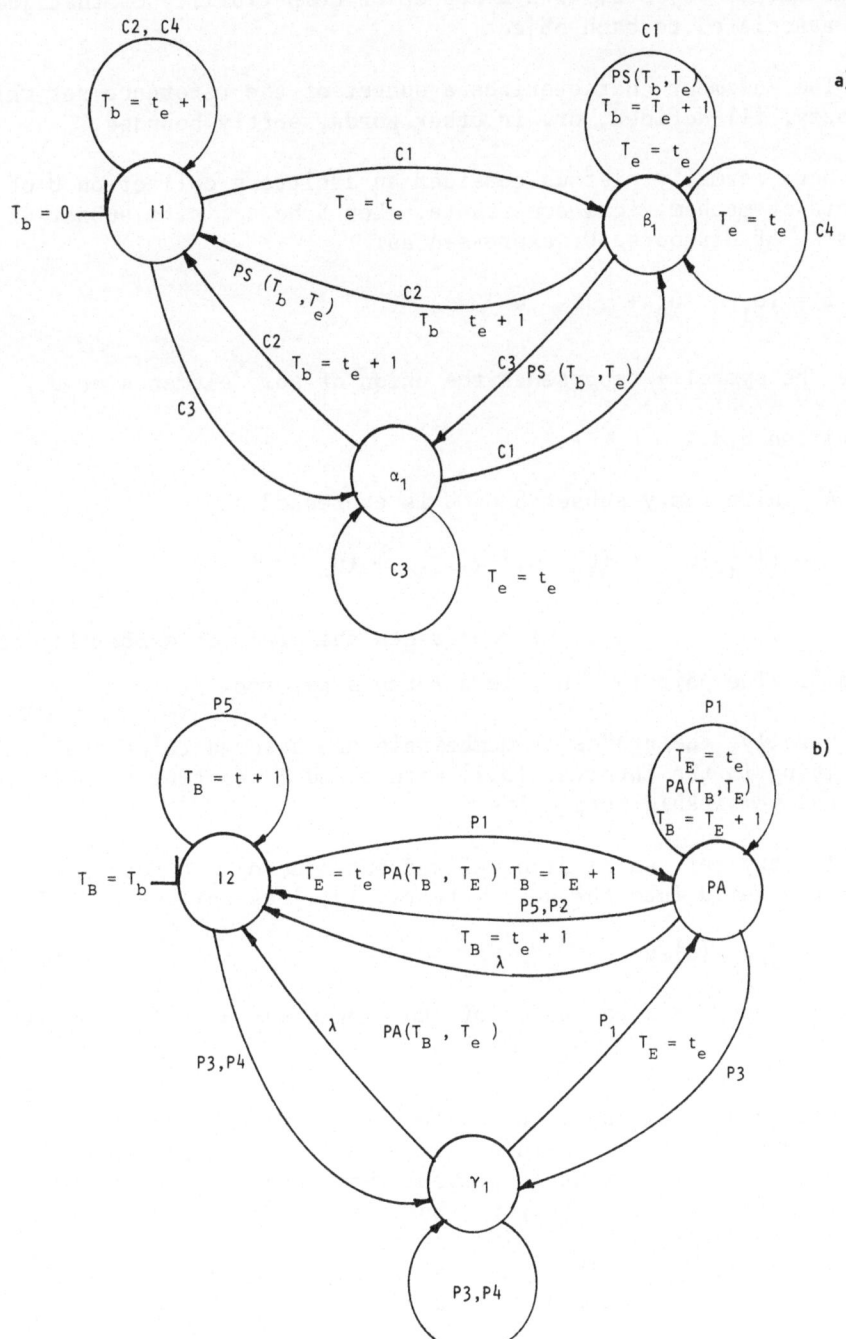

Fig. 3.9 Automata for the recognition of PS (a) and PA (b)

One such scale for the attribution of a judgment may be a reference set of objects, with a degree of compatibility of that judgement associated to each object.

The judgment characterizes a subset of the reference set which is fuzzy, ill-defined, or, in other words, softly bounded.

More formally, let us consider an arbitrary collection U of objects or mathematical constructs. Let S be a finite subset of the universe of discourse U, expressed as:

$$S = \{u_1 + u_2 + \ldots + u_n\}$$

where the symbol + represents the union of some elements of U.

Definition 3.4.1

A finite fuzzy subset $\tilde{S}$ of U is expressed as:

$$\tilde{S} = \{(\mu_1 /u_1) + (\mu_2 /u_2) + \ldots + (\mu_n /u_n)\}$$

where μ_i (i = 1, 2, ..., n) represents the *grade of membership* of u_i in $\tilde{S}$. The pair (μ_i, u_i) is a *fuzzy singleton*.

Usually, the grades of membership are assumed to be real numbers lying in the interval [0,1] with 0 and 1 denoting *no* membership and *full* membership respectively.

A fuzzy set can be also defined introducing a function $\mu_S (u)$ mapping a set U into the unit interval [0,1] as follows:

$$\tilde{S} = \int_U \mu_S (u)/u \tag{3.2}$$

where $\int_U$ represents the union of the singletons $\mu_S (u)/u$ for all $u \in U$.

Membership values are not probabilities; they represent different concepts. Let us consider, for example, a collection of types of accidents. If we want to prevent them we are not only interested in their frequency of occurrence but also in their degree of danger. In many countries, electrical accidents have little frequency of occurrence but, generally, they have a high degree of danger, a high membership in the scale of danger.

3.4.2 Operations on fuzzy sets

Definition 3.4.2

If $\tilde{S}_1$ and $\tilde{S}_2$ are fuzzy subsets of U, their *union* is defined as follows:

$$\tilde{S}_1 \cup \tilde{S}_2 \overset{\Delta}{=} \int_U (\mu_{S_1}(u) \vee \mu_{S_2}(u))/u \qquad (3.3)$$

where $\overset{\Delta}{=}$ means "by definition" and $\vee$ is an operator which takes the maximum of $\mu S_1(u)$ and $\mu S_2(u)$ (*max* operator).

Definition 3.4.3

The *intersection* of two fuzzy subsets $\tilde{S}_1$ and $\tilde{S}_2$ of U is defined as follows:

$$\tilde{S}_1 \cap \tilde{S}_2 \overset{\Delta}{=} \int_U (\mu_{S_1}(u) \wedge \mu_{S_2}(u))/u \qquad (3.4)$$

where $\wedge$ is the *min* operator (i.e., an operator which takes the minimum of the operands).

Definition 3.4.4

The complement $\neg\tilde{S}$ of a fuzzy sub-set $\tilde{S}$ of U is defined by:

$$\neg\tilde{S} \overset{\Delta}{=} \int_U (1 - \mu_s(u))/u \qquad (3.5)$$

Definition 3.4.5

If $\tilde{S}$ is a fuzzy subset of U, then a *t-level set of* $\tilde{S}$ is a non-fuzzy hard set, denoted by S_t, which contains all the elements of U whose grade of membership in $\tilde{S}$ is greater than or equal to t:

$$S_t = \{u/\mu_S(u) \geqslant t\} \qquad (3.6)$$

Definition 3.4.6

A *fuzzy algebra* is the system:

$$FA = <Z, +, *, \bar{}>$$

where Z is a set of at least two distinct elements and $\forall x, y, z \in Z$, system FA satisfies the following set of axioms:

3.1) *Idempotency*

$$x + x = x ; \qquad x * x = x ;$$

3.2) *Commutativity*

$$x + y = y + x ; \qquad x * y = y * x ;$$

3.3) *Associativity*

$$(x + y) + z = x + (y + z); \quad (x * y) * z = x * (y * z);$$

3.4) *Absorption*

$$x + (x * y) = x; \qquad x * (x + y) = x$$

3.5) *Distributivity*

$$x + (y * z) = (x + y) * (x + z); \quad x * (y + z) = x * y + x * z$$

3.6) *Complement*

If $x \in Z$ then there is a unique complement $\bar{x}$ of x such that $\bar{x} \in Z$ and $\bar{\bar{x}} = x$. The complement of x is denoted $\bar{x}$ rather than $\neg x$ for the sake of simplicity.

3.7) *Identities*

$(\exists! \, e_+)$ $(\forall x)$ such that:

$$x + e_+ = e_+ + x = x$$

$(\exists! \, e_*)$ $(\forall x)$ such that

$$x * e_* = e_* * x = x$$

3.8) *De-Morgan Laws*

$$\overline{x + y} = \bar{x} * \bar{y}; \quad \overline{x * y} = \bar{x} + \bar{y}$$

We shall be using a particular fuzzy algebra defined by the system:

$$FA = < [0,1], +, *, ^- > \tag{3.7}$$

where the operators $(+, *, ^-)$ are interpreted as *max, min* and *complement* $(\bar{x} = 1 - x, \forall x \in [0,1])$, respectively.

Notice that the Boolean algebra is also a fuzzy algebra, but not every fuzzy algebra is Boolean.

3.4.3 Fuzzy restrictions

A fuzzy set $\tilde{S}$ can be used to define the extent to which an element $u \in U$ possesses a certain property X. This property can be represented by a variable taking values in U. The property X may define a binary valued restriction on U; in this case a variable takes value 1 for every element of U having the property X, 0 otherwise. A property X may induce a *fuzzy restriction* defined over U; in this case the restriction is represented by a fuzzy subset $\tilde{S}$ of U acting as an elastic constraint on the elements of U which may possess the property X.

Definition 3.4.7

A *fuzzy restriction* is the assignment of a linguistic value X to an element $u \in U$ with a degree of compatibility μ_S (u) according to the following definition:

$$X = u: \mu_S (u) \tag{3.8}$$

Expression (3.8) can be extended over elements of U defining a fuzzy subset of U. Variables with associated fuzzy restrictions may be words or phrases of a natural language; the associated restrictions represent their *meaning*. These variables may be combined to form more complex compounds, using, for example, the linguistic connectives <and> (conjunction) and <or> (disjunction). A complex compound is again a fuzzy restriction represented by a fuzzy set, obtained by combining the fuzzy sets of the component restrictions with the $\cap$ operator when an <and> is encountered and with the $\cup$ operator when an <or> is encountered. Thus:

$$\tilde{S}_1 \text{ <and> } \tilde{S}_2 \overset{\Delta}{=} \tilde{S}_1 \cap \tilde{S}_2 \tag{3.9}$$

$$\tilde{S}_1 \text{ <or> } \tilde{S}_2 \overset{\Delta}{=} \tilde{S}_1 \cup \tilde{S}_2 \tag{3.10}$$

By these definitions <and> and <or> are implied to be *non-interactive* in the sense that there is no "trade-off" between their operands. When this is not the case, <and> and <or> are denoted by and and or, respectively, and are defined in a way that reflects the nature of the trade-off.

For example, we may have:

$$\tilde{S}_1 \text{ and } \tilde{S}_2 \overset{\Delta}{=} \int_U (\mu_{S_1} (u) \cdot \mu_{S_2} (u))/u \tag{3.11}$$

$$\tilde{S}_1 \text{ or } \tilde{S}_2 \overset{\Delta}{=} \int_U (\mu_{S_1} (u) + \mu_{S_2} (u) - \mu_{S_1} (u) \cdot \mu_{S_2} (u))/u \tag{3.12}$$

where + denotes here the arithmetic sum and . denotes the arithmetic product.

In general, the interactive versions of and and or do not possess the simplifying properties of the connectives defined for the non-interactive case, e.g. associativity, commutativity, distributivity etc.

Often a fuzzy set, representing a fuzzy restriction, is selected as a point of reference. One such set could be the definition of "young" as a restriction over the universe of possible ages. We may modify this set and use these modifications in a very rough way as multiples of a unit of measurement. Such modifications can be: "very young", "old", "very old", "more or less young", etc. So some modifiers like "very", "more or less" are introduced.

In order to show how may they change the basic fuzzy set "young", the definition of power of a fuzzy set is introduced.

Definition 3.4.8

If a is a real number, then $\tilde{S}^a$ is defined by:

$$\tilde{S}^a \triangleq \int_U (\mu_S (u))^a /u \qquad\qquad (3.13)$$

This operation may be used to approximate, very roughly, the effects of some linguistic modifiers. For example one may define:

$$\text{very } \tilde{S} \triangleq \tilde{S}^2 \qquad\qquad (3.14)$$

$$\text{more or less } \tilde{S} \triangleq \tilde{S}^{1/2} \qquad\qquad (3.15)$$

$$\text{old} \quad \triangleq \neg \tilde{S}$$

3.4.4 Possibility distributions

Definition 3.4.9

Let X be a variable taking values in U and let $\tilde{S}$ represent a fuzzy restriction, R(X) associated with X. Then the proposition "X is $\tilde{S}$" which translates into:

$$R(X) = \tilde{S}$$

associates a possibility distribution π_X with X which is postulated to be equal to R(X).

A possibility distribution function π_X is associated with X and is defined as:

$$\pi_X (u) = \mu_S (u) \qquad \forall u \in U \qquad\qquad (3.16)$$

$\pi_X (u)$ is the *possibility* that X = u.

Possibility relates to the perception of the degree of feasibility whereas probability is associated with the concept of frequency of occurrence. What is possible may not be probable and what is
improbable need not to be impossible.

Since what is impossible is necessarily improbable, possibility
values must always be greater than probability values.

Even if possibilities are non-statistical in nature, this does
not prevent one from using statistics in the estimation of membership
functions. But this estimation does not require necessarily as large
a number of experiments as the estimation of a probability density.
Given a set of statistical data in the form of a histogram, the induction of a probability density can be made by reducing to unity
the surface of the histogram. On the contrary the induction of a
possibility density can be largely influenced by the a priori knowledge of human experience or models of the phenomenon to be analyzed
and is less constrained.

In many applications, a possibility distribution for a feature
f, which may assume different values of a base variable u, is simply
established in the following way. For those values of u which are
assumed by f and only by it, the possibility distribution takes
value 1. For those values of u which are never assumed by f, the
possibility distribution takes value 0. Usually, the possibilities
take values 1 and 0 in two intervals of u. These two intervals can
be known to some extent by the a priori knowledge. Between them the
possibility distribution can be approximated with an increasing
function varying from 0 to 1.

In investigating models for speech understanding, we are interested in evaluating the possibility that a feature f is in a pattern
p, i.e.,

Poss (f is in p)   (3.17)

3.4.5 A simple example

In Section 3.3 a grammar has been introduced for controlling
the detection of the feature VOCALIC in continuous speech. It has
been shown that the phonetic feature VOCALIC is detected when the
following acoustic features appear simultaneously in the "vocalic
pattern feature" VPF:

- high ratio between low and high frequency energies denoted as X_1 = high (mR(VPF));

- X_2 = long (dur(VPF));

- X_3 = high $(M(VPF))$.

The evidence of the feature VOCALIC in the portion of the acoustic data contained in the time interval (T_b, T_e) of a VPF can be defined by the following possibility:

Poss (VOCALIC is in VPF (T_b, T_e))

The above possibility is expressed by the following disjunction:

VOCALIC = high $(mR(VPF))$ <and> long $(durVPF))$<and> high $(M(VPF))$

The three variables can be defined as fuzzy restrictions on the universes of the acoustic measurements u_1: R, u_2: duration of VPF and u_3: maximum amplitude of the peaks, $M(VPF)$. After collecting histograms of these parameters for about 500 vowels and using some notions from experimental phonetics about the nature of vowels and their duration, the fuzzy restrictions shown in Figs. 3.10 a,b,c have been induced.

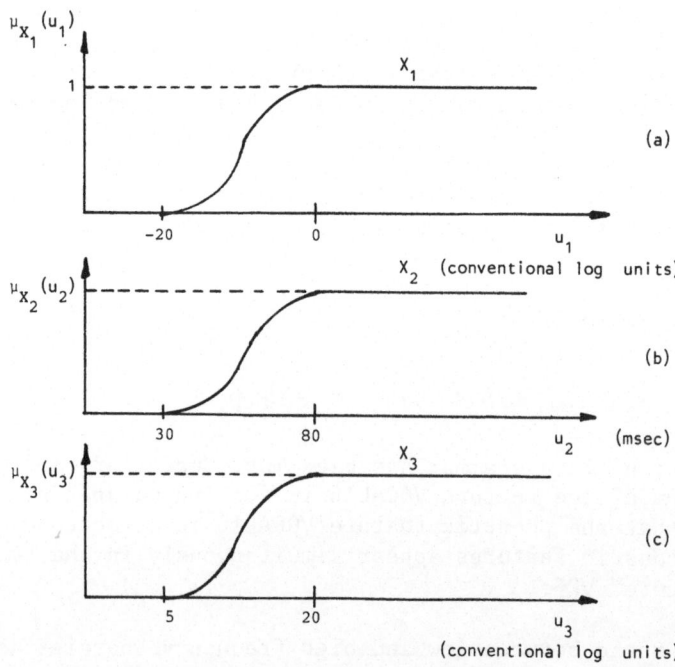

Fig. 3.10   Fuzzy sets defining the meanings of "high" for the three acoustic measurements characterizing a Vocalic Pattern Feature.

The evidence of the feature VOCALIC can be evaluated as follows:

$$\text{Poss (VOCALIC is in VPF}(T_b, T_e)) = \mu_{X_1}(u_1) \, \lambda \, \mu_{X_2}(u_2) \tag{3.18}$$

$$\wedge \, \mu_{X_3}(u_3).$$

Where u_1, u_2, u_3 are the measurements taken in a VPF (T_b, T_e).

The fuzzy sets in Fig. 3.10 have been obtained by simply assigning membership 1 to the values of u assumed only by the vocalic segments, membership 0 to the values of u assumed only by the consonants, and interpolating with a continuous function between the two intervals.

Example 3.4.1

The following data have been measured after the features PS and PA have been detected concurrently in an acoustic pattern:

$u_1 = 5$ c.u. (conventional units)

$u_2 = 70$ ms

$u_3 = 15$ c.u. (conventional units)

The corresponding memberships are:

$$\mu_{X_1}(u_1) = 1; \; \mu_{X_2}(u_2) = 0.8; \; \mu_{X_3}(u_3) = 0.7,$$

$$\text{Poss (VOCALIC is in VPF}(T_b, T_e)) \quad 1 \wedge 0.8 \wedge 0.7 = 0.7$$

3.5 Fuzzy Relations and Languages

In this section, fuzzy relations and languages will be introduced in order to provide the necessary background for describing the algorithms used to generate and evaluate phonetic and phonemic hypotheses.

3.5.1 Fuzzy relations

Definition 3.5.1

If $\tilde{S}_1$, $\tilde{S}_2$, ..., $\tilde{S}_n$ are fuzzy subsets of U_1, U_2, ..., U_n, then the *cartesian product* of $\tilde{S}_1$, ..., $\tilde{S}_n$ is a fuzzy subset of $U_1 \times U_2 \times ... \times U_n$ defined by:

$$\tilde{S}_1 \times \tilde{S}_2 \times ... \times \tilde{S}_n = \int_{U_1 \times U_2 \times ... \times U_n} \mu_{S_1}(u_1) \wedge ... \wedge \mu_{S_n}(u_n) / (u_1 ... u_n)$$

(3.19)

Definition 3.5.2

Let $X = \{x\}$ and $y = \{y\}$ be collections of objects denoted generically by x and y, then a *fuzzy relation* R in XᴜY is a fuzzy subset of X×Y characterised by a membership function $\mu_R(x,y)$ which associates with each pair (x,y) its "grade of membership" in R.

Definition 3.5.3

The *domain* of a fuzzy relation R, dom R is a fuzzy set defined by:

$$\mu_{\text{dom } R}(x) = \bigvee_y \mu_R(x,y) \qquad \forall x \in X.$$

Definition 3.5.4

The *support* of R, denoted S(R), is the non-fuzzy subset of X×Y over which $\mu_R(x,y) > 0$.

Definition 3.5.5

If $R \subset X \times Y$ and $Q \subset Y \times Z$ are two fuzzy relations, their max-min *composition*, denoted $R \circ Q$ is defined by:

$$\mu_{R \circ Q} = \bigvee_y (\mu_R(x,y) \wedge \mu_Q(y,z))$$

(3.20)

$$\forall x \in X, z \in Z$$

Note that the operation of composition has the associative property:

$$A \circ (B \circ C) = (A \circ B) \circ C$$

So far we have introduced a binary fuzzy relation; an n-ary relation is defined as a fuzzy subset of $U_1 \times U_2 \times ... \times U_n$, the cartesian product of n non-fuzzy sets.

Definition 3.5.6

The *projection* of a fuzzy subset of $U_1 \times U_2 \times ... \times U_n$ on

$U_{i_1} \times \ldots \times U_{i_k}$, where $(i_1, \ldots, i_k)$ is a subsequence of $(1, \ldots, n)$.

is a relation in $U_{i_1} \times \ldots \times U_{i_k}$ defined by:

$$Proj\ R\ on\ U_{i_1} \times \ldots \times U_{i_k} \triangleq \int_{(u_{j_1}, \ldots, u_{j_1})} \bigvee \mu(u_1 \ldots u_n)/(u_1 \ldots u_n)$$
$$U_{i_1} \times \ldots \times U_{i_k}$$

$$(3.21)$$

where $(j_1, \ldots, j_1)$ is the sequence complementary to $(i_1, \ldots, i_k)$ (e.g. if $n = 6$, then $(1,3,6)$) is complementary to $(2,4,5)$ and

$$(u_{j_1}, \ldots, u_{j_1})$$

denotes the supremum over $U_{j_1} \times \ldots \times U_{j_1}$.

Definition 3.5.7

If R is a fuzzy subset of $U_{i_1}, \ldots, U_{i_k}$, then its *cylindrical extension* $\bar{R}$ in $U_1 \times \ldots \times U_n$ is a fuzzy subset of $U_1 \times \ldots \times U_n$ defined by:

$$\bar{R} = \int_{U_1 \times \ldots \times U_n} \mu_R (u_{i_1}, \ldots, u_{i_k})/(u_1, \ldots, u_n) \qquad (3.22)$$

3.5.2 The extension principle

The *extension principle* for fuzzy sets is in essence a basic identity which allows the domain of the definition of a mapping or a relation to be extended from points in U to fuzzy subsets of U.

Definition 3.5.8

Suppose that f is a mapping from U to V, and $\tilde{S}$ is a fuzzy subset of U expressed as:

$$\tilde{S} = \mu_1/u_1 + \mu_2/u_2 + \ldots + \mu_n/u_n$$

The *extension principle* asserts that:

$$f(\tilde{S}) = \mu_1/f(u_1) + \mu_2/f(u_2) + \ldots + \mu_n/f(u_n) \qquad (3.23)$$

3.5.3 Fuzzy languages

The names of fuzzy restrictions like "young", "old", "very young" etc., may be seen as *linguistic values* which may be assumed by a *base variable* such as "age".

All the linguistic values which may be assumed by the base variable define a *set of terms* T.

Definition 3.5.9

A *fuzzy language* L is a fuzzy relation from a set of terms T to a universe of discourse U and is characterized by a membership function:

$$\mu_L : T{\times}U \rightarrow [0,1] \qquad\qquad (3.24)$$

Let us consider a particular element $y_0 \in U$; then $\mu_L(x/y_0)$, $\forall x \in T$, defines a fuzzy set $D(y_0)$ in T which is called *descriptor* and serves to characterize the extent to which each term in T describes a given element y_0 of U.

Let $d \in T$, $u \in U$; d may be a *description* of a pattern u. In such a case $\mu_L(d,u)$ expresses the *vagueness* of the description d of u.

In the general case the term set T may contain an infinite number of elements. This infinite number of elements may be *structured*. This means that an element of the set may be generated algorithmically by concatenating, for example, the terms of an alphabet V_T, applying the rewriting rules of a grammar.

Definition 3.5.10

To each element X' of T can be associated a *fuzzy linguistic variable* X characterized by a triple $(X',U,R(X',u))$ in which X' is the name of the variable, $u \in U$ is a generic element of the universe of discourse U, and $R(X';u)$ is a fuzzy restriction on the values of U imposed by the name X'.

In order to obtain the possibility of each term in T, a *semantic rule* has to be associated with each syntactic rule. The semantic rules relate the possibility of the left-hand side of a rewriting (syntactic) rule and the possibilities of the terms at the right-hand side.

Let us consider now a case in which pattern classes are defined by pattern grammars and the terminal symbols are the primitive features of the pattern.

Let p be an unknown pattern. In practical cases, and particularly in speech decoding, a feature f is very often not well evident in the pattern. Thus, several features are extracted from the pattern with a degree of evidence expressed by Poss {f is in p}. Strings of detected features may be seen as terms of a fuzzy language whose universe of discourse U is that of all the possible patterns. A term may belong to many term sets and many terms of different term sets may describe an unknown pattern. We are interested in the descriptors of the pattern p for every term set generated by pattern grammars. Strings of features are parsed under the control of pattern grammars, yielding syntax trees whose leaves are primary terms, modifiers, and markers, which serve as aids to parsing. Starting from the bottom, the primary terms are assigned their meaning and, using the semantic rules, the meaning of nonterminals connected to the leaves are computed.

Repeating this process, the meanings of the terms associated with the roots are computed, giving the possibility that the interpretation corresponding to a root applies to the unknown pattern p.

In many cases, and speech is one of these, even syntactic rules are not well known. The semantics of fuzzy languages can be extended to incorporate also this type of imprecision.

For this purpose, the concept of *fuzzy structured language* is introduced.

Let G_T be a grammar capable of generating the term set T, with a set of syntactic rules S_T, a nonterminal alphabet V_n, and a terminal alphabet V_T.

Definition 3.5.11

A fuzzy structured language is a quadruple $L(U, S_T, V'_T, S_N)$, where V'_T denotes the set of all finite strings which can be extracted as pattern descriptions, these strings are composed elements of V_T, S_N is the set of semantic rules associated to S_T, and the term set T is assumed to be a fuzzy subset of V'_T. The function of the syntax S is to provide a set of rules for generating strings in the support V' of T with their grade of membership in T. A particular form of fuzzy grammar can be obtained by generalizing the notion of a phrase-structure grammar to represent a case in which the knowledge is imprecise, but the input string is unambiguous. Let:

$$G = (V_T, V_N, S, P)$$

be a grammar where $S \in V_N$ is a starting symbol standing for the syntactic category characterizing a given pattern class and P is a finite set of fuzzy productions of the form:

$$\alpha \xrightarrow{\rho} \beta$$

α and β are strings composed of elements of $V_T \cup V_N$, and $\rho \in [0,1]$ represents the grammaticality of the rule.

Definition 3.5.12

If A and B are two strings in $(V_T \cup V_N)^*$ and there exist strings $\alpha_1, \alpha_2, \ldots, \alpha_{n-1}$ in $(V_T \cup V_N)^*$ such that;

$$A \xrightarrow{\rho_1} \alpha_1 \xrightarrow{\rho_2} \alpha_2 \xrightarrow{\rho_3} \ldots \xrightarrow{\rho_n} B,$$

then B is said to be derivable from A via the derivation chain $(A, \alpha_1, \alpha_2, \ldots, B)$.

The *strength* of this chain is defined as follows:

$$\text{strength} (A, \alpha_1, \alpha_2, \ldots, B) = \rho_1 \wedge \rho_2 \wedge \ldots \wedge \rho_n \qquad (3.25)$$

The *strength* of the relation between A and B is defined to be the strength of the strongest chain between A and B. Let ρ be such a strength. Then we write:

$$A \xrightarrow{\rho} B.$$

A terminal string x is in T if it is derivable from S. Its grade of membership in T is the strength of the relation between S and x.

For many purposes it is convenient to express the productions in an algebraic notation. This is obtained by representing the following rules:

$$A \xrightarrow{\rho_1} B$$

$$A \xrightarrow{\rho_2} C$$

by the equation

$$A = \rho_1 B + \rho_2 C \qquad (3.26)$$

with the following associated semantic rule:

$$\mu(A) = \rho_1 \wedge \mu(B) \vee \rho_2 \wedge \mu(C) \qquad (3.27)$$

The most general case is the one in which both the knowledge is imprecise and the input is a lattice of ambiguous hypotheses.

3.6 Use of Fuzzy Algorithms for Feature Hypothesization

3.6.1 Fuzzy algorithms

Definition 3.6.1

A *fuzzy recognition algorithm* of the hypothesis H is a collection of steps acting on a given object $u \in U$ which, upon execution, yields the degree to which u is compatible with an hypothesis H.

An hypothesis H may have the form of a branching questionnaire Q, in which both the constituent questions and the answers are allowed to be fuzzy in nature.

A question Qi in Q may be either *classificational* or *attributional*.

In the case of classificational questions, Qi is concerned with the grade of membership of an element $u \in U$ in a fuzzy set representing H, or with the truth-value of a predicate which corresponds to H.

For example, Qi may be; "Does a pattern p contain a sonorant consonant?" The answer could be a degree of compatibility of the phonetic feature "sonorant consonant" in p.

In the case of attributional questions, Qi relates to the value of an attribute of the subject. For example, an instance of Qi may be "What is the evidence of a word in the acoustic pattern p" and the answer could be "high".

The totality of the questions in Q constitutes a basis for Q or the "fuzzy concept" defined by Q.

Definition 3.6.2

An *atomic* question is a triple $Q \triangleq (U,B,A)$, where U is the set of objects to which Q applies; B, the body of Q, is a label of either a class or an attribute; A, the *answer set* is a set of admissable answers to the question. B may be an element of a fuzzy language.

A *composite* question is a collection of constituent questions which may be composite or atomic.

Following Zadeh (1976), consider a composite classificational question $Q = B?$, whose constituents are classificational questions $Q_1 = B_1$?, $Q_2 = B_2$?, ... $Q_n = B_n$? in which B_i ($i \in [1,n]$) is a fuzzy subset of the universe U_i.

A composite question like "Does a pattern p contain a sonorant consonant?" can be represented as an explicit function between B_i (sonorant consonant) and the bodies B ($i \in [1,n]$) of component questions like Q_i = "Is R low in the consonant?" (B_i is "low").

Such an expression for B as a function of B_i ,..., B_n will be referred to as *analytic representation* of B.

An equivalent formulation of the problem leading to an easier, even if less elegant interpretation of the relations consists in expressing Q_i as "How is R?" and admitting answers like "low", "high", etc., which are fuzzy sets defined over the universe of the values of R. This allows the same component question to be used for different composite questions characterized by different relations between the body and the answers to atomic questions. These relations may be the structural rules of a fuzzy language with associated semantic rules for obtaining the meaning of the answers obtained by composition.

It is generally possible to represent a composite question in terms of the union of all sequence of answers to atomic questions followed by the consequent answers to the composite question.

This is a representation of a *branching questionnaire* in algebraic form.

Fuzzy grammars are similar to attributed grammars (see You and Fu, 1979, for definitions).

The use of semantic rules associated with the syntactic ones allows one to combine in a simple way evidences of terminal symbols and degrees of worthiness of the rules. Attempting the same task in a probabilistic framework would be more complex without substantial advantages in many applications.

The generation of hypotheses according to the *hypothesize-and-test paradigm* is in our case the execution of a fuzzy algorithm because the goal here is that of evaluating the evidence of a hypothesis H in a pattern p belonging to the universe of discourse.

A natural way to describe the hypothesis formation process is through a branching questionnaire; it just consists in asking questions about an unknown pattern.

The approach followed in this book is that of expressing the bodies of the hypotheses, following an algebraic representation, and using the formalism of structured fuzzy languages, because it allows modelling the knowledge imprecision (fuzzy rules) as well as the vagueness of the features extracted (the primitives, i.e., the terminal alphabet, are composed of fuzzy linguistic variables) and taking advantage of the system knowledge.

The semantic rules are defined according to the fuzzy algebra defined in Sec. 3.4; with such an approach, the operations can be performed easily and fast; furthermore, no other algebra leading to better performances for this problem has been proposed so far.

3.6.2 An example of application

The generation of the phonetic hypotheses "sonorant" or "non-sonorant" is described here in detail as an example of application of fuzzy algorithms. Such a process belongs to *precategorical classification*.

Precategorical classification consists primarily in the assignment of phonetic features to non-vocalic segments. The main purpose of such a classification is for segmentation of continuous speech into pseudo-syllable segments (PSS) and for driving a context-dependent extraction of more detailed features.

The spectra are processed in order to obtain these "gross spectral features":

- S : the total energy of a spectrum,

- B : the energy in the 200-900 Hz band,

- F : the energy in the 5 - 10 kHz band,

- A : the energy in the 3 - 5 kHz band,

- R : the ratio between B and F.

The gross features are described by the language presented in Section 3.3 and the description in terms of these features, denoted DGF, is used for precategorical classification, denoted PC; PC is then used to segment continuous speech into syllabic units, as described in Chapter 4.

The results of the precategorical classification may drive the extraction of detailed spectral features, such as formant evolutions, characteristics of frication noise, burst, etc.

Precategorical classification has also been postulated in human perception (see Stevens, 1973, for a discussion on this item).

The approach proposed in this chapter is based on the detection of phonetic categories related to some of the distinctive features proposed by Hughes and Hemdal (1965).

A phonetic feature is a syntactic category related to the lower level description DGF of the acoustic cues by a "branching question-naire" and is assigned to a speech segment after the answer to a composite classificational question $Q \stackrel{\triangle}{=} B$ is obtained, where the bodies of the component questions Q_j ($j = 1,2,...,n$) are fuzzy sets B_1, B_2, ..., B_n involved in an analytic representation of B. The fuzzy sets B_1, B_2, ..., B_n are linguistic variables defined over the range of the previously defined acoustic measurements.

Let Q_i ($i = 1,2,...,K$) be the composite question related to the i-th phonetic feature to be hypothesized. The answers to component questions Q_{ij} of the composite question Q_i are fuzzy linguistic variables (e.g. high, medium, low, more or less high, more or less low) whose membership functions are computed from diagrams stored in the syllabic source of knowledge of the system and are defined over the universe of an acoustic measurement.

The answers to the composite question Q_i are related by syntactic rules to the answers to the component questions Q_{ij}; thus to each answer to Q_i can be associated a membership function whose value is computed under the control of semantic rules from the values of the membership functions associated to each answer to the component questions.

This use of fuzzy algorithms models to some extent the fact that most of the acoustic-phonetic properties of speech sounds are only known with a degree of vagueness, e.g.: the signal energy is high for vowels, nonsonorants have high frequency components, in unvoiced plosives there is an interval of silence followed by some noise.

To a fuzzy linguistic variable, representing a judgement that can be expressed after the inspection of some acoustic parameters, is associated a fuzzy restriction.

For the sake of simplicity, the memberships (compatibility functions) of each restriction will be labelled by abbreviations of adjectives of phonetic features.

Fig. 3.11 shows the compatibility function of the fuzzy vari-

able X = "high consonant duration" over the universe of the possible
values of u, which represents time in this example.

L_{14} and H_{14} are the so called "break-points" of the possibility
distribution; they define the interval in which the membership values are neither 0 nor 1.

The fuzzy variables form a basis for a composite question Q.
An answer to Q may be interpreted as a specification of the grade of
evidence of a phonetic feature in a speech segment.

This grade is a function of the grades of membership of the
speech segment with every fuzzy variable.

The membership functions associated with each restriction were
established subjectively after inspection of the distribution of
acoustic measurements made on a large number of sound samples.

Let Σ be the alphabet of the linguistic variables that may be
answers to the component questions of the "questionnaire" used to
characterize the phonetic feature (FF). Obviously $X \in \Sigma$. Let p
be an acoustic pattern represented by its description in terms of
acoustic cues. We are interested in the following possibility:
Poss {(FF) is in p}. Let $D \in \Sigma *$ be a string of linguistic variables;
the string D identifies a composite fuzzy restriction $R_p(D)$ defined
on some of the acoustic attributes of p. There may be many descriptions of the same pattern.

The possibility that a feature FF is present in a pattern p is
obtained by the following composition:

$$R_p(FF) = R_p(D) \circ R((FF), D) \qquad (3.28)$$

and is computed as follows:

$$\text{Poss } \{(FF) \text{ is in } p\} = \bigvee_{D \in \Sigma*} \{\mu_p(D) \wedge \mu((FF), D)\} \qquad (3.29)$$

where $\vee$ is the *max* operator, $\wedge$ is the *min* operator, $\mu_p(D)$ is the
compatibility of the string D with the pattern p and $\mu((FF), D)$ is
the compatibility of the name (FF) with the string D.

The classification of a consonant as "sonorant" or "nonsonorant"
is performed after the detection of vocalic intervals and is obtained
as an answer to a composite question: Q_1 = How is the interval corresponding to the acoustic pattern p?

The question Q_1 has six component questions:

- Q_{11} = how is R?

- Q_{12} = how is the minimum dip of R with respect to the values of R in the preceding and following vowels?

- Q_{13} = how is the minimum value of S with respect to the values assumed by S on the silences?

- Q_{14} = how is the duration of the signal dip?

- Q_{15} = how is the minimum dip of the signal?

- Q_{16} = how is the maximum dip of R?

Each question admits two possible answers: low or high. These answers will be indicated as:

$$l_{1i} , h_{1i}$$

where the subscripts $1i$ (i = 1,2...,6) refer to the answer to the i-th question; it is assumed that:

$$l_{1i} = \overline{h}_{1i} .$$

The answers to such questions are fuzzy linguistic variables defined over the ranges in which the parameters they refer to may vary. Such parameters are defined as follows:

- u_{11} = R

- u_{12} = min $(R_p - R_c : R_f - R_c)$

- u_{13} = $\min_{S \in dip}$ $(S - S_{sil})$

- u_{14} = consonant duration

- u_{15} = min $(S_p - S_c ; S_f - S_c)$

- u_{16} = max $(R_p - R_c ; R_f - R_c)$

where the subscript p refers to the detected vowel preceding the interval in which a nonsonorant feature is being sought; the subscript f refers to the detected vowel following the interval, the subscript c refers to the consonant interval and S_{sil} is the level of the signal energy in the silence.

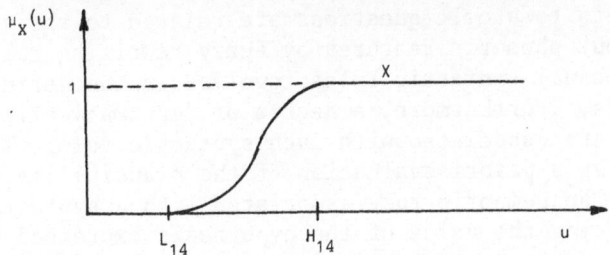

Fig. 3.11 Compatibility function of the fuzzy variable x : = high
         consonant duration

The membership functions were defined after consideration of
the range of the above parameters for each consonant in every context, assigning values different from 1 and from zero to $\mu_{h_{1i}}$ and

$\mu_{1_{1i}}$ (i = 1,2,...,6) only in the range where sonorant and nonsonorant

sounds may coexist.

Notice that $\mu_{1_{1i}}$ is for μ_{low} (u_{1i}) and $\mu_{h_{1i}}$ is for μ_{high} (u_{1i});
furthermore $\mu_{1_{1i}}$ = 1 - $\mu_{h_{1i}}$.

The fuzzy restrictions obtained from the experiments and corresponding to the linguistic values *high* have the diagram shown in
Fig. 3.11; the break points, indicated as L_{14} and H_{14} in Fig. 3.11
are the bounds of the interval where the memberships are neither 0
nor 1.   In general, the break points for the variable h_{ji} will be
indicated as L_{ji} and H_{ji}; these parameters assume different values
depending on the questions, as listed in Table 3.2.

TABLE 3.2

Variable	L_{ji}	H_{ji}	Dimension
u_{11}	-15	7	dB
u_{12}	- 3	21	dB
u_{13}	2	20	dB
u_{14}	50	130	ms
u_{15}	6	23	dB
u_{16}	0	24	dB

The answers to atomic questions are related to the values of hypotheses about phonetic features by fuzzy rewriting rules. These rules are the usual syntactic rules involved in the definition of phrase grammars. Furthermore, a degree of "grammaticality" and a semantic rule are associated with each syntactic rule. The grammaticality is an a priori evaluation of the plausibility of the syntactic rule. The semantic rule associated with a syntactic one is used for computing the value of the hypothesis expressed by the left-side member of the syntactic rule as a function of the membership functions of the right-side components.

A set of fuzzy rules has been inferred for defining the syntactic categories "sonorant" and "nonsonorant" that correspond to the answer to Q_1. The inferred rules are given by the following:

$$\Pi_1$$

<nonsonorant> $\xrightarrow{1}$ <dip of S>   (P11)

<nonsonorant> $\xrightarrow{1}$ <dip of R>   (P12)

<dip of S> $\xrightarrow{0.85}$ $l_{13}P_S$ (P13)

<dip of S> $\xrightarrow{0.9}$ $h_{14}P_S$ (P14)

<dip of S> $\xrightarrow{1}$ $h_{15}P_S$ (P15)

<dip of S> $\xrightarrow{1}$ $h_{14}l_{13}P_S$ (P16)

The notation <dip of S> $\xrightarrow{0.85}$ l_{13} means that l_{13} (a low difference between the minimum of S and the silence level) allows us to generate the hypothesis that there is a dip of S with 0.85 plausibility.

The semantic rule associated with the syntactic definition of <dip of S> is.

$$\mu_{\text{<dip of S>}} = P_S \wedge \{(0.85 \wedge \mu_{l_{13}}) \vee (0.9 \wedge \mu_{h_{14}}) \vee$$
$$\vee \mu_{h_{15}} \vee (\mu_{l_{13}} \wedge \mu_{h_{14}}) ;$$ (3.30)

P_S is a boolean variable defined as follows:

$$P_S = \begin{cases} 1 \text{ if } (S-S_{sil}) \le 8 \text{ dB} \\ 0 \text{ if } (S-S_{sil}) > 8 \text{ dB} \end{cases}$$

$$<\text{dip of R}> \xrightarrow{0.6} 1_{11} \tag{P19}$$

$$<\text{dip of R}> \xrightarrow{0.9} h_{12} \tag{P110}$$

$$<\text{dip of R}> \xrightarrow{0.9} h_{16} \tag{P111}$$

$$<\text{dip of R}> \xrightarrow{1} 1_{11} h_{12} \tag{P112}$$

$$<\text{dip of R}> \xrightarrow{1} 1_{11} h_{16} \tag{P113}$$

$$<\text{dip of R}> \xrightarrow{1} h_{12} h_{16} \tag{P114}$$

The semantic rule associated with the definition of $<\text{dip of R}>$ is:

$$\mu_{<\text{dip of R}>} = (0.6 \wedge \mu_{1_{11}}) \vee (0.9 \wedge \mu_{h_{12}}) \vee (0.9 \wedge \mu_{h_{16}}) \vee$$
$$\vee (\mu_{1_{11}} \wedge \mu_{h_{12}}) \vee (\mu_{1_{11}} \wedge \mu_{h_{16}}) \vee (\mu_{h_{12}} \wedge \mu_{h_{16}}). \tag{3.31}$$

The symbols $\vee$ and $\wedge$ represent the *max* and the *min* operators respectively.

The definition of the syntactic category $<\text{sonorant}>$ could be simply established as: $<\text{sonorant}> = \overline{<\text{nonsonorant}>}$.

Nevertheless, it has been found that this phonetic feature is better characterized in terms of fuzzy rules involving the answers to component questions that chiefly characterize the sonorant consonants. These elements are the complements of the elements appearing in the definition of nonsonorant.

Π_2

$$<\text{sonorant}> \xrightarrow{0.9} <\text{SRV}> \tag{P21}$$

$$<\text{sonorant}> \xrightarrow{0.9} <\text{SS}> \tag{P22}$$

$$<\text{sonorant}> \xrightarrow{1} <\text{SRV}> <\text{SS}> \tag{P23}$$

$$<\text{SRV}> \xrightarrow{0.85} h_{11} \tag{P24}$$

$$<\text{SRV}> \xrightarrow{0.85} 1_{12} \tag{P25}$$

$$<\text{SRV}> \xrightarrow{1} 1_{16} \tag{P26}$$

$$<SRV> \xrightarrow{1} h_{11} \, l_{12} \tag{P27}$$

$$<SS> \xrightarrow{1} l_{14} \tag{P28}$$

$$<SS> \xrightarrow{1} l_{15} \, p_S \tag{P29}$$

The semantic rules associated with the definition of <SRV> and <SS> are given as follows:

$$\mu_{<SRV>} = (0.85 \wedge \mu_{h_{11}}) \vee (0.85 \wedge \mu_{l_{12}}) \vee \mu_{l_{16}} \vee (\mu_{l_{12}} \wedge \mu_{h_{11}})$$

$$\mu_{<SS>} = \mu_{l_{14}} \vee (p_S \wedge \mu_{l_{15}}).$$

The measure of the hypotheses <sonorant> and <nonsonorant> are computed considering <dip of S>, <dip of R>, <SRV>, <SS> as descriptions of the acoustic pattern p and the grammaticalities of Π_1 and Π_2 as memberships of the fuzzy relations between the left side and the right side phrases of the rules. One gets:

$$\text{Poss } \{<\text{sonorant}> \text{ is in } p\} = \mu_{<\text{dip of S}>} \vee \mu_{<\text{dip of R}>} \tag{3.32}$$

$$\text{Poss } \{<\text{nonsonorant}> \text{ is in } p\} = (\mu_{<SRV>} \wedge 0.9) \vee (\mu_{<SS>} \wedge 0.9) \vee$$

$$\vee (\mu_{<SRV>} \wedge \mu_{<SS>}) \tag{3.33}$$

Example 3.6.1

The following values are assumed by the parameters u_{1i} ($i = 1,2, \dots, 6$) for the consonant /n/ of an utterance of the pseudo-syllable /oni/ (See Fig. 3.12):

$$u_{11} = 15.5 \quad \text{dB}$$

$$u_{12} = -20 \quad \text{dB}$$

$$u_{13} = 10.4 \quad \text{dB}$$

$$u_{14} = 100 \quad \text{ms}$$

$$u_{15} = 7.45 \quad \text{dB}$$

$$u_{16} = 4.5 \quad \text{dB}$$

Using the diagram of Fig. 3.11 with the break points given by Table 3.2 one gets:

$$\mu_{h_{11}} = 1, \qquad \mu_{l_{11}} = 0,$$

$$\mu_{h_{12}} = 0, \qquad \mu_{l_{12}} = 1,$$

$$\mu_{h_{13}} = 0.55, \qquad \mu_{l_{13}} = 0.45,$$

$$\mu_{h_{14}} = 0.7, \qquad \mu_{l_{14}} = 0.3,$$

$$\mu_{h_{15}} = 0.1, \qquad \mu_{l_{15}} = 0.9,$$

$$\mu_{h_{16}} = 0.2, \qquad \mu_{l_{16}} = 0.8,$$

Applying the semantic rules one gets:

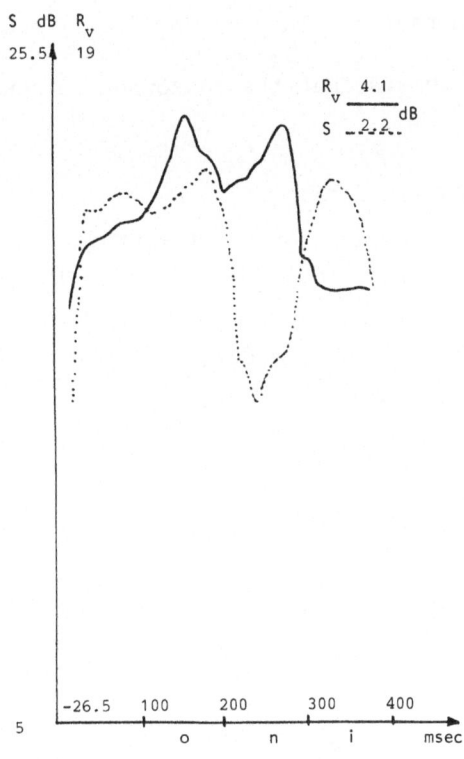

Fig. 3.12 Time evolutions of the parameters S and R for the pseudo-syllable-segment /oni/

$\mu_{<dip\ of\ S>}$ = Min {0, [Max(Min(0.85,0.45),Min(0.9,0.7),
                        Min (0.45,0.7), 0.1]} = 0

$\mu_{<dip\ of\ R>}$ = Max {Min(0.6,0),Min(0.9,0),Min(0.9,0.2),
                        Min(0,0),Min(0,0.2),Min(0,0.2)} =
                     = Max(0,0,0.2,0,0,0) = 0.2

$\mu_{<SRV>}$ = Max {Min(0.9,1),Min(0.9,1),Min(1,1),0.8} =
                   = Max(0.9,0.9,1,0.8) = 1

$\mu_{<SS>}$ = Max(0.3,0.9) = 0.9

Poss {sonorant is in p} = Max {Min(0.85,1),Min(0.9,0.9),
                              Min(1,0.9)} = Max(0.85,0.9,0.9) =
                           = 0.9

Poss {<nonsonorant> is in p} = Max(0,0.2) = 0.2

Thus the hypothesis that the consonant is sonorant is more consistent.

Example 3.6.2

Using the same approach as for the preceding example, for the pseudo-syllable /edi/, the following values have been measured (see Fig. 3.13);

u_{11} = -13 dB $\mu_{h_{11}}$ = 0.1, $\mu_{1_{11}}$ = 0.9

u_{12} =   0 dB $\mu_{h_{12}}$ = 0.8, $\mu_{1_{12}}$ = 0.2,

u_{13} =   0 dB $\mu_{h_{13}}$ = 0, $\mu_{1_{13}}$ = 1,

u_{14} = 100 ms $\mu_{h_{14}}$ = 0.7, $\mu_{1_{14}}$ = 0.3,

u_{15} = 16.7 dB $\mu_{h_{15}}$ = 0.6, $\mu_{1_{15}}$ = 0.4,

u_{16} = 5.6 dB $\mu_{h_{16}}$ = 0.4, $\mu_{1_{16}}$ = 0.6,

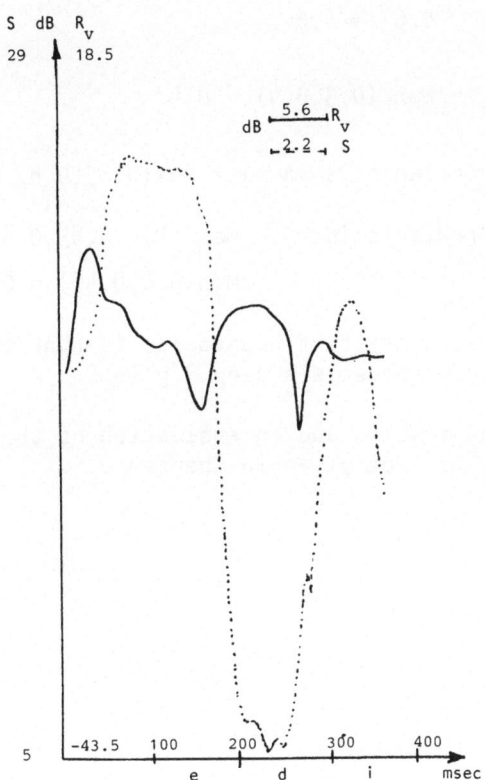

Fig. 3.13 Time evolutions of the parameters S and R for the
          pseudo-syllable-segment /edi/

$$\mu_{<dip\ of\ S>} = Min\ \{1,\ [Max(Min(0.85,1),Min(0.9,0.7),$$
$$Min(1,0.7),0.6)]\} = 0.85$$

$$\mu_{<dip\ of\ R>} = Max\ \{Min(0.6,0.9),Min(0.9,0.8),Min(0.9,0.4),$$
$$Min(0.9,0.8),Min(0.9,0.4),Min(0.8,0.4)\} =$$
$$= Max(0.6,0.8,0.4,0.8,0.4,0.4) = 0.8$$

$$\mu_{<SRV>} = Max\ \{Min(0.9,0.1),Min(0.9,0.2),Min(0.2,011),$$
$$0.6\} = 0.6$$

$$\mu_{<SS>} = Max\ (0.3,0.4) = 0.4$$

Poss $\{<nonsonorant>$ is in $p\} = Max(0.85,0.8) = 0.85$

Poss $\{<sonorant>$ is in $p\} = Max\ \{Min(0.85,0.6),Min(0.9,0.4),$
$$Min(0.6,0.4)\} = 0.6$$

Thus the most consistent hypothesis is that the consonant is nonsonorant because there is a deep dip in S.

Experimental results and an evaluation of the performances of these algorithms will be given in Chapter 5.

3.7   References

Fuzzy sets were first introduced by Zadeh (1965). A good presentation of the theory of fuzzy sets can be found in a book by Kaufmann (1973). One of the most recent bibliographies on fuzzy set theory, with more than 500 references, is contained in a paper by Kandel and Byatt (1978). The concepts of fuzzy linguistic variables, fuzzy restrictions, fuzzy relations, and fuzzy languages are described in papers by Zadeh (1973, 1975).

Recent books by Chorayan (1979), Dubois and Prade (1980) and Bezdek (1981) report advances in fuzzy set theory and applications.

The application of branching questionnaires to the definition of complex and imprecise concepts is described by Zadeh (1976). The introduction of fuzzy grammars for interpreting continuous speech is described in a paper by De Mori and Laface (1980).

An excellent discussion on Problem Reduction Representation, State Space Representation, equivalence between AND/OR graphs and context-free grammars, non-directional parsers can be found in the Ph.D. Thesis by Stockman (1977) at the University of Maryland.

Spectrogram reading experiments are reported by Klatt and Stevens (1973) and in the book edited by Cole (1980).

Journals that often contain material related to syntactic pattern recognition and fuzzy set theory include IEEE Transactions on Pattern Analysis and Machine Intelligence, IEEE Transactions on System, Man and Cybernetics, Pattern Recognition, Fuzzy Sets and Systems, Information and Control, Information Sciences, Communications of the ACM, Journal of the ACM, Mathematical System Theory, International Journal of Man-Machine Studies.

4. DESIGN PRINCIPLES FOR CONTROLLING

THE USE OF STRUCTURAL RULES FOR SEGMENTATION

4.1 The Meaning of the Meaning

Once the possibility that a feature hypothesis $H \triangleq f$ is in the pattern p has been computed, a precise number $\mu_H(p)$ can be obtained. But, during the computation of this number, many subjective assumptions have been made, i.e., the definitions of primary fuzzy sets, the assignment of grammaticalities, the use of a particular fuzzy algebra for defining the semantic rules. Thus, because the procedure is affected by imprecision, it would be inappropriate to say that the evidence of a hypothesis is a precise number. Evidence values should be affected by the same order of imprecision as the procedure used for computing them. This is obtained by allowing the values of a judgement of evidence to be fuzzy subsets of the interval [0,1] on which the possibilities of feature hypotheses are defined.

These fuzzy sets with fuzzy memberships are called type-2 fuzzy sets. This definition can be generalized.

Definition 4.1.1

A fuzzy set is of *type-n* (n>2) if its membership function ranges over fuzzy sets of type n-1.

The membership function of a membership value in our approach is defined as a *fuzzy number*, i.e., a continuous, normal, convex fuzzy subset of the interval [0,1].

A fuzzy number is a "bell-shaped" fuzzy set (continuous, convex) and the maximum value of its membership function is 1. The

algebra of fuzzy numbers (see Dubois and Prade, 1978) offers definitions extended to fuzzy numbers of the usual operations of real-
algebra.

The membership distribution of a membership value can be approximated with standard fuzzy sets defining restrictions of fuzzy
variables representing the elements of the answer set of a branching questionnaire.

This is equivalent to modeling the process of hypothesis generation by a branching questionnaire of composite attributional
questions, such as: "How evident is the hypothesis 'sonorant' in the
data?" and an answer set containing, for example: "The hypothesis
is very likely to be in the pattern". Here "sonorant" is the body
of the composite question and the answer set contains qualitative
evidence judgments like "very likely", "uncertain", etc.

These approximations allow one to express the evidence of a
hypothesis by the linguistic value which is the label of the fuzzy
variable that best approximates the fuzzy number built on the membership value computed by fuzzy algorithms.

The vagueness of this linguistic value represents the vagueness
of the entire process executed for getting it and is the meaning of
the meaning represented by a membership value.

Control strategies based on inference rules can be designed
with these linguistic values and an *Approximate Reasoning* can be
made with them. This item will be treated in Chapter 8.

Another interesting aspect is that "a posteriori" *linguistic
probabilities* can be obtained from imprecise data and rules.

The idea of introducing linguistic values of evidences or
probabilities comes from the following motivations.

Fuzzy algorithms are essentially a tool for reducing the dimensions of a problem, giving a single measure of trustworthiness
for a complex hypothesis generated by interpreting a certain
amount of unclear data under the control of an imprecise knowledge.
For example, a phonetic feature is related to many acoustic cues
detected with imprecision; furthermore, the relations are often
imprecise rules.

Acceptable statistics of these imprecisions are practically
not feasible and fuzzy algorithms are the proper tool for dealing
with this kind of vagueness, as such imprecisions can never be expressed by precise numbers.

Once the dimensions of a problem are reduced and a degree

of compatibility μ_H of a hypothesis has been obtained, the a posteriori probability:

$$Pr\ (H\ |\ \mu_H) = \frac{Pr(\mu_H|H)\ Pr(H)}{Pr(\mu_H)} \tag{4.1}$$

could be computed with some approximations.

The reasons for the imprecision of such a calculation are in
the imprecision of μ_H and of the knowledge of the a priori proba-
bilities, $Pr(\mu_H|H)$ which are usually estimated from histograms of a
limited volume of data. Notice that $Pr(\mu_H)$ may be computed by sum-
ming all the a priori probabilities, namely the probability of μ_H,
given H, and of μ_H, given all the hypotheses competing with it.

Furthermore, we are really interested in the total probability
that H is in the pattern, but we can only estimate the probability
that H is in the pattern, given μ_H.

For all these reasons, because the probabilities we need cannot
be computed precisely, linguistic probability values, defined as
fuzzy sets over the interval [0,1], are introduced; the imprecision
of an a posteriori probability is represented by a fuzzy number which
is described by the linguistic probability whose restriction best
approximates such a fuzzy number.

Eventually, we do not need precise numbers for measuring the
evidence. Qualitative judgments are adequate and suitable for designing a control strategy based on inference rules.

Often researchers on Artificial Intelligence have raised the
problem of how to measure evidences and combine them when the data
are so complex that there are limited samples for each combination
and acceptable statistics cannot be obtained.

One of the best solutions proposed to solve this problem is
that of using interval functions for combining evidences and performing an Approximate Reasoning when a system has to decide how to
continue an interpretation or a deduction process.

The approach proposed here is more general, because a fuzzy
set can be seen as a collection of intervals (every t-level set
identifies an interval) and a fuzzy algebra is a more powerful tool
than an interval algebra.

These aspects of high level control strategy will be discussed
in Chapter 8.

4.2   The Control Problem in the
Segmentation Process

In this section, a realization of the principles introduced in the previous section will be described in detail.

One of the operations which should be performed by a Speech Understanding System in a "bottom-up" way is the segmentation of the acoustic pattern into Pseudo-Syllable-Segments (PSS). The data used for this operation are phonetic features that have been hypothesized by the use of context-independent rules. Let us call these features Context-Independent Features (CIF). The *result* of segmentation is a collection of segments.

These segments may overlap in time, reflecting ambiguities in the hypothesization of CIFs in the acoustic pattern. Each segment contains a single sequence of CIFs that will be used for driving a more detailed extraction of acoustic cues and a further hypothesization of phonetic and phonemic features. Coarticulation effects are expected only inside syllabic segments.

The structural knowledge used by the segmentation process is a grammar whose rules define all the possible phonetic compositions of a PSS and all the relations between each phonetic feature and some acoustic cues to be detected in the pattern. Let us denote as SKS (Segmentation Knowledge Source) such a structural knowledge.

The procedural knowledge for using the SKS is a parsing algorithm characterizing a Segmentation Control Knowledge Source (SCKS).

The SKS and the SCKS will be described in detail in Section 4.4, later in this chapter.

The execution of the segmentation algorithm is scheduled by the General Controller (Strategy KS) after the acoustic pattern has been obtained and the hypotheses about the feature VOCALIC have been generated.

Whether the execution of these processes is strictly sequential or is concurrent, provided that some synchronization rules are respected, is an implementation problem which is not treated here.

The execution of the segmentation algorithm is not a parsing in the classical sense, because the input is only partially available when the syntactic rules are applied.

Rather than for recognition, the rules are used for *predicting* possible CIFs in a PSS.

A prediction results in a request to the Acoustic Control KS

(ACKS) to extract some Acoustic Cues (AC).

This situation is sketched in Fig. 4.1.

Once the acoustic cues and their degrees of evidence have been written into the blackboard and a synchronization message has been sent to the segmentation controller, SCKS, the segmentation algorithm has to evaluate the evidence of a phonetic feature and to decide whether to use it for establishing PSS bounds or to discard it because its evidence is too poor. If the decision criteria are too severe, then the right hypothesis will often be discarded, resulting in a high rate of segmentation errors; if the decision criteria are too generous, then there will be too many ambiguities.

Because segmentation is one of the preliminary processes of the system, ambiguities can be tolerated, but missing the right hypothesis should be a very rare event. Based on these considerations, the following control statement (CS1) turns out to be a good rule for deciding when a hypothesis has to be discarded:

A PHONETIC HYPOTHESIS HAS TO BE DISCARDED IF IT IS EXTREMELY UNLIKELY TO BE IN THE PATTERN.

The phonetic knowledge used for segmentation can be represented by a grammar which incorporates a binary predicate whose truth depends on the application of the control statement CS1 to the evidence of a particular feature. The introduction of such a predicate will be informally discussed now, whereas a formal presentation of the segmentation grammar and of its parsing algorithm will be given in Section 4.4.

Fig. 4.2 shows a portion of the state diagrams of the knowledge used for analyzing an interval between two successive segments for which the label VOCALIC has been hypothesized.

Let this intervocalic segment me indicated as IVS (t_i, t_j).

For each IVS (t_i, t_j), the segmentation algorithm, or the SCKS, schedules the execution of a process that, after requesting to and obtaining from the acoustic controller, ACKS, the memberships of the acoustic cues introduced in Sec. 3.6, applies the rules $\prod_1$ and $\prod_2$ of Sec. 3.6, obtaining the possibilities

Poss {<sonorant> is in p (t_i, t_j)} and

Poss {<nonsonorant> is in p (t_i, t_j)} .

The pattern p (t_i, t_j) is made of the time evolutions of S,A,R

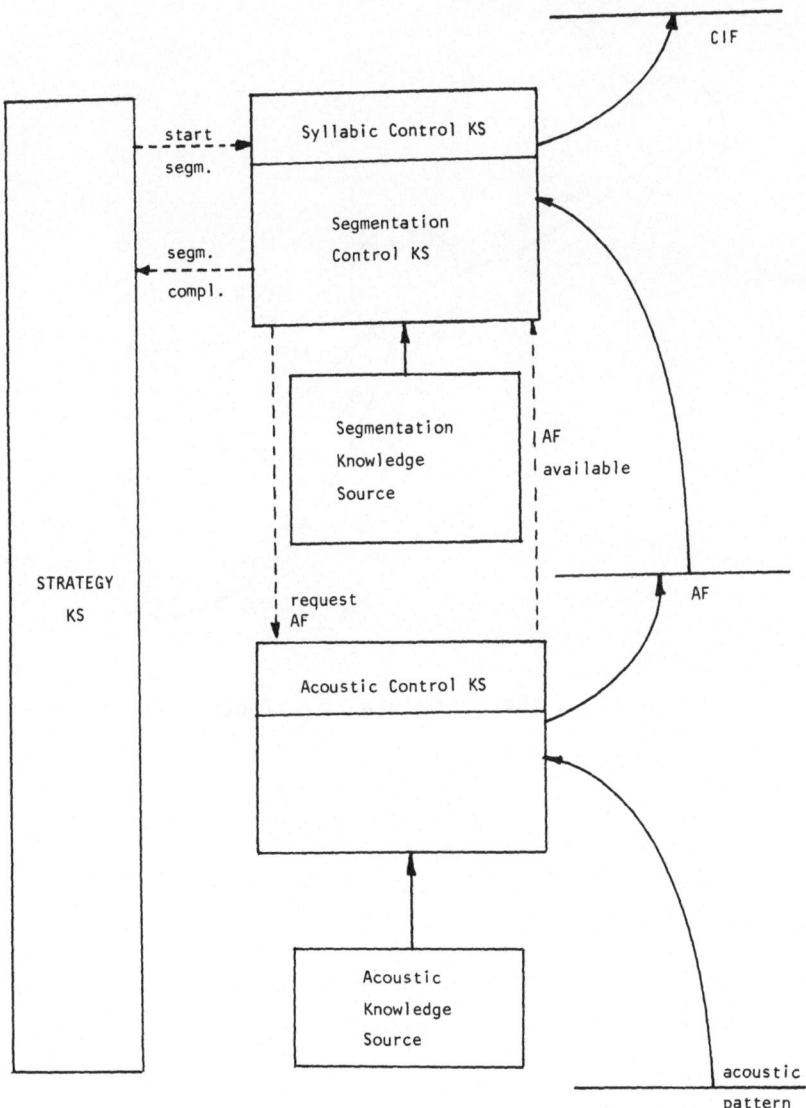

Fig. 4.1 Scheme for the extraction of Context-Independent Features

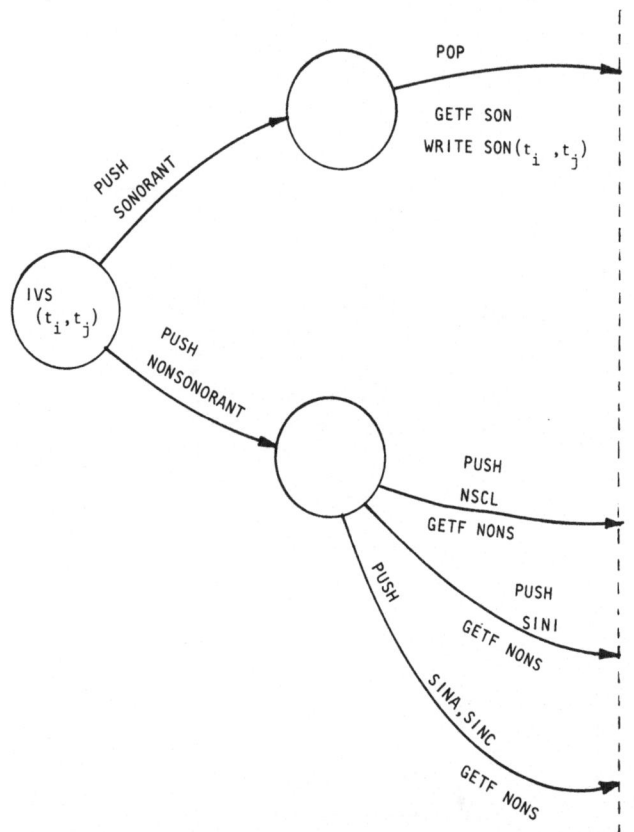

Fig. 4.2 A fragment of the procedural knowledge used for segmen-
         tation

(see Sec. 3.6 for definition) in the interval (t_i, t_j). For the
sake of brevity, let us indicate the above possibilities as μ_S and
μ_N respectively.

According to Fig. 4.2, the SCKS generates the hypothesis that
the whole IVS (t_i, t_j) is "sonorant", performing the action: "write
SON (t_i, t_j)", or it may accept the hypothesis that there is a nonsonorant segment in IVS (t_i, t_j).

In the latter case, SCKS will look to see whether IVS (t_i, t_j)

contains a consonant cluster NSCL with at least a nonsonorant consonant or if IVS (t_i, t_j) is a single-intervocalic-nonsonorant-consonant
which may be affricate (SINA), continuant (SINC) or interrupted
(SINI).

The possibility of following an arc in Fig. 4.2 depends on the
value of the binary variables GETF SON and GETF NONS, indicated just
below the arcs.

The binary variable GETF SON is false only if the hypothesis
"sonorant" respects the control statement CS1; analogously, the binary variable GETF NONS is false only if the hypothesis "nonsonorant"
respects CS1. As the two phonetic features are defined by two fuzzy
grammars whose semantic rules are not complementary, the binary
variables can be both true, yielding a request of parallel execution
of the processes invoked by the phonetic labels of the arcs in Fig.
4.2, which have GETF SON and GETF NONS as conditions. This is a
case of "beam-search" on a "problem space", a fragment of which is
represented in Fig. 4.2.

The remaining part of this section will be devoted to a discussion of the assignment of truth values to GETF SON and GETF NONS,
based on the control statement CS1.

Fig. 4.3 shows the a priori probability densities of the hypotheses:

$H_S \triangleq$ feature "sonorant" is in p

and

$H_N^{\cdot} \triangleq$ feature "nonsonorant" is in p

defined over the parameter:

$x = \mu_S - \mu_N$; obviously:

$-1 \leq x \leq 1$.

These distributions are approximate and we are interested in
defining over the interval $[-1,1]$ some fuzzy sets representing the
judgments of evidence for the features "sonorant" and "nonsonorant".

The set of labels of these fuzzy sets is the answer set of
composite attributional questions of the type "How evident is the
phonetic feature sonorant in the pattern p?".

As we cannot avoid imprecisions in establishing the evidence
of a hypothesis, let us now consider in detail how they force us to

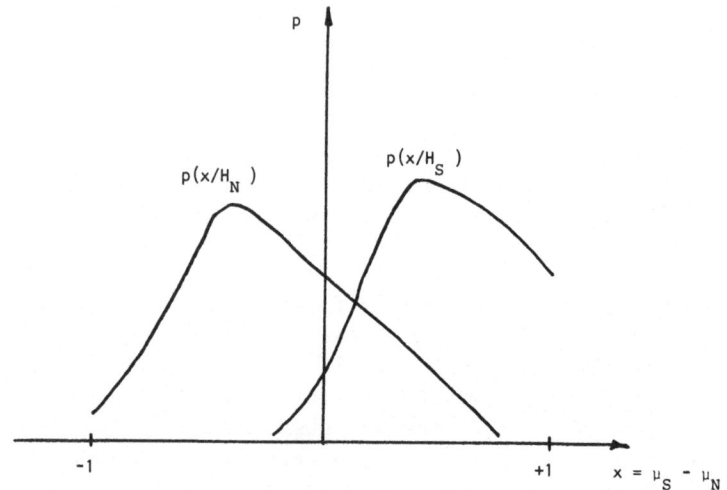

Fig. 4.3   Probability densities for the hypotheses H_S and H_N as functions of $x = \mu_S - \mu_N$

express an evidence with a qualitive, ill-defined judgment.

Given an intervocalic segment, the memberships μ_S and μ_N are computed and a value x^* of x is obtained. This value is an imprecise measure of evidence; this imprecision is accounted for by considering a possibility distribution around x^* and saying that this possibility distribution is a fuzzy set labelled "close to x^*", abbreviated as "x^*". As the actual goal is that of applying the control statement CS1, based on the evidence of the hypothesized features, a correspondence between the linguistic probability value EXTREMELY UNLIKELY and the fuzzy set "x^*" has to be established in order to identify some intervals along the x-axis where GETF SON and/or GETF NONS are true.

This gives the opportunity of introducing the concept of linguistic probability, providing new tools for Approximate Reasoning.

4.3 Computation with Linguistic Probabilities

Definition 4.3.1

A *linguistic probability* is a fuzzy linguistic variable defined over the interval [0,1], where probability values may vary. The label of the restriction is a probability term and the restriction is a possibility distribution over probability values.

Fig. 4.4 shows the definition of some linguistic probabilities with the corresponding fuzzy restrictions on the unit interval $[0,1]$ of the axis of numerical probability curves. Curve 1 is the possibility distribution of the base variable *likely*, curve 2 is for *not likely*, curve 3 is for *unlikely*, curve 4 is for *extremely unlikely* which can be assumed to be "very, very, very ......very unlikely".

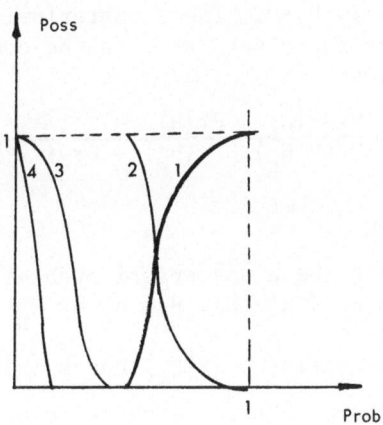

Fig. 4.4 Definitions of some linguistic probabilities

Definition 4.3.2

According to Zadeh (1975), a *term-set* T_p *of linguistic probabilities*, all defined over a universe U_p, can be expressed as follows:

T_p = likely + not likely + unlikely + very likely + more or less likely + very unlikely + ...,

where the operator + indicates the union.

U_p may be either the unit interval $[0,1]$ or the finite set:

U_p = {0 + 0.1 + ... + 0.9 + 1};

the fuzzy linguistic variable *likely* plays the role of a base variable and is defined by a possibility distribution over U_p as shown in Fig. 4.4. Let $p \in U_p$ be a numerical variable, let $\mu_{likely}(p)$ be the possibility distribution of "likely"; the following definitions hold:

$$\mu_{not\ likely}(p) = 1 - \mu_{likely}(p) \qquad (4.2)$$

$$\mu_{\text{unlikely}} (p) = \mu_{\text{likely}} (1 - p) \tag{4.3}$$

Given a measure of evidence $x^* = \mu_S - \mu_N$, $-1 \le x^* \le + 1$, the imprecision of the calculation of μ_S and μ_N is represented by a possibility distribution $\pi_{x^*} (x)$ across x^*. This possibility distribution may be equal to the membership function of the fuzzy set "close to x^*" or, briefly "x^*". The a posteriori probability of a hypothesis H, given the fuzzy set "x^*", can be computed by applying Bayes' Theorem as follows:

$$Pr (H|"x^*") = \frac{Pr ("x^*"|H) \quad Pr (H)}{Pr ("x^*"|H_S) \; Pr (H_S) + Pr ("x^*"|H_N) \; Pr (H_N)} \tag{4.4}$$

where H may be H_S or H_N.

In order to compute the a posteriori probability in (4.4), the a priori probabilities $Pr ("x^*"|H), (H = H_S$ or $H_N)$ have to be computed.

The probability densities $p (x|H_S)$ and $p (x|H_N)$ are not known precisely; we shall assume that for each value of x an interval is known, inside which it is very likely to find the true probability density value $p (x|H)$. By joining the extremes of such intervals the two curves $p - (x|H)$ and $p + (x|H)$ are obtained, as shown in Fig. 4.5 together with the fuzzy set "x^*".

Given now a value t of $\pi_{x^*} (x)$ and the corresponding t-level-set π_t, the a posteriori probability:

$$Pr^+ (H| \; x \in \pi_t)$$

can be obtained by the following relations:

$$Pr^+ (H_S| \; x \in \pi_t) = \frac{1}{1 + ry} \tag{4.5}$$

$$Pr^+ (H_N| \; x \in \pi_t) = \frac{1}{1 + \dfrac{1}{ry}} \tag{4.6}$$

$$r = \frac{Pr (H_N)}{Pr (H_S)} ; \; y = \frac{Pr^+ (x \in \pi_t|H_N)}{Pr^+ (x \in \pi_t|H_S)} \tag{4.7}$$

$Pr^+ (x \in \pi_t|H)$ is obtained by integrating $p^+ (x|H)$ over π_t. In a similar way $P_r^- (H|x \in \pi_t)$ can be obtained.

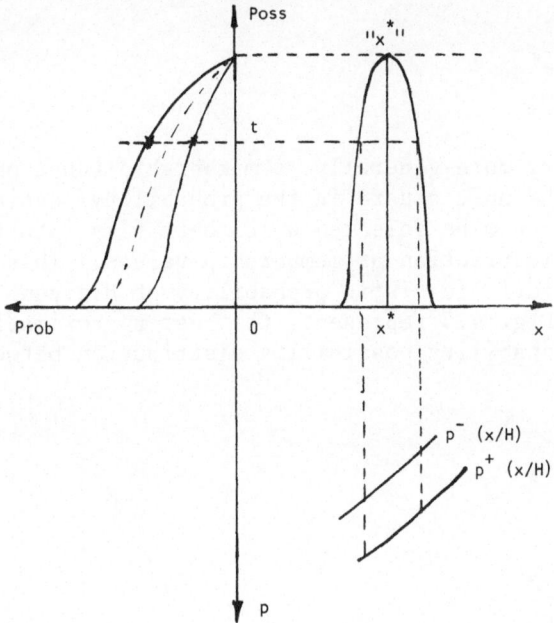

Fig. 4.5 Example of calculation of probability/possibility
          distributions

From the statistics of the consonants, it has been found that:

   Pr (H_S) = 0.4; PR (H_N) = 0.6; r = 1.5 .

By applying the extension principle, the possibility value t can
be assigned to the a posteriori probabilities Pr^+ $(H|x \in \pi_t)$ and
Pr^- $(H|x \in \pi_t)$, which may be seen as functions of t. If more than
two probability densities are considered, instead of only p - $(x|H)$
and p + $(x|H)$, and a degree of possibility is assigned to each dis-
tribution, more than two points are obtained in the possibility/
probability plane and each one of these can be assigned the possibility of the probability density curve which contributed to obtaining it.

     Now, for each x*, letting t vary between 0 and 1, two proba-
bility/possibility distributions named:

Pr^+ $(H|''x*'')$

and

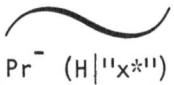

$$Pr^- (H|''x*'')$$

are obtained, or, more generally, a membership function defined on
all points of the unit square in the probability/possibility plane.
A term in Tp can now be selected which best fits into the two
curves or the distribution of membership values. This term will
be the a posteriori linguistic probability of H given "x*". The
dashed line in Fig. 4.5 represents the best approximation of the
a posteriori probability/possibility distribution between the two
curves:

$$Pr^+ (H|''x*'')$$

and

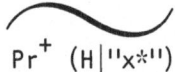

$$Pr^- (H|''x*'')$$

All the values of x* for which the curve of EXTREMELY UNLIKELY
is the a posteriori linguistic probability of a hypothesis H identify a subset of the x-axis for the points of which a control
statement like CS1 can be applied.

By repeating the same operation for other linguistic probability values, other subsets of the x-axis can be obtained and fuzzy
subsets of the x-axis can be built, having the obtained subsets of
points as level sets for some value of t.

Fig. 4.6 shows a choice of fuzzy subsets built on level sets
obtained from a posteriori linguistic probabilities.

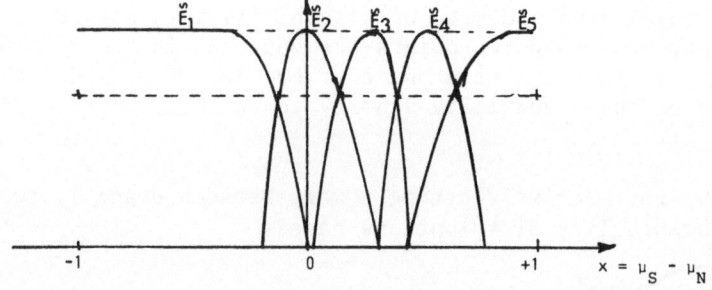

Fig. 4.6 Linguistic evidences represented by fuzzy subsets

These fuzzy subsets can be considered as linguistic values of
evidence. Table 4.1 summarizes the correspondence between linguistic a posteriori probability values, intervals in the [-1, +1] interval of the x-axis, and labels for fuzzy subsets of evidence of
the phonetic features "sonorant" and "nonsonorant". These data have
been obtained by De Mori and Saitta (1979).

The only constraint in building convex, normal evidence fuzzy
sets is that they must have a well-defined t-level set for a value
of t. As the freedom involved in this choice is high, one may try
to assign one of them a basic evidence judgment; for example one may
establish that the label "evident" corresponds to the linguistic a
posteriori probability "extremely likely". This label can be assumed as a primary term on which a term set T_E of evidences can be

built. The other evidence fuzzy subsets can be drawn from t-level
sets in order to approximate as much as possible some fuzzy restrictions corresponding to terms in T_E. This can be useful for perform-

ing Approximate Reasoning and for assigning priorities to processes
activated by evidence hypotheses. Following the approach introduced
in this section, the intervals in which GETF SON and GETF NONS are
true are defined as follows:

TABLE 4.1

linguistic probability	interval on the x-axis	fuzzy variable of evidence
Extremely unlikely H_S Extremely likely H_N	-1, -0.12	$\tilde{E}_1$
Uncertain H_S Uncertain H_N	-0.12, 0.12	$\tilde{E}_2$
Likely H_S Unlikely H_N	0.12, 0.4	$\tilde{E}_3$
Very likely H_S Very unlikely H_N	0.4, 0.65	$\tilde{E}_4$
Extremely likely H_S Extremely unlikely H_N	0.65, 1	$\tilde{E}_5$

I {GETF   SON = true}   = (-0.12, 1)

I {GETF   NONS = true}   = (-1, 0.65)

Using these intervals, a large number of experiments performed on data different from that used for learning gave an error rate of less than 1%. An error is a missing of the right hypothesis. Almost all the errors are due to a wrong extraction of acoustic cues or to the fact that the acoustic pattern was extremely unclear. Only higher levels of knowledge can recover these errors.

In spite of the width of the interval where GETF SON and GETF NONS are both true, it has been found that in more than 50% of the cases only one of these binary control features is true and a single path is followed in the problem space.

The approximations involved in obtaining the a posteriori linguistic probabilities cannot be avoided because the data are often very sparse and imprecise. For these reasons, performances with little degradation can be obtained if an expert designer decides level sets like the ones shown in Fig. 4.6. His decisions can be based on the inspection of the a priori probability density estimations and the knowledge of the absolute probabilities of the competing hypotheses. He should be a little generous in overlapping the intervals for which GETF SON and GETF NONS are both true. In this case he will keep negligible the rate of missing errors, possibly increasing the competition of hypotheses with respect to an optimal choice.

Some concluding remarks are in order at the end of this description of an application of fuzzy set theory. The approach that has been followed is *algebraic* rather than *numerical*.

The reason for this choice is that the knowledge about speech is properly represented by a logical language rather than by pure numbers.

The use of a fuzzy algebra is a choice for obtaining similarity measures between imprecise knowledge structures and corrupted descriptions. A recent paper by Simon et al. (1980), contains a discussion of this problem, showing that from the algorithmic point of view there is no deep difference between the fuzzy set and the usual probability density approach for computing similarities. The former approach is simpler and more flexible, incorporating expert knowledge.

Because of the imprecisions and approximations involved in the evaluation of similarities, fuzzy sets with linguistic labels are more suitable than pure numbers for representing evidences, which are similarity evaluations between data and hypotheses.

In the example shown in this section, a conditioned action associated to an arc like

PUSH NSCL/GETF NONS

may be seen as a process whose execution on a specific time interval of the speech data is requested by the fact that the feature "non-sonorant" is detected in the interval with enough evidence, while the hypothesis that there is a nonsonorant consonant cannot be discarded. The decision on the activation of the new process is made on the basis of the a priori probabilities of the evidences and the attempt to keep the error probability small. As these probabilities may only be known with imprecision, the decision is based on linguistic probabilities rather than on numerical ones, but this is just a different way to describe a problem solution which is available in statistical decision theory. The introduction of linguistic probabilities or, more generally, of linguistic evidences is more useful for assigning priorities to the new processes. Priorities may be based on the linguistic evidences of the features detected in the creating processes and on other control features which are known with imprecision.

A simple and effective way for obtaining priority classes based on premises expressed by linguistic judgments in a framework of Approximate Reasoning will be presented in Chapter 8.

Let us now move to the segmentation problem and the bottom-up generation of phonetic and phonemic hypotheses. For this generation, all the activated processes have the same priority because the sub-lexical hypotheses, which can be generated only from the data, have to be completed before the activation of the lexical knowledge.

4.4 Segmentation of Continuous Speech into Pseudo-Syllabic Nuclei

4.4.1 Introduction

As has been pointed out in Chapter 1, the discrete sequence of articulatory commands generates an acoustic pattern in which the acoustic correlates of these commands exhibit continuous evolutions in time as a consequence of the so-called "coarticulation effects". Based on the hypotheses of phonetic features which can be detected with context-independent rules, a segmentation algorithm will be introduced. This algorithm allows one to hypothesize the bounds of pseudo-syllabic segments (PSS), where coarticulation is expected to predominate.

For each PSS, detailed features can be further hypothesized, using context-dependent rules whose contextual constraints cannot be extended outside the PSS.

The importance of segmenting continuous speech into syllabic nuclei has been pointed out by Fujimura (1974), De Mori (1974), Mermelstein (1975).

Mermelstein (1975) and De Mori (1974) have also proposed grammars for relating syllables to acoustic cues.

A slightly different approach has been tried by Nakatsu and Kohda (1978) and Kasuya and Wakita (1978) for Japanese.

They proposed methods for extracting "inter-syllable segments", which are speech intervals between syllable centers in two successive syllables. Syllable centers may be vowels or easily detectable phonemes. A similar approach is also followed by Ruske and Shotola (1977) for German.

It will be easy to verify that the segmentation grammar described in the following is also capable of generating inter-syllable segments defined as Pseudo-Syllabic-Segments.

In spite of the interesting content of the above mentioned papers, it appears that very little has been done so far for characterizing the structure of syllables; robust efficient structural representations have been proposed for the high levels of task-dependent knowledge such as syntax and semantics.

In order to understand the types of structures of the Pseudo-Syllable-Segments, a grammar G_S for generating phonemic transcriptions of them is now introduced. This grammar has been proven useful for isolating coarticulation instances of Italian. Phoneticians expert in other languages can modify this grammar according to their knowledge. As this grammar can also apply to English, a set of phonetic classes containing the phonemes introduced in the book by Flanagan (1972) will be considered as belonging to the terminal alphabet V_{TS} of the segmentation grammar, and a segmentation example will be given for English. The segmentation grammar G_S is defined as follows:

$$G_S = \{V_{TS}, V_{NS}, P_S, PSS\}$$

where:

V_{TS} = {VOC, VFRIC, UFRIC, VSTOP, USTOP, NAS, SEMIV, GLIDES}

V_{NS} = {PSS, INTERVSON, PREVSON, POSTVSON, UN, SINIL, VSGA, VSGB, VSG, VSGBA}

The terms in V$_{TS}$ are related to the phonemes of the English
Language as follows (comments are given in square brackets):

VOC: {i, I, e, ε, æ, a, ʌ, u, ʊ, o, ɔ, ɑ} [vowels];

VFRIC: {v, ð, z, ʒ} [voiced fricatives];

UFRIC: {f, θ, s, ʃ, h} [unvoiced fricatives];

VSTOP: {b, d, g} [voiced stops];

USTOP: {p, t, k} [unvoiced stops];

NAS: {m, n, ŋ}   [nasals];

SEMIV: {l, r}   [semivowels];

GLIDES: {w, j}   [glides].

Each terminal symbol corresponds to one of the phoneme classes
described by Flanagan (1972). The phoneme symbols come also from
Flanagan's book (1972) where, for each phoneme, the configuration
of the vocal tract is given, showing the place of articulation, i.e.,
the point of maximal constriction along the vocal tract; furthermore,
examples are given of words beginning with each phoneme.

The terminal symbols in V$_{TS}$ are related to the nonterminal ones
in V$_{NS}$ by rules which are partially "context-dependent". Some of
them, belonging to P$_S$, are now introduced informally for the sake of
simplicity.

Rules in P$_S$:

UN is VFRIC or UFRIC or USTOP or VSTOP in a consonant cluster
or any sequence of them;

INTERVSON is any sequence, between two vowels, of NAS, SEMIV
and GLIDES;

PREVSON is any sequence, after a sequence labelled UN and before a vowel, of NAS, SEMIV and GLIDES;

POSTVSON is any sequence, before a sequence labelled UN and after a vowel, of NAS, SEMIV and GLIDES.

SINIL is a VSTOP between two vowels.

The other rules of P$_S$ are the following:

1) PSS → UN VSGA

2) PSS   →   VOC   VSGB

3) VSGA   →   VSG

4) VSGA   →   PREVSON   VSG

5) VSGB   →   VSG

6) VSGB   →   VSGBA   VSG

7) VSG   →   VOC

8) VSG   →   VOC   POSTVSON

9) VSGBA →   INTERVSON

10) VSGBA →   SINIL

Semivowels are often called liquid or oral consonants.

Examples of derivations using P_S, with the index of the applied rule associated to each arrow, are:

PSS $\overset{1}{\Rightarrow}$ UN   VSGA $\overset{4}{\Rightarrow}$ UN   PREVSON   VSG $\overset{8}{\Rightarrow}$ UN   PREVSON

VOC   POSTVSON $\Rightarrow$ ... $\Rightarrow$ USTOP SEMIV VOC NAS $\Rightarrow$ ... $\Rightarrow$ tran.

PSS $\overset{1}{\Rightarrow}$ UN   VSGA $\overset{3}{\Rightarrow}$ UN   VSG $\overset{8}{\Rightarrow}$ UN VOC   POSTVOC $\Rightarrow$ ...

... $\Rightarrow$ (UFRIC) $\overset{2}{\Rightarrow}$ VOC   SEMIV $\Rightarrow$ ... $\Rightarrow$ sfer.

By concatenating the two PSSs obtained by the above derivations, the word "transfer" can be generated.

For segmenting continous speech, the rules of P_S introduced informally have to be written in such a way that they can be used even if two successive PSSs have a vowel in common; the input is a lattice of acoustic cues, and the output may be a lattice of partially overlapping PSS hypotheses. The possibility of having a vowel in common between two PSSs is required by the fact that in a sequence like VOC NAS VOC NAS VOC, the vowel between the two nasals affects both of them by coarticulation. These problems are solved by modifying the *Segmentation Grammar*, which is a structural KS, and by introducing a *parser* especially designed for this particular application.

Such a parser represents the procedural knowledge according to the scheme of Fig. 4.7; this scheme is similar to the one of Fig. 4.1, except that the Segmentation Parser uses the features VOCALIC, which have to be hypothesized before applying any rule, and schedules the requests for the generation of hypotheses about context-

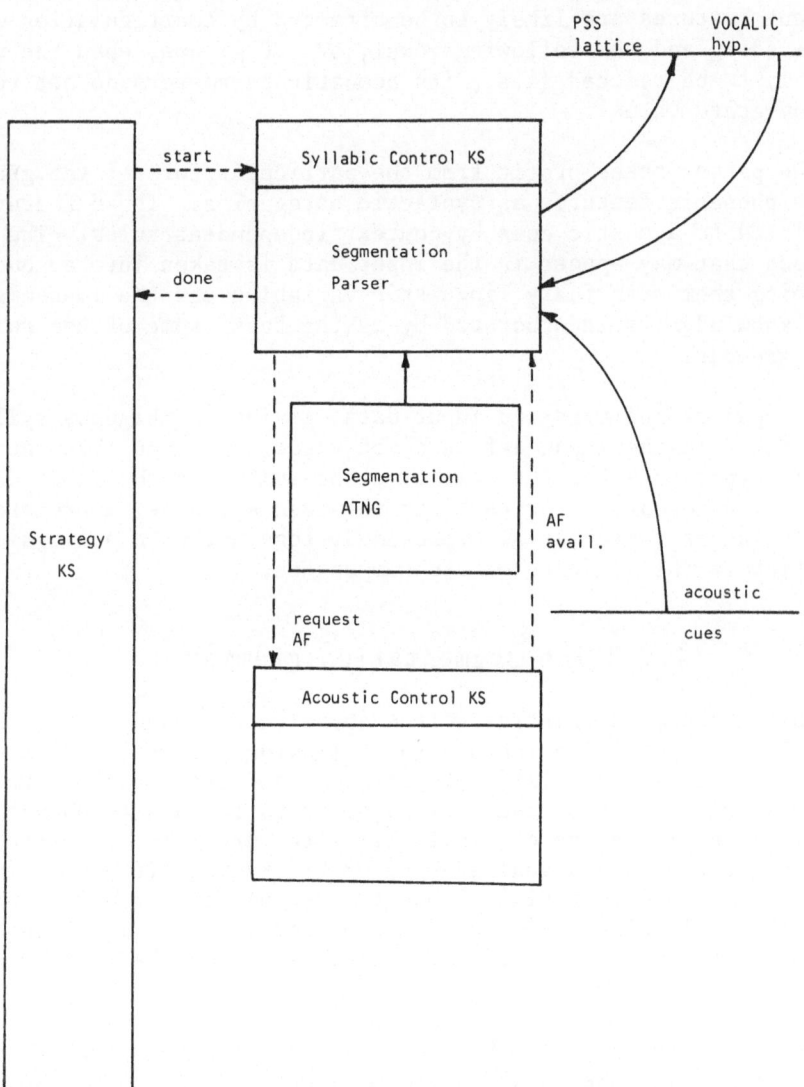

Fig. 4.7 Scheme for segmentation

independent features, for which also the detection of some acoustic cues (AF) may be required.

The Segmentation Grammar is represented in LISP-like form by a simple Transition Network, which defines several types of syllabic structures, including Vowel-Consonant-Vowel (VCV) groups, when the consonant features are likely to be affected by coarticulation with the preceding and the following vowel, or CVC groups, when the vowel is likely to be reduced (i.e., its acoustic parameters do not reach a steady-state value).

The parser takes profit from the particular form of the grammar and has phonetic features as syntactic categories. These features are related to acoustic cues by context independent rules. The vagueness that may appear in the input data is taken into account by describing them with fuzzy linguistic variables and the imprecision of the knowledge is incorporated by making fuzzy some of the rules of the grammar.

The parser analyzes the input data, giving unambiguous syllabic hypotheses when the input data are not vague, and more than one syllabic hypothesis for the same time interval when the input data are not clear enough. For each hypothesized syllable, a precategorical classification is also provided, together with a linguistic evaluation of the evidence of its components.

4.4.2 The segmentation grammar

The main characteristics of the approach proposed in this book for syllable hypothesization are the following. First, an integrated syllabic knowledge, consisting in an Augmented Transition Network Grammar (ATNG), relates descriptions of acoustic cues to phonetic and phonemic interpretations of a syllable structure. The acoustic cues can be valleys of the signal energy, or "energoids" (frequeny intervals of energy concentration of burst spectra), or formant transitions. The syllable structure can be expressed in terms of phonetic features like nasality or, at the highest level of abstraction, by a phonemic transcription.

Secondly, the ATNG can be used for several purposes by properly setting the conditions on its arcs and specifying the subnetworks to be used for a given task. The ATNG can be used essentially in a data-driven way for segmentation and for generating preconditions for the lexical knowledge source.

The latter task is achieved by generating rough descriptions of the phonetic structure of the syllable, leaving to a later phase a detailed extraction of phoneme components and complex consonant clusters. Eventually, the ATNG can be used in a model-driven way to obtain phonemic transcriptions of a syllable, looking for possible complex consonant clusters that cannot be effectively hypothesized purely bottom-up.

Carefully selected rules make it possible to reduce the number of ambiguities in hypothesizing syllables and to obtain well defined bounds for them before starting the interpretation of their structure.

Thirdly, the ATNG mostly contains structural rules about features to be detected in short time intervals of the speech patterns. Thus, in most cases, the subnetworks can be used for interpreting the input data in a simple left-to-right manner, avoiding the conversion of the structural representation into a procedural one.

A portion of the syllabic ATNG is used for segmentation. Only this portion will be introduced in this section.

An ATNG is composed of states, indicated here as $Q_j (j = 1,2, ..., N_j)$, and arcs connecting the states.

The ATNG contains a main network and some subnetworks. The starting state of the main network corresponds to the starting symbol of the grammar.

Once the final state of the main network has been reached, the input is recognized as belonging to the language generated by the grammar.

An arc leaving a state has the following general form:

(<arc-type> / <test> <action> <destination>)

The <arc-type> may be a CAT followed by a terminal symbol of the grammar, a PUSH followed by a non-terminal symbol indicating a subnetwork which has to be completed in order to make the arc accepted, a JUMP indicating a branch that may be conditioned by the content of the test item, a POP indicating the successful completion of a subnetwork. Networks with only these types of arcs, without tests and actions, may generate any context-free language. The <test> associated to an arc may concern the truth of a binary feature, such as GETF X, evaluated on a variable X, or every test of the L|SP Language.

The <action> sets auxiliary registers used to condition the consumption of some arcs. The <destination> specifies the next state in the subnetwork; it is absent when the arc is of the POP type. The details of the ATNG are given in the following. Most of the actions concern the setting and resetting of a register TIMER, which controls the time-advance of a window of flexible bounds. The window defines the time interval in which the input date are analyzed.

When TIMER = NIL, passing from a state to one of its successors requires the detection of some new features in the same time interval;
when TIMER = T, features have to be looked for by moving the time
window forward without a gap between the new and the old positions.
The register RP controls if a vowel is shared by two syllables; in
such a case it is set to T.

The starting symbol of the network is SEN (sentence) and the
first rule of the segmentation ATNG establishes that a sentence is
a sequence of PSSs ending with a PAUSE. The phonetic features used
for recognition can be related to those of G_s; the motivation for
their use is that their detection is relatively easy and reliable.
All the phonemes in the class VOC contain the feature VOCALIC.
Among the consonants, a "sonorant" segment can be a PREVSON, a
POSTVON or an INTERVSON. This feature "sonorant" is related to
acoustic cues.

A "nonsonorant" segment is a UN or a SINIL. The VSTOP sounds
are identified by the detection of the features nonsonorant-
interrupted-lax; the USTOP have the features nonsonorant-
interrupted-tense; the VFRIC have the features nonsonorant-
continuant-lax; the UFRIC have the features nonsonorant-continuant-
tense.

Combinations of the phonemes like tʃ and tθ are characterized by the detection of the features nonsonorant-affricate-tense,
while dʒ and dð are characterized by the detection of the features nonsonorant-affricate-lax.

Table 4.2. summarizes the meaning of the phonetic arc labels
of the ATNG.

TABLE 4.2

Phonetic arc labels

VOCALIC,

PREVSON: prevocalic sonorant consonant in a cluster,

POSTVSON: postvocalic sonorant consonant in a cluster,

INTERVSON: intervocalic sonorant consonant,

NSCL: nonsonorant consonant in a cluster,

SINS: single intervocalic nonsonorant consonant,

SINI: an interrupted SINS,

SINA: an affricate SINS,

SINC: a continuant SINS,

SINIL: a lax SINI,

SINIT: a tense SINI.

In order to complete an arc labelled with one of these symbols, a path of a subnetwork with arcs labelled by acoustic cues has to be accepted. The relations between phonetic and acoustic cues are imprecise, reflecting the vagueness of our knowledge, and the evidence of the acoustic cues in the acoustic patterns is very often uncertain. For these reasons, the rules relating phonetic and acoustic cues are fuzzy rules with neither conditions nor actions associated and the terminal alphabet of the acoustic cues consists of fuzzy linguistic variables, defined over the universe of certain acoustic parameters.

The rules relating pseudo-syllable-segments and phonetic features are listed in Table 4.3. In order to make easier their understanding, a graphical representation of them is given in Fig. 4.8.

TABLE 4.3

```
(SEN      (PUSH PSS/T        (TO Q1)))
(Q1       (PUSH PSS/T        (TO Q1))
          (CAT PAUSE/T       (TO QP1)))
(QP1      (POP))
(PSS      (PUSH UN/T         (SETR TIMER T)(TO Q2))
          (PUSH VE/T         (SETR TIMER T)(TO Q3)))
(Q2       (PUSH VSGA/T       (TO QP2)))
(Q3       (PUSH VSGB/T       (TO QP2)))
(QP2      (POP))
(VSGA     (PUSH VSGAA/T      (SETR TIMER T)(TO Q4))
          (PUSH VSG/T        (TO QP3)))
(Q4       (PUSH VSG/T        (TO QP3)))
(QP3      (POP))
(VSGB     (PUSH VSGBA/T      (SETR TIMER T)(TO Q4))
          (PUSH VSG/T        (TO QP3)))
(VSG      (PUSH VBG/T        (TO QP4))
          (PUSH VR/T         (SETR TIMER T)(TO Q5)))
(Q5       (PUSH VSGC/T       (TO QP4)))
(QP4      (POP))
(VBG      (PUSH VOCALIC/T    (SETR TIMER NIL)(SETR RP T)(TO QP5)))
(VE       (JUMP/RP           (SETR TIMER NIL)(SETR RP NIL)(TO QP5)))
(QP5      (POP))
```

```
(VR       (PUSH VOCALIC/T      (SETR TIMER NIL)(TO QP6)))
(VSGAA    (PUSH PREVSON/T      (SETR TIMER NIL)(TO QP7)))
(QP6      (POP))
(QP7      (POP))
(VSGC     (PUSH POSTVSON/T     (SETR TIMER NIL)(TO QP8)))
(QP8      (POP))
(UN       (PUSH SINS/T         (SETR TIMER NIL)(TO Q6))
          (PUSH NSCL/T         (SETR TIMER NIL)(TO QP12)))
(Q6       (PUSH SINI/T                (TO Q7))
          (PUSH SINA/T                (TO QP12))
          (PUSH SINC/T                (TO QP12)))
(Q7       (PUSH SINIT/T               (TO QP9)))
(QP9      (POP))
(QP12     (POP))
(VSGBA    (PUSH INTERVSON/T    (SETR TIMER NIL)(TO QP13))
          (PUSH SINS/T         (SETR TIMER NIL)(TO Q8)))
(Q8       (PUSH SINI/T                (TO Q9)))
(Q9       (PUSH SINIL/T               (TO QP14)))
(QP13     (POP))
(QP14     (POP))
```

Some of the relations between phonetic and acoustic features
have been described in Section 3.5; other relations will be introduced in successive chapters of this book. They are not given here
for the sake of simplicity.

Let us now compare the use of ATNGs for analyzing a written
text of a natural language and its use for the generation of syllabic hypotheses. In the former case, the input is a written string
of symbols analyzed from left to right; in the latter case, the input is a pattern which may be analyzed along the time axis. A time
interval of the speech pattern may have more than one structural
description, reflecting imprecision in feature extraction and knowledge representation.

For these reasons, a special parser has to be designed for
segmenting the pattern into time intervals suitable for performing
a structural analysis inside them.

For example, the grammar described in Section 3.3 is used for
detecting the segments where the Vocalic Pattern Feature (VPF) does

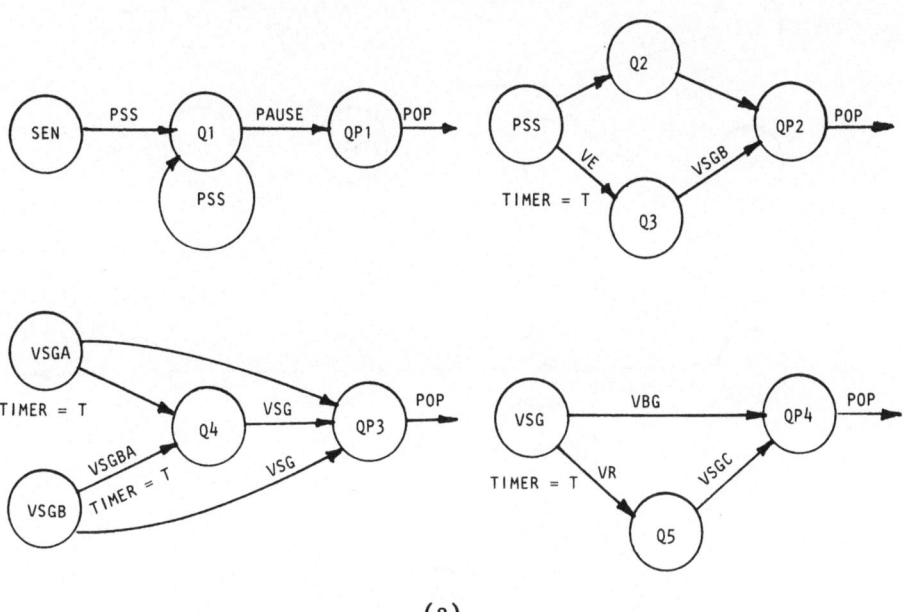

(a)

Fig. 4.8 Graphical representation of the segmentation grammar.

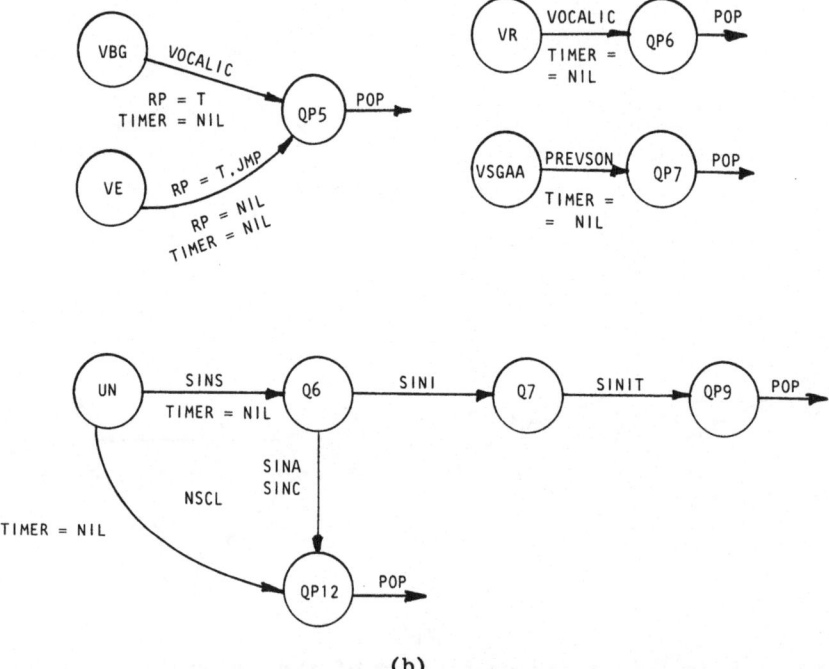

(b)

Fig. 4.8 continued

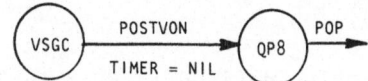

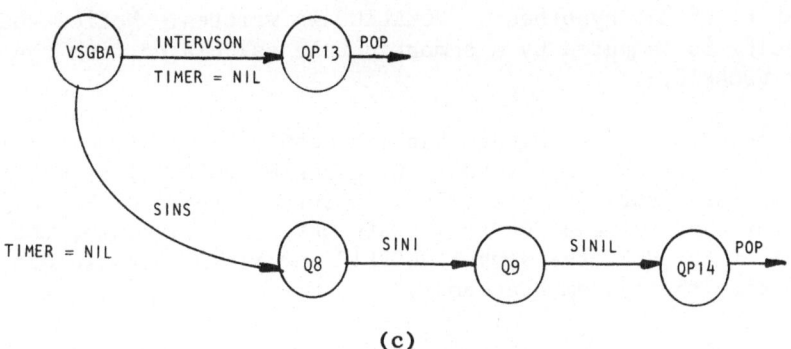

(c)

Fig. 4.8 continued

exist. For each one of these segments, the rules of the segmentation grammar are applied in order to generate syllabic hypotheses.

Let us consider, as an example, the subnetwork VBG. A move
from the state VBG to state QP5 can be performed if the subnetwork
VOCALIC can be completed. From QP5 a POP is executed.

The structure of the subnetwork VOCALIC can be derived from
Rule (3.18) of Example 3.4.5 by inserting the evaluation of evidences as actions associated with the arcs as follows:

```
(VOCALIC    (CAT    X1/T         (SETR EX1 MU(X1))(TO QV1)))
(QV1        (CAT    X2/T         (SETR EX2 MU(X2))(TO QV2)))
(QV2        (CAT    X3/T         (SETR EX3 MU(X3))(TO QV3)))
(QV3        (JMP                 (SETR EV  MU(VOCALIC))(TO QPV)))
(QPV        (POP    (GETF V))))
```

X1, X2, X3 are the linguistic variables defined in Example 3.4.5.
The action $(SETR\ EX_j\ MU(X_j)(j = 1,2,3)$ writes into register X_j the
membership $MU(X_j)$ of the acoustic parameters of a VPF in the linguistic variable X_j. EV is the register in which the possibility
MU(VOCALIC) of the hypothesis "VOCALIC" is written. Notice that
MU(VOCALIC) is computed by a semantic rule associated with the subnetwork VOCALIC.

If MU(VOCALIC) is higher than a threshold delimiting the interval in which the hypothesis VOCALIC, given MU(VOCALIC) is extremely
unlikely, the binary feature GETF V is set to true and the condition
is met for performing a POP from state QPV. In this case, the time
interval of the VPF, the symbol VOCALIC and its membership are
written into the system blackboard.

The subnetwork VOCALIC and its use in the way described above
are invoked whenever the syntactic pattern recognition procedure
described in Section 3.3 detects a VPF. This invocation is performed by the following control rule of the SCKS:

IF a VPF is found THEN PUSH VOCALIC

Other procedural rules of the same type are used by the segmentation parser which is described in the next subsection.

The segmentation network has controls on the binary features
GETF X associated with POP arcs of each subnetwork corresponding to
a phonetic feature, rather than distributed over other arcs as shown
in the introductory example of Fig. 4.2.

TABLE 4.4

Phonetic feature	Binary control feature
VOCALIC	V
INTERVSON (intervocalic sonorant)	INTERVSON
PREVSON (prevocalic sonorant)	PREVS
POSTVSON (postvocalic sonorant)	POSTVS
SINS (single intervocalic sonorant)	BSINS
NSCL (nonsonorant in a cluster)	BNSCL
SINI (interrupted SINS)	BSINIT
SINIL (lax SINI)	BSINIL
SINA (affricate SINS)	BSINA
SINC (continuant SINS)	BSINC

The reason for this is efficiency. Table 4.4 shows the binary features corresponding to the phonetic features used in the Segmentation ATNG.

There are many possible sets of phonetic features which can describe a language. The choice of the features used here depends on the fact that it has been possible to establish relations between these features and acoustic cues which have been proven very effective in the model for speech understanding.

Notice that the features described in this book and their relations with acoustic cues have been introduced for the Italian Language. Nevertheless, they have a lot of similarities with the phonetic features and their relations with acoustic cues investigated by Hughes and Hendal (1965) for English.

4.4.3 The segmentation algorithm

The procedural Knowledge Source builds up on the blackboard a State Space Representation of the interpretations by writing hypotheses with acceptable evidences. This is a parallel process, because many nodes of the ATNG can be simultaneously active in calling the attention of the procedural KS.

In order to make efficient the parallel generation of hypotheses, it is convenient in some cases to proceed in a "non-directional" manner, trying to solve first the easier and less complex problems which correspond to actions associated with some arcs, and use the results of their solution, in case of success, to delimit the portions of the acoustic pattern where the solution of more complex and difficult problems is attempted.

This is the case of segmentation, for which an equivalent representation of the ATNG has been derived as a Problem Reduction Representation.

Figure 4.9 shows the PRR of the segmentation ATNG. A sequence of PSS problems has to be solved until a PAUSE is found (i.e., a PAUSE sub-problem is solved).

Each labelled sub-problem is solved when the corresponding network is successfully completed.

There are two types of AND nodes; the ones represented by lines relate sub-problems which have to be solved in the *same* interval; AND nodes represented by dashed lines relate sub-problems which have to be solved in *contiguous sequential* time intervals.

The sub-problems UN, VSG and VE are the only ones which determine the time bounds of an hypothesis of a Pseudo-Syllabic Segment. Here VE is connected by a dashed line to VBG because, when VBG is solved, VE is automatically solved by taking the ending portion of the vocalic segment. These different relations between nodes make the PRR a relational graph rather than an AND/OR graph. A partial ordering followed by the function SEGMENT of the parser, described below, is indicated by the integers associated with certain terminal sub-problems (1 precedes 2, etc.).

The sub-problems labelled by an ordering integer are the only ones which need to be solved in order to generate hypotheses about Pseudo-Syllabic-Segments.

Their generation does not necessarily mean that a PSS problem has been solved. Rather, it signifies that the focus of attention of the system is called to schedule suitable actions for solving the remaining sub-problems (indicated by dashed arrows coming out from nodes in Fig. 4.9) of an hypothesized PSS. This happens when PSS is "solved" for the sub-problems represented in Fig. 4.9. As the solution of these sub-problems has to be performed inside the PSS, the invoked sub-networks of the ATNG can be used in a strictly *left-to-right* manner without any need to introduce further procedural rules. The sub-problems which do not have a phonetic significance are not labelled in the figures, for the sake of simplicity.

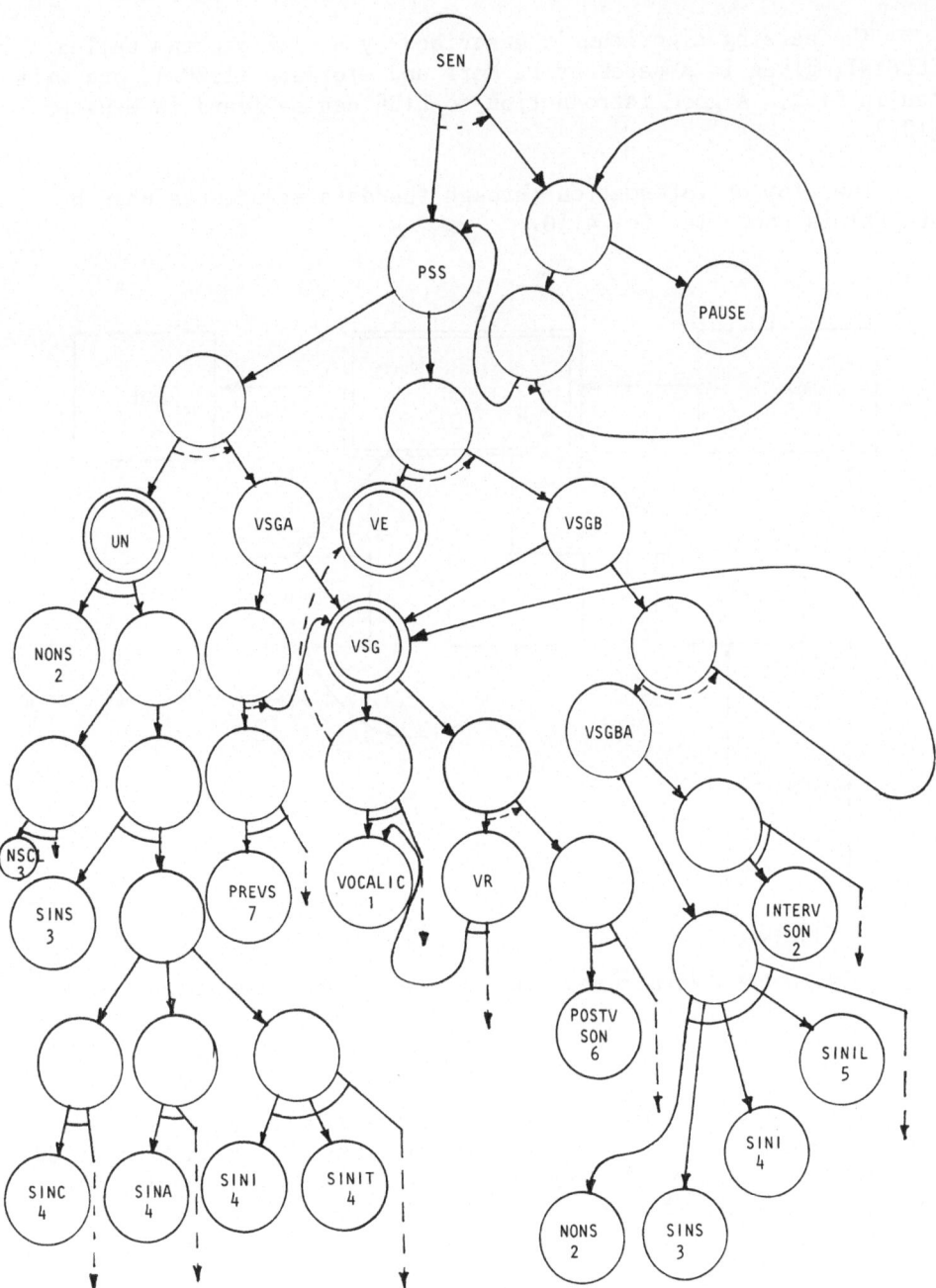

Fig. 4.9 Problem Reduction Representation of the segmentation
        grammar

Notice that a Problem Reduction Representation is used for deriving control rules to be used in the segmentation parser.

The parsing algorithm is described by a LISP program called SEGMENT, given in a paper by De Mori and Giordano (1980a), and written in LISP. A good introduction to LISP can be found in Winston (1977).

The flow of information through the data structures used by SEGMENT is shown in Fig. 4.10.

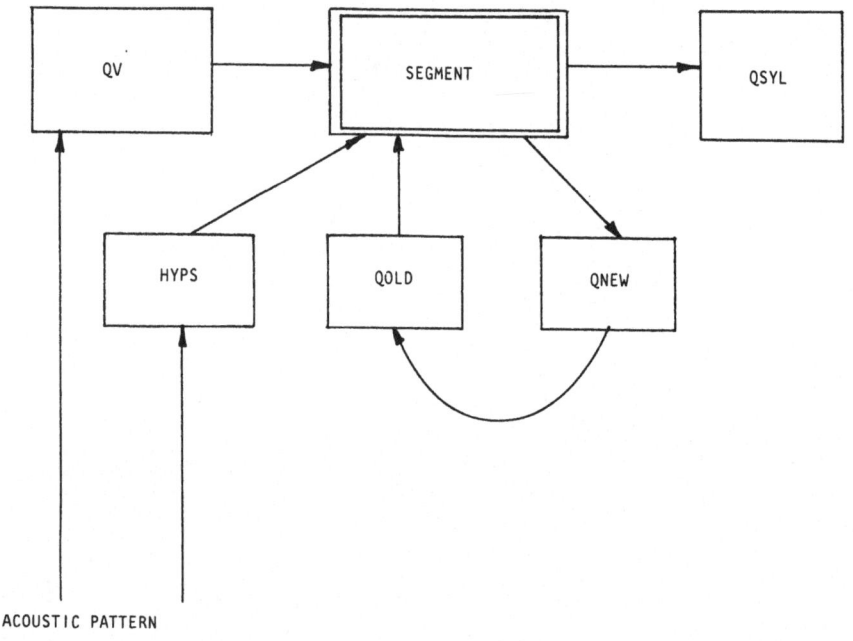

Fig. 4.10 Data structures used by the segmentation algorithm

The vowels are hypothesized using the syntactic procedure described in Chapter 3 and stored into the list QV. Each item of QV contains the times of the beginning and the end of a vocalic hypothesis and a membership expressing the evidence with which the feature "VOCALIC" has been detected in the above-mentioned interval of the acoustic pattern.

The list DD contains all the possible descriptions of intervocalic segments (consonants or clusters of consonants) for every possible pair of consecutive vocalic hypotheses.

Such a description is obtained by the program PRESYL and is

made of symbols belonging to a subset VN1 of the non-terminal
alphabet:

VN1 = (UN, SON),

where:

SON = INTERVSON + PREVSON + POSTVSON + SINIL

The analysis of an intervocalic tract follows a Problem
Reduction Representation (PRR) of the complex problems, consisting
in interpreting the content of every consonantal interval (CNTINT)
between two vowels. For this purpose, the problem reduction scheme
in Fig. 4.11 has been introduced.

Above the dashed line are indicated the problems which have
to be solved by the segmentation parser. Once a consonantal
interval has been hypothesized between two vowels, the problem
solver looks to see whether all the consonants are sonorant (problem INTERVSON) or there is at least one sonorant (problem NONS).

This implies the extraction of some acoustic parameters in the
consonantal interval, their description with fuzzy linguistic variables, and the use of the rules of the subnetworks INTERVSON and
NONS. The application of these rules leads to the computation of
the evidences MU(INTERVSON) and MU(NONS).

In the case of a nonsonorant in a cluster, the possibility of
having a postvocalic sonorant consonant (problem POSTVSON) preceding the nonsonorant is looked for.

Once problems have been solved in specific time intervals, the
function PHONETICS recognizes as UN segments the portions of NSCL
possibly following a POSTVSON segment or single intervocalic NA
(Nonsonorant Affricate), NC (Nonsonorant Continuant), or NIT
(Nonsonorant Interrupted Tense) segments, and recognizes as SON
(VSGBA) all the INTERVSON and the single intervocalic NIT segments.
The details of the rules for relating these phonetic features to
acoustic cues will be given in Chapters 5 and 6.

The extraction of acoustic cues is usually performed by syntactic methods.

Each fuzzy rule of the segmentation grammar has associated a
semantic rule for computing the plausibility of an hypothesis represented by a left-hand symbol, given the evidences of the right-
hand symbols. The semantic rules use the MAX operator for logical
disjunction and the MIN operator for logical conjunction. The
evidence of the feature F is represented by MU(F). Essentially,

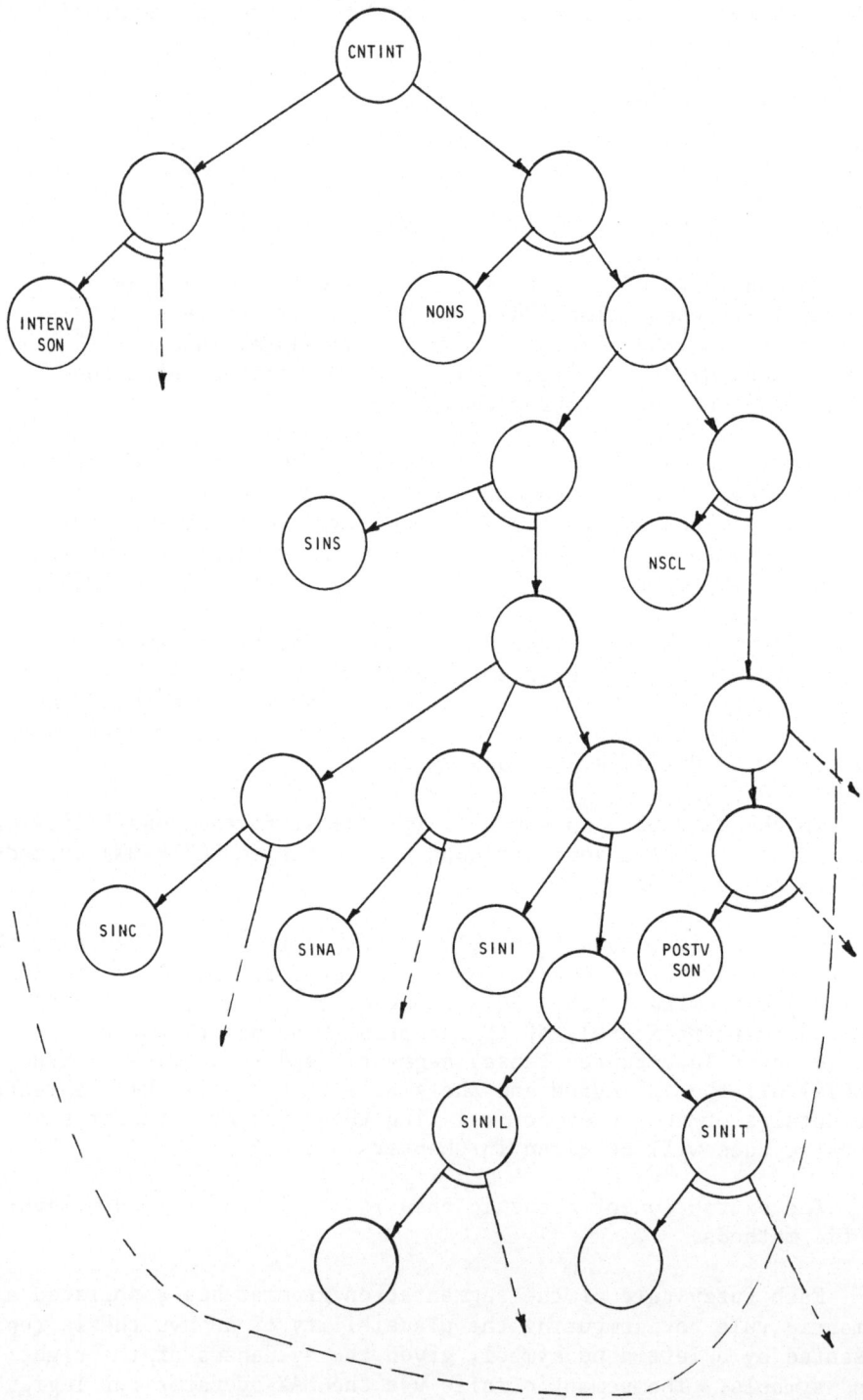

Fig. 4.11 Problem Reduction Representation of a consonantal
         interval

the program PHONETICS rewrites phonetic features into UN or SON
only if the evidence of the phonetic features is higher than a giv-
en threshold. This fact makes some binary features like (GETF
INTERVSON) associated to a feature (i.e., INTERVSON), assume the
value T (true).

Figure 4.10 shows the data structures used by the program
SEGMENT; this program is defined as an expression EXPR which depends (LAMBDA stands for function) on QV, the queue of hypothesized
VOCALIC segments, the descriptions of the signal S with the language L_S, of the signal A with the language L_A, and of the ratio
(denoted in this program DRV); these descriptions of acoustic patterns form phonetic hypotheses written into HYPS.

Details of SEGMENT are discussed below; the instructions of
SEGMENT are written after the LISP primitive PROG. V is a list
which will contain the attributes of any hypothesized vowel popped
from QV; each syllabic hypotheses is built starting from the
current vowel hypothesis written in V.

The key of the PARSER, which does not proceed in a strictly
left-to-right manner, is a prediction of possible syllabic structures following a time interval under analysis and a verification
of the predictions in the interval under analysis.

Predictions are stored in a list called QNEW. When an interval has been analyzed, the list QNEW is renamed QOLD and a new
QNEW is created. The hypotheses in QOLD are refined and written
into QSYL. This refinement produces new hypotheses for the succeed-
ing time-segment (whose bounds are not known) and a rough description of them is written into QNEW. At the beginning, the hypothesis
(UN VSGA) is written into QOLD with the initial time set to 0 and
the ending time set to X.

A typical node in QOLD has the following structure: $(t_b, t_e$
(rough syllable hypothesis)); for example: (0 X (UN VSGA)). As
usual, b is for beginning and e is for end.

QNEW is a queue similar to QOLD, containing all the partial
syllabic hypotheses generated during the refinement of a single
syllabic hypothesis. Once this refinement is completed, the
syllabic hypothesis is written into QSYL; when the entire content
of QOLD has been analyzed, the content of QNEW is pushed into QOLD.

A node in QSYL has the following components: $(t_b, t_e$ (structure of the syllable in terms of the symbols V, UN, SON)):

The function EXTR-AC-PAR extracts the acoustic parameters necessary for emitting the phonetic hypotheses of Table 4.2. The function PHONETICS reads these parameters, generates phonetic hypotheses like those of Table 4.2, evaluates evidences, and checks whether subnetworks like UN can be completed or not. If they can be completed, the nonterminal symbols UN or SON, depending on the subnetworks successfully completed, are stored into HYPS without any evidence judgement.

UNPROC refines a hypothesis in QOLD when an UN has been produced by PHONETICS in the time segment under analysis; VCDSYL performs a similar operation when no UN has been found between two vowels in the segment under analysis; LAHEAD connects two consonantal hypotheses across a vowel of uncertain evidence. COMPLETE performs certain adjustments at the end of the sentence.

In order to recognize the features POSTVSON and PREVSON the following rules are used:

PREVSON (P5) → s((DS), (LPA)) (P5)

where   DS → C4
         DS → C4   DS
         SA → P1
         SA → P2   P1
         SA → P1   SA
         SA → P2   P1   SA

s is a predicate for "simultaneous" and LPA is a SA with a low absolute maximum of energy.

(P5) POSTVSON → (P5)   s((AS), (LPA))

         XS → C3
         XS → C3   XS
         AS → XS
         AS → C1

These rules are applied whenever a nonsonorant segment is hypothesized. Because of the type of the rules, a cluster of nonsonorant consonants is always recognized as nonsonorant and never as an interrupted lax, thus it is relabelled with UN.

4.4.4 Examples

Figure 4.12 shows the data produced by PHONETICS for a sentence.

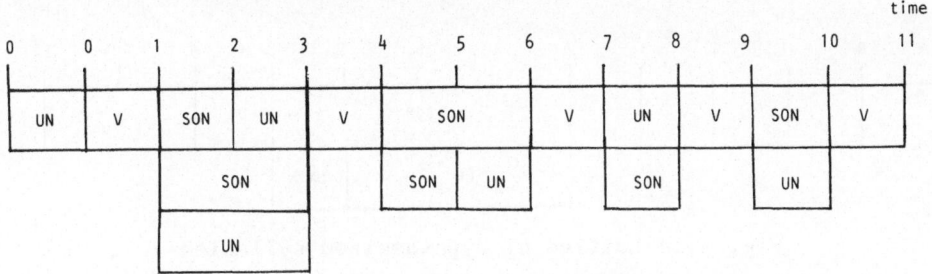

Fig. 4.12 Example of segmentation lattice

For the sake of simplicity, the time bounds are indicated by integer numbers. At the beginning of the sentence, a silence, lab-elled UN, is assumed to cover the interval 0-0.

Only the vowel between times 6 and 7 has uncertain evidence.

The parser reads the data in Fig. 4.12, stored into the lists QV and HYPS, generates the lattice of syllables given in Fig. 4.13 and stores then into QSYL.

Because of the uncertainty of the vowel in the time interval (6 - 7) the procedure LAHEAD adds to the lattice of Fig. 4.12 the portion shown in Fig. 4.14.

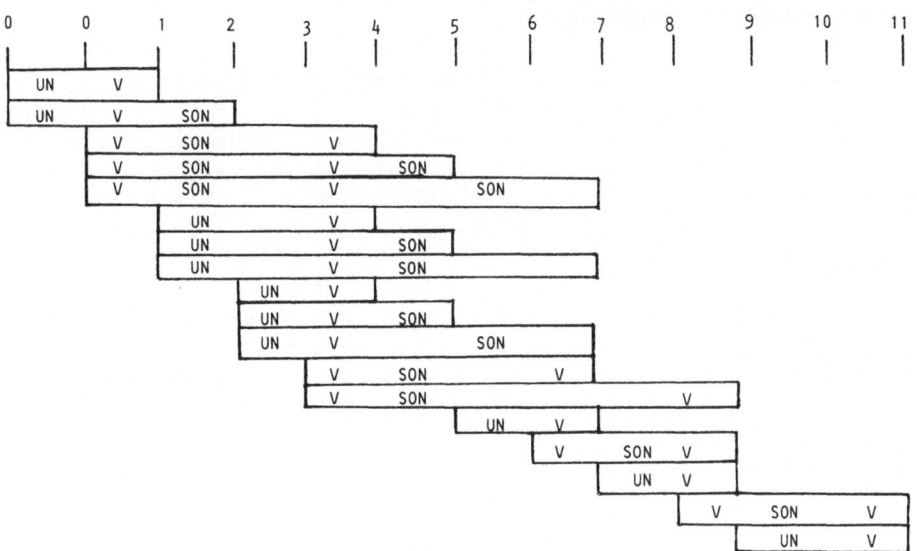

Fig. 4.13 Lattice of the phonetic features added because of the uncertain evidence of a VOCALIC hypothesis

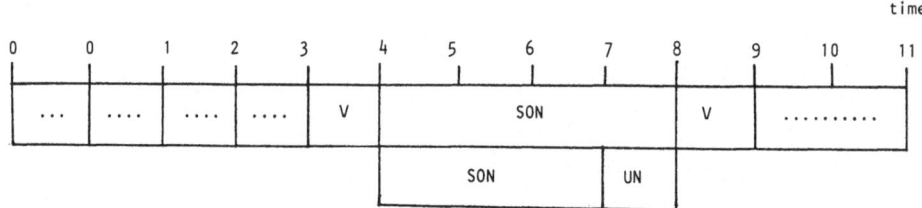

Fig. 4.14 Lattice of hypothesized syllables

4.5 A Parallel Processing Model for Generating Phoneme Hypotheses

The parser described in the previous section is suitable for execution in a classical sequential machine.

Simpler and more efficient algorithms can be conceived for a parallel computation model. This model will be described in the following. Rules introduced in Chapters 5 and 6 can be used both in sequential or parallel models.

Distributed processing is now practical (Lesser and Corkill, 1981). This makes it attractive to consider very complex tasks such as speech understanding or speech recognition in a framework of distributed problem solving.

With this view, decoding is performed by a pluralism of inter-pretation activities performed by a society of experts according to the scientific community metaphor (Kornfeld and Hewitt, 1981). The task of hypothesis generation is decomposed into a number of sub-tasks performed by a society of basic experts capable of using knowledge sources and dynamically creating several instantiations of them.

Instantiations are computing agents which cooperate by ex-changing messages. They are generated by a basic expert, use its knowledge and communicate with other experts to obtain all the in-formation they need to achieve their task.

There are two main motivations for using this model.

- The first one is that phonetic and acoustic knowledge ac-quired by human experts about speech sounds is scattered in a variety of propositions and conjectures. This makes it natural to allow generating multiple interpretation hypotheses which grow up concurrently.

- A second motivation is that a distributed model of an expert phonetician is coherent with the models of other components of a speech understanding system when they are conceived as distributed problem solvers because the system has to work close to real-time, facing ambiguous data and a very large variety of possible solutions.

Many valuable speech recognition systems do not use rule-based experts for phoneme hypothesization. In spite of this, an attempt to model human knowledge in interpreting speech patterns may contribute to progress in the speech scientific community for the following reasons:

1 - Conceiving models and making experiments with them may help in improving our knowledge about speech perception;

2 - Automatic speech recognition systems with complex tasks, such as voice activated typewriters, require highly accurate phoneme hypothesization;

3 - Rule-based experts for phoneme recognition are promising for speaker-independent, task-independent and, perhaps, language-independent applications, provided that suitable rules are given to the system.

4 - Learning can be performed by updating or modifying the rules.

An expert receives requests for the generation of hypotheses from other experts.

Hypotheses are data-structures written into a Short Term Memory (STM) associated to each expert.

Experts are grouped into Societies according to their level of expertise.

The task-independent knowledge is structured on two levels, corresponding to the Auditory Expert Society (AES) and the Syllabic Expert Society (SES).

The AES has the task of extracting acoustic cues. Another task of the AES is that of interpreting and executing the requests from the SES. These requests may concern the extraction of acoustic cues for phonetic features like VOCALIC, PLOSIVE-CONSONANT etc.

The behaviour of an expert is shown in Fig. 4.15. To each expert EXPj is associated a KSj stored into a Long Term Memory LTMj and a data-base stored into STMj which contains hypotheses. The STM of an expert society may be partitioned into a set of disjoint parts SMTj (j = 1,2,..., J).

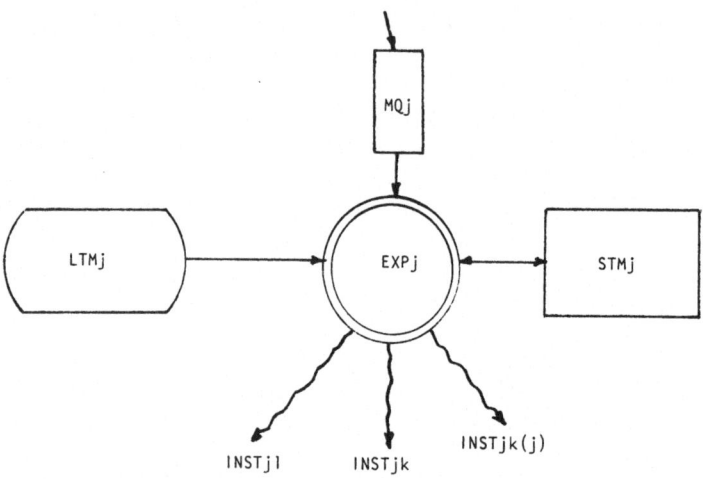

Fig. 4.15 Expert structure

To each expert EXPj is also associated a message queue MQj containing the requests made to it from other experts. EXPj reads
sequentially these requests. If a request concerns some information
which has not been requested before, then EXPj creates an instantiation of KSj. Let INSTj1, INSTj2,...,INSTjk,...,INSTjK(j) be the
instantiations created at a given time tk.

An instantiation is a computing agent that may create other instantiations of KSj or send requests to other experts or send answers
to the experts which have made requests to EXPj. In other words, an
instantiation INSTjk can send a message MESSjk1 to other experts.

Messages for the experts can be stimuli coming from lower level
experts or verification requests from higher level experts or commands
from a strategy KS.

When an instantiation has performed its task, it terminates and
leaves the system.

The auditory experts use some of the rules of AKS for extracting
acoustic cues, such as dips or peaks of energies in certain bands, or
the frequency values of the zones of energy concentrations in a segment for which the phonetic hypothesis VOCALIC has been previously
generated.

Figure 4.16 shows the experts of the auditory (below) and syllabic (above) societies.

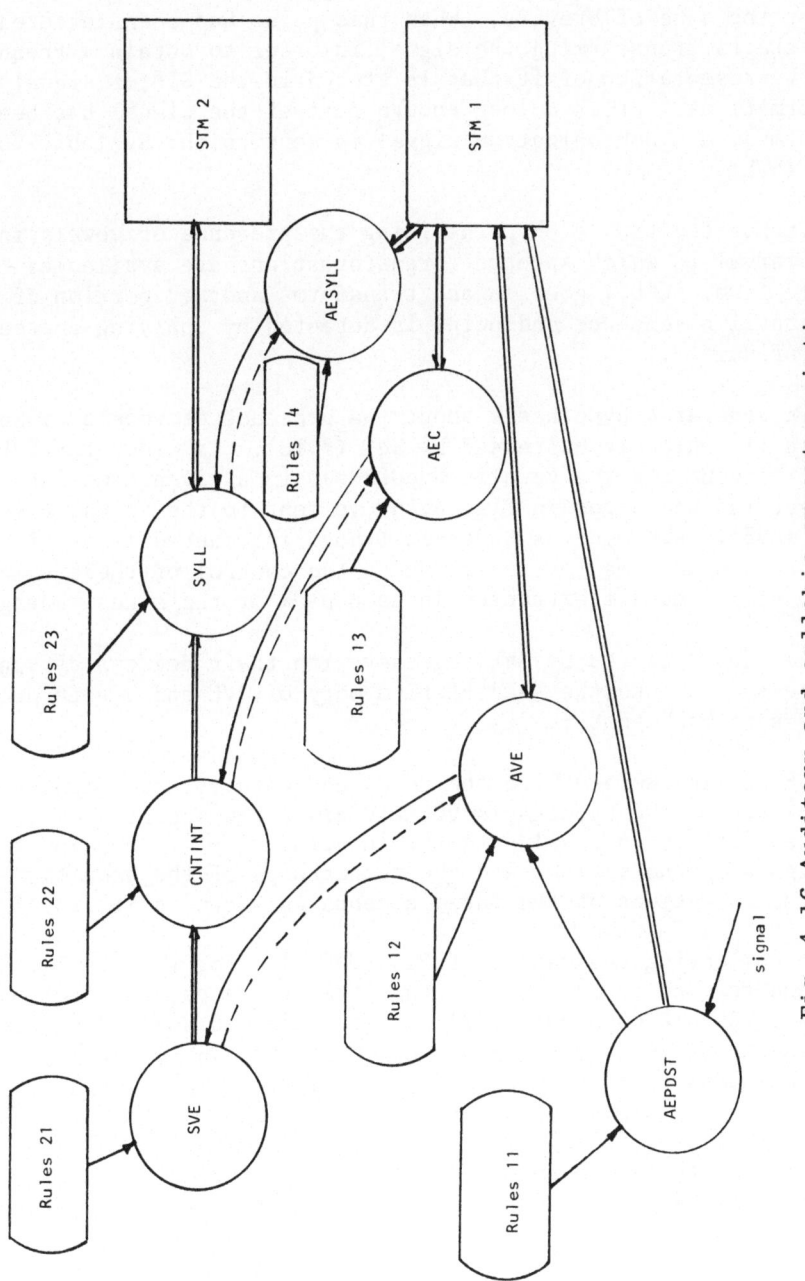

Fig. 4.16 Auditory and syllabic expert societies

The speech signal is sampled, quantized, and transformed by an expert called "Auditory Expert for End-Points Detection and Signal Transformation" (AEPDST). AEPDST looks for the starting point of a sentence by using a subset of Rules-11, whose details are omitted here for the sake of brevity. When this point has been detected, AEPDST starts transforming the signal in order to obtain a frequency-domain representation of it that is stored in the STM of signal transformations. After a long enough part of the signal has been transformed, a synchronization signal is sent to the Syllabic Vocalic Expert (SVE).

SVE has the task of hypothesizing the presence of vowels in a time interval in which spectral transformations are available. At the same time, AEPDST goes on and transforms another portion of the signal until a sentence end-point is detected by applying another subset of Rules-11.

SVE generates hypotheses about the presence of vowels by applying Rule-21, which is expressed by Eq. (3.18). In order for SVE to be able to generate the vocalic hypotheses in a given time interval, a request (dashed arrow in Fig. 4.16) is sent to the Acoustic Vocalic Expert (AVE). AVE applies Rules-12 (those introduced in section 3.3) in the requested time interval. Under the control of these rules, the acoustic cues are extracted in each peak of the signal energy.

The cues detected by AVE together with their degrees of vagueness are stored into the short-term-memory of AVE and a reference to them is sent back to SVE.

SVE applies Rules-21 to the received features and computes the plausibility of the hypothesis VOCALIC given the acoustic cues. This plausibility is a value in the interval [0 - 1] computed from the weights of the strings and the memberships of the acoustic cues, using the operations of the fuzzy algebra as shown in Eq. (3.18).

By collecting the values of Poss(VOCALIC) for the vocalic segments and for the other peaks of the total energies, the probability density p(VOCALIC|Poss(VOCALIC)) can be estimated and a decision criterion based on the values of Poss(VOCALIC) can be established in order to ensure that the probability of missing the right hypothesis is kept lower than a desired threshold.

Based on the decision criterion, vocalic hypotheses are generated for some peaks of signal energy. As soon as these hypotheses are generated, they are sent to another syllabic expert called CNTINT.

CNTINT considers every possible pair of vowels and assigns some consonantal features to the intervals between pairs of vowels. In order to perform these operations, CNTINT sends requests to the

Acoustic Expert for Consonantal intervals (AEC) which sends back references to acoustic cues stored into its short-term-memory. CNTINT
applies Rules-22 to every set of acoustic cues it receives and generates hypotheses about some phonetic features. Rules-22 are those
introduced in Section 3.6.

Notice that Rules-22 relate phonetic features with acoustic cues
in a context-independent way; furthermore the rules are affected by
degrees of imprecision and CNTINT is capable of combining the degrees of rule imprecision with the vagueness with which the acoustic
cues are detected.

Rules-13 control the extraction of the acoustic cues used for
precategorical classification (see subsection 3.6.2). Their details
are omitted here for the sake of brevity. Hypotheses about consonantal features are sent as messages from CNTINT to the SYLLabic
classification expert (SYLL). SYLL uses a subset of Rules-23 from
the segmentation grammar represented in Table 4.3 and the messages
from CNTINT for generating hypotheses about the bounds of PSS. The
Short Term Memory is organized in a data structure of syllabic hypotheses as shown in Fig. 4.14. This data structure grows according
to the scheme of Fig. 5.1.

In parallel with the generation of PSS hypotheses, the SYLL expert attempts to generate more detailed hypotheses about the already
detected Pseudo Syllabic Segments. For this purpose it sends
requests to the Acoustic Expert for Syllabic features (AESYL) which
uses Rules-14 for extracting acoustic cues according to the requests
it receives. References to these cues are sent back to SYLL. SYLL
uses a subset of Rules-23 which is invoked by the pattern of the
consonantal features representing the PSS for which more detailed
hypotheses are sought. Rules-14 and Rules-23 will be described in
Chapters 5 and 6.

SYLL can also be requested to generate or verify syllabic hypotheses from the Society of Lexical Experts. It reacts to these
requests by establishing dialogs with AESYL.

For each message received by an expert, an instantiation of its
knowledge is created.

Fig. 4.17 shows a scheme of actor creation inside an expert subsystem. A controller, which is a permanent actor, decodes the messages from MQj. For each message representing a request of information not yet available in STMj, a set of four actors is created by
the controller.

Rule-applier is an actor which sends messages to other experts
based on the rules it has to apply.

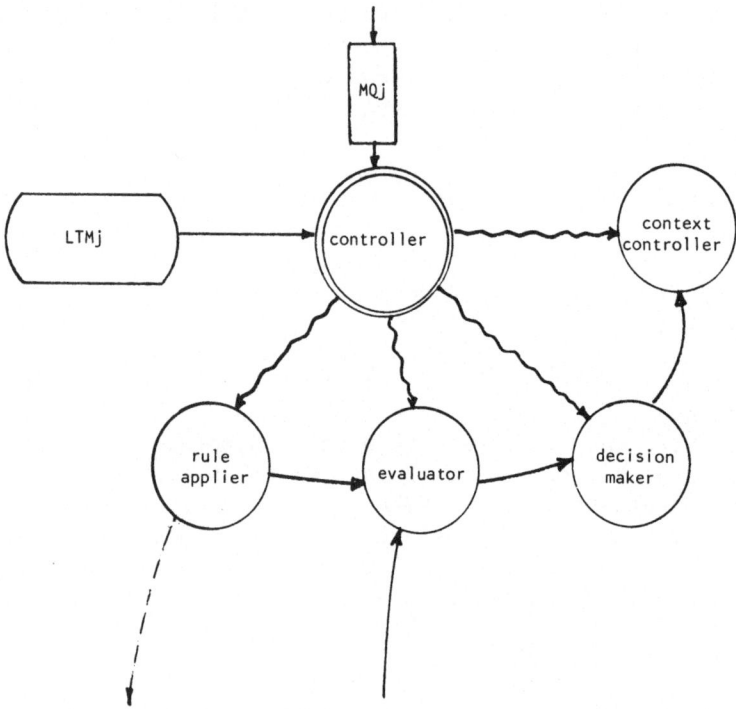

Fig. 4.17 Scheme for actor creation

Evaluator is an actor which receives the answers to the questions advanced by the Rule-applier and performs evidence composition.

Decision-Maker receives evidence qualifications from Evaluator and decides whether it is worthwhile to apply again the rules to other data by creating new sets of actors.

Context-controller controls the access and the arrival of new hypotheses from Decision-Maker.

When it is worthwhile for Decision-Maker to continue the problem solving activity, it may create one or more triplets of the first three actor-types introduced above. Decision-Maker creates more than one triplet when it discovers a disjunction of possible continuations.

Simulation results have shown that with an average parallelism degree of 12 the whole process can be done in real-time using standard multi-microprocessor architectures. Instantiations are expert activities; most of them can be carried out in parallel, improving the speed of the system.

4.6 A Review of Previous Work
on Phoneme Recognition

The next two chapters will be devoted to a description of the fuzzy algorithms used for generating a detailed description of the composition of each Pseudo-Syllable-Segment. As the approach proposed in this book is rather new and is partially described in very recent publications of the author, a brief review of other approaches taken in the past is given in this section. The reader interested in more details on this subject may refer to an excellent work recently published by Sakai (1980).

The number of phonemes in a language is not very high. Usually there are less than 100 phonemes, but the variety of the speech patterns in which they may appear is high and difficult to characterize because of the following factors.

Speech patterns of a phoneme depend on the manner of speech production. Speed, loudness and prosody have an influence on the phoneme patterns that is often not negligible. Other modifications due to coarticulation effects and speaker dependencies make the variety of phoneme realizations larger.

A rule-based approach should overcome these obstacles, allowing the structure of the pattern which represents the essence of the phonemes to be extracted.

Most of the approaches try to match a segment of the speech pattern to a prototype. Segments may be obtained synchronously (for example every 10 ms) or asynchronously.

The former approach, also called the "centisecond approach" is discussed by Bakis (1977); the latter approach is described in great detail by Goldberg (1975) in his Ph.D. thesis. More recently, Shikano and Kohda (1978) have compared both theoretically and experimentally four distance measures in matching the spectral envelopes of a reference and an unknown pattern, when the spectra are computed from linear prediction coefficients.

These distance measures are called the likelihood ratio measure (or linear predictive residual), the normalized residual power measure the COSH measure, the LPC cepstrum distance. The differences among these measures seem to be very slight, although the COSH measure has given the best comparative performances.

Its definition is given below:

$$COSH = \frac{v}{u\,R_f} \sum_{k=p}^{-p} A_k\,s_k + \frac{u}{v\,R_g} \sum_{k=p}^{-p} B_k\,r_k - 2 \qquad (4.8)$$

where u is the power, r_k are the correlation coefficients,

$$A_k = \sum_{i=|k|-p}^{p} a_i \, a_{i-|k|} \, ,$$

a_i are the prediction coefficients, R_f is the normalized residual and p is the order of the model; all these symbols refer to the prototype spectrum; v is the power, s_k are the correlation coefficients,

$$B_k = \sum_{i=|k|-p}^{p} b_i \, b_{i-|k|}$$

b_i are the prediction coefficients and R_g is the normalized residual of the input spectrum.

Dixoh and Silverman (1976) have compared spectral classification methods using power spectra obtained by Fast Fourier Transformation.

They obtained the best performances using a quadratic distance with a linear mean correction.

Similar results have been obtained by Nakagawa (1976), using the power spectrum obtained from a filter bank.

The quadratic discriminant function based on Bayes decision rules gave the most powerful performances in vowel recognition. In accordance with the behaviour of the auditory system, logarithmic values of energies give better spectral distances than absolute values.

Creation of reference patterns involves decisions on how many templates should be created for each phoneme, how a reference pattern should be created from samples, and how samples should be extracted.

A reference function can be created in a probabilistic framework from the statistics of many reference samples. The statistics are usually obtained by regarding the distribution of feature parameters as a normal distribution. The reference pattern is usually made of the arithmetic mean of each component of the sample set if all of them have the same number of components. If the samples have different length or different number of components, a dynamic programming technique can be used (Niimi, 1978).

Reference patterns can be also obtained by clustering techniques (Tanaka 1978, Rabiner 1978).

R. Nakatsu and M. Kohda (1978) have applied a dynamic programming technique for the recognition of VCV syllables.

The input speech is represented by a sequence of vectors s_1 ... s_i ... s_m of correlation coefficients and the reference pattern is represented by a succession of the maximum likelihood spectral vectors A_1 ... A_j ... A_n . The similarity L between these two successions is given by:

$$L = \max_f l(i, f(i)) \qquad\qquad (4.9)$$

where:

$$f(1) = 1, \ f(m) = n, \ f(i) - f(i-1) = 0, 1, \text{ or } 2$$

and:

$$l(i,j) = -\log \sum_{k=1}^{K} s_{ik} \ A_{jk}$$

K is the dimension of each vector.

As $f(1) = 1$ and $f(m) = n$, the matching is constrained to start at the beginning and to stop at the end of the unknown pattern. This is called *edge-fixed* DP matching.

As in continuous speech it is rather difficult to detect the beginning and the end of a syllable precisely, the above mentioned constraint is relaxed by making

$$1 \leq f(1) \text{ and } f(m) \leq n$$

obtaining an *edge-free* DP matching.

Usually all the phoneme recognition methods provide a *lattice* of phoneme hypotheses. This lattice is a collection of phoneme hypotheses and time intervals.

Nakagawa (1976) has found that better performances are achieved if phoneme hypotheses are scored by a degree of confidence in the input data. Furthermore, he found that omission errors in segmentation are the only ones that often become fatal for word recognition.

Results of spectral matching may be represented by *phones* instead of phonemes.

A phone may correspond to a particular realization of a phoneme in a given context.

Usually a phoneme may correspond to many phones, and phone recognition precedes phoneme recognition.

Nakagawa (1976) has proposed a preclassification of words based on sequences of acoustic features like "silence", "stationary part of a vowel", "central part of a voiced consonant", "vowel-consonant transition", "consonant-vowel transition". These features could be directly detected on the speech waveform.

Another important problem in phoneme recognition is that of the *speaker differences*.

From a perceptual point of view, there is experimental evidence that the vocal tract parameters contain more characteristics of the speaker than the glottal wave (Tabata and Sakai, 1977).

Methods for the normalization of speaker differences have been proposed, based on the assumption that vocal tract configurations of the speakers differ only in length.

Fujisaki et al. (1970), propose a normalization method based on a vocal tract representation with polar coordinates (r, θ, ϕ). These coordinates are related to the first three formants F1, F2, F3 as follows:

$$r = (F_1^2 + F_2^2 + F_3^2)^{1/2}$$

$$\theta = \tan^{-1} (F1/F2)$$

$$\phi = \tan^{-1} ((F_1^2 + F_2^2)^{1/2}/F3)$$

(4.10)

Only r is assumed to be speaker-dependent.

Wakita (1977) proposes a normalization of formants F and band-with B using an estimated vocal tract length (L) and a standard vocal tract length (L_s) as follows:

$$\bar{F}_i = F_i \frac{L}{L_s} \quad ; \quad \bar{B}_i = B_i \frac{L}{L_s}$$

Sambur and Rabiner (1975) proposed for digit recognition a self-normalization of the threshold made during the articulation of the words to be recognized.

As the problem of speaker normalization has not yet been satisfactorily solved, many systems still use speaker-dependent reference patterns.

In the approach followed in this book speaker differences are assumed to be elements that "fuzzify" the knowledge sources. In order to reduce these effects as much as possible, relative measures are taken whenever this is feasible.

4.7 References

The foundations of Approximate Reasoning with fuzzy relations and the computation with linguistic probabilities are described in a paper by Zadeh (1975).

Definitions and operations on fuzzy numbers, fuzzy integrals, and applications to linguistic probabilities, fuzzy logic and approximate reasoning are contained in a report by Dubois and Prade (1978) and their book (1980). An interesting book on fuzzy algorithms in thinking processes has been published in Russian by Chorayan (1979).

The use of linguistic probabilities with Bayes' theorem and their application to scheduling KS instantations in a Speech Understanding System was proposed by De Mori and Saitta (1979).

The grammar for segmenting continuous speech into Pseudo-Syllable-Segments was proposed by De Mori (1974) and the Segmentation Parser is described in a paper by De Mori and Giordano (1980a).

The place and manner of articulation for the phonemes of American English can be found in a book by Flanagan (1972). Other details can be found in the book on phonetics by Abercombie (1967). Recent investigations of perception and production of speech are reported in a book edited by Cole (1980).

5. RULES FOR CHARACTERIZING
SONORANT SOUNDS

5.1 A Fragment of the Structural
Knowledge Source for Pseudo-Syllables

5.1.1 Generalities

The purpose of this chapter is to introduce and describe the rules which may be applied when the feature "sonorant" is hypothesized for a segment.

The sonorant sounds are generated by exciting the vocal tract with a quasi-periodic sequence of glottal pulses. The corresponding spectrograms exhibit narrow energy peaks whose central frequency changes continuously in time because of the coarticulation effects.

Most of the quasi-stationary portions of the time evolutions of energy concentrations correspond to vowels, which, in many cases, can be used as anchor points for segmentation and for the interpretation process. In specific contexts, detected and taken into account by the segmentation process, the evolutions of energy concentrations do not reach a quasi-stationary configuration corresponding to the target values of the formant frequencies of a vowel. This is the so-called "vowel reduction" effect.

When nasal sounds are pronounced, the velum is open and the resonances of the nasal cavity give rise to many lines of energy concentrations in the spectra; these lines can be partially masked by the antiresonances, due to the vocal tract which is closed. In this case, a good model of the speech production system should have a transfer function with both poles and zeros, whereas, for the other sonorant sounds, a production model with an all-pole transfer function is adequate, at least if the sounds are not nasalized.

Research is in progress for investigating new speech analysis systems based on vocal-tract models and with an acceptable computational complexity.

For this purpose, the works by Nakajima (1976) and Shirai and Honda (1976) are noteworthy.

Concerning Speech Understanding Systems, it will be shown in Section 5.5, devoted to a discussion on the recognition of nasal sounds, that the antiresonance effects can be described in terms of formant patterns, provided that formants are extracted according to certain rules. For this purpose, Section 5.2 will be devoted to the problem of formant tracking. Recall that formants are the time-evolutions of the frequencies of the three main energy concentrations in the 0.1 - 3.5 kHz band and that their values can be related to such features as place and manner of articulation (see Flanagan, 1972).

Section 5.3 will be devoted to the recognition of vowels, Section 5.4 will deal with the rules for a detailed hypothesization of the phonetic features "liquid" (characterizing semivowels) and "nasal", Section 5.5 will present the detailed rules for the recognition of nasal sounds.

Most of the experiments reported in the following refer to Pseudo-Syllable-Segments of the type Vowel-Consonant-Vowel (VCV) because they exhibit complex coarticulation effects. Knipper (1980) has investigated similar types of rules for Consonant-Vowel segments. These rules can be applied in the approach proposed in this book when the consonant is a "PREVOCALIC SONORANT" (PREVSON).

The phonetic knowledge described in this chapter is incorporated into the Syllabic Grammar, which is an extension of the Segmentation Grammar. The Syllabic Grammar contains the conditions named SEGM and LABEL; these conditions are set by the controller; when SEGM is true, the Segmentation ATNG is activated, when LABEL is true, the rules for phoneme recognition are applied.

This means that new arcs and states have to be added to the Segmentation ATNG in order to obtain the whole Syllabic ATNG. The rules of the resulting version of the syllabic ATNG are reported in Table 5.1. The portion of the ATNG dealing with a more detailed generation of hypotheses about the sonorant sounds will be discussed in Subsection 5.1.2, below.

One more condition, which has not been exhibited previously, is added to the ATNG.

This condition is called LEX and is set under request of the lexical component for "model driven" hypotheses formation or verification. In this way the system generates, in a "data-driven" manner, all the hypotheses it can. When the system has detected difficult cases of dipthong consonant clusters, the "data-driven" generation of hypotheses does not advance up to the generation of phonemic hypotheses; rather it stops at the level of phonetic features, waiting for specific verification requests from the lexical component.

Examples of such cases are the clusters of sonorant and non-sonorant consonants.

5.1.2 Generation of hypotheses about sonorant sounds

The ATNG, whose details are given in Table 5.1, is described in a LISP-like notation.

Each set of characters following a parenthesis at the left is the label state. Each state label is followed by one or more arc descriptions. Each arc description starts with an arc-type description (PUSH, CAT, JUMP, POP) which may be followed by a terminal or a nonterminal symbol. For example

(Q1   (PUSH   PSS/T ....

indicates that Q1 is a state which can be left by consuming the PUSH PSS arc.

The PUSH PSS arc is consumed if the subnetwork PSS is completed and one of its POP arcs is consumed. The PUSH arcs may be followed by a condition written after a slash "/". This condition has to be satisfied before an attempt is made to consume the arc.

The symbol T (True) means that a condition is always verified. The conditions associated to the POP arcs are not preceded by a slash but follow directly the arc-label. The actions associated to each arc are listed between parentheses after the conditions. A typical action is SETR (set register); it is followed by the name of the register and the content to be written into it. After the actions, the destination of the arc is a state name, indicated after the word TO.

Sometimes comments are written in lower case letters at the right.

The fragment of the network reported in Table 5.1 relates pseudo-syllabic segments to phonetic and phonemic features.

The relations between these features and the acoustic ones will be described in other sections, in an analytic or *algebraic form*. The reason for this choice is that these relations do not require conditions or actions to be associated with the arcs, making an

TABLE 5.1

```
(SEN      (PUSH PSS/T
          (TO Q1)))
(Q1       (PUSH PSS/T
          (TO Q1))
          (CAT PAUSE/T
          (TO QP1)))
(QP1      (POP))
(PSS      (PUSH UN/T
          (SETR TIMER T)
          (TO Q2))
          (PUSH VE/T
          (SETR TIMER T)
          (TO Q3)))
(Q2       (PUSH VSGA/T
          (TO QP2)))
(Q3       (PUSH VSGB/T
          (TO QP2)))
(QP2      (POP))
(VSGA     (PUSH VSGAA/T
          (SETR TIMER T)
          (TO Q4))
          (PUSH VSG/T
          (TO QP3)))
(Q4       (PUSH VSG/T
          (TO QP3)))
(VSGB     (PUSH VSGBA/T
          (SETR TIMER T)
          (TO Q4))
          (PUSH VSG/T
          (TO QP3)))
(QP3      (POP))
(VSG      (PUSH VBG/T
```

```
                    (TO QP4))
                    (PUSH VR/T
                    (SETR TIMER T)
                    (TO Q5)))
(Q5                 (PUSH VSGC/T
                    (TO QP4)))
(QP4                (POP))
(VBG                (PUSH V/T
                    (SETR TIMER NIL)
                    (SETR RP T)
                    (TO QP5)))
(VE                 (JUMP/RP
                    (SETR TIMER NIL)
                    (SETR RP NIL)
                    (TO QP5)))
(VR                 (PUSH VOCALIC/T
                    (SETR TIMER NIL)
                    (TO Q6)))
(Q6                 (POP/(SEGM))
                    (PUSH VRED/(LABEL)
                    (TO QP5)))
(QP5                (POP))
(VSGAA              (PUSH PREVSON/T
                    (SETR TIMER NIL)
                    (TO Q7)))
(Q7                 (POP))
                    (PUSH PREVS/(LABEL)
                    (TO QP6)))
(QP6                (POP))
(VSGBA              (PUSH INTERVSON/T
                    (SETR TIMER NIL)
                    (TO Q8))
                    (PUSH SINS/T
```

```
                    (SETR TIMER NIL)
                    (TO Q9)))                    (details in Sec. 6.1)
      (Q8          (POP/(SEGM))
                    (PUSH SISON/(LABEL)
                    (TO QP7)))
                    (PUSH MSON/(AND LABEL LEX)
                    (TO QP7)))
      (Q12         (POP/(SEGM))
                    (PUSH POSTVS/(LABEL)
                    (TO QP8)))
      (QP8         (POP))
      (Q13         (POP/(SEGM))
                    (PUSH SV/(LABEL)
                    (TO QP9)))
                    (PUSH DIFTONG/(LABEL)
                    (TO QP9)))
      (QP9         (POP))
      (SISON       (PUSH SILIQ/T
                    (TO QP14))                    single-intervo-
                    (PUSH SINAS/T                 calic-sonorant
                    (TO Q15)))
      (Q14         (PUSH SILIQF/T
                    (TO QP10)))
      (Q15         (PUSH SINASF/T
                    (TO QP10)))
      (QP10        (POP))
      (SILIQF      (PUSH/L//T                     single-intervo-
                    (TO QP11))                    calic-liquid
                    (PUSH/R//T
                    (TO QP11))
                    (PUSH/GL//T
                    (TO QP11)))
      (QP11        (POP))
```

```
(SINASF        (PUSH/N//T              single-intervocalic
               (TO QP12))
               (PUSH/M//T
               (TO QP12))
               (PUSH/GN//T
               (TO QP12)))
(QP12          (POP))
(SV            (PUSH VB/T               single-vowel
               (TO Q27))
               (PUSH VF/T
               (TO Q28))
               (PUSH VC/T
               (TO Q29)))
(Q27           (PUSH VBF/T
               (TO QP22)))
(Q28           (PUSH VFF/T
               (TO QP22)))
(Q29           (PUSH VCF/T
               (TO QP22)))
(QP22          (POP))
(VBF           (PUSH/O//T              back-vowel
               (TO QP23))
               (PUSH/U//T
               (TO QP23)))
(QP23          (POP))
(VFF           (PUSH/E//T              front-vowel
               (TO QP24))
               (PUSH/I//T
               (TO QP24)))
(QP24          (POP))
(VCF           (PUSH/A//T              central-vowel
               (TO QP25)))
(QP25          (POP))
```

algebraic representation more concise than the equivalent LISP-like description.

The syllabic ATNG is similar to the segmentation ATNG up to state Q6 of the subnetwork VR.

Here, if SEGM is true and GETF V is true (the vocalic feature has enough evidence), a move is made on a POP after the state QP5 has been reached. If LABEL and GETF V are both true, then the control is switched to the subnetwork VRED, whose details are omitted here for the sake of brevity; VRED should recognize the reduced vowels. In a similar way, if LABEL and GETF V are both true and the state Q13 of the subnetwork VSGBA has been reached, the subnetwork SV is used for performing a detailed recognition of the vowels. In a similar way, the other subnetworks, like PREVS (Prevocalic Sonorant), POSTVS (Postvocalic Sonorant), and SISON (Single Intervocalic Sonorant), are used for generating hypotheses about liquid and nasal sounds. For the sake of brevity only the details of SISON are given, the other two subnetworks being very similar.

The description of the subnetwork MSON is also omitted here for the sake of brevity; it contains the rules for the interpretation of clusters of sonorant consonants.

The subnetworks SILIQ and SINALS contain the rules for the recognition of the phonetic features "liquid" and "nasal".

Some details of these rules will be given in Section 5.4. The subnetworks invoked in SILIQF and SINASF contain rules for the recognition of phonemes. The subnetworks VB, VF and VC contain rules for the recognition of a back-vowel, a front-vowel, and a central-vowel and VBF, VFF, and VCF invoke subnetworks containing the rules for the detailed recognition of vowels. The details of the subnetwork DIFTONG for the recognition of dipthongs are also omitted for the sake of brevity.

The subnetwork UN as well as the portion of the network following the state Q9 will be described in Chapter 6.

The Syllabic Network can be used in a "bottom-up" way for emitting all the hypotheses it may, provided they have enough evidence, or in a "top-down" way for verifying a hypothesis; in this case it gives the evidence of that hypothesis, provided that enough evidence is obtained through the invoked subnetworks.

The blackboard at the syllabic level is organized with a directory of syllables accessed by increasing values of the time instants Tb, the beginning of a syllable, and by the ending time Te for those syllables having the same Tb. The directory contains, for every entry, pointers to a data structure where the various hypotheses are linked together. Figure 5.1 shows an example of such an organization.

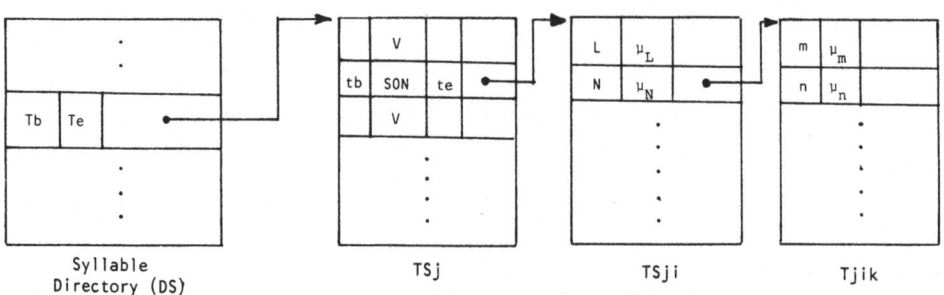

Fig. 5.1 Example of data structure at the syllabic level

Each entry of the syllabic directory DS has a pointer to a table TS_j (j=1,2,..., number of syllables), where the segments used for segmentation, such as V, SON, V are described. For each segment, the time of the beginning, t_b, and of the end, t_e are given. To each segment is associated a pointer to another table (TS_{ji} where more detailed phonetic hypotheses about the segment components are described (i=1,2,..., number of segments of the j-th syllable). In order to fill in this table, other, more detailed, acoustic cues have to be extracted.

This operation is performed when the Syllabic Control KS decides to know something more about a syllable. After reading TS_j the syllabic controller decides what kind of request has to be sent to the acoustic controller KS.

The k-th entry in T_{ji} has pointers to another table T_{jik} containing phonemic hypotheses, their membership values, and other information. A pointer field may contain two pointers if a segment is further split into two time-adjacent intervals containing different sounds.

The acoustic parameters used for generating more detailed hypotheses about a PSS segment are stored in the blackboard at the acoustic level. Links may be provided between corresponding syllable segments in the two directories.

The items of the tables in Fig. 5.1 are written by actions associated to the syllabic ATNG. The details of these actions have been omitted for the sake of simplicity.

The data structures written into the blackboard by the segmentation ATNG activate states of the syllabic ATNG on specific time intervals of the speech pattern.

A scheduler inside the syllabic control KS orders a "ready list" of interpretation processes in a computational model suitable to be implemented on a sequential computer.

Each one of such interpretation processes is described by a table containing an indication of the state of the ATNG which has been activated, together with the time interval of the speech pattern to be analyzed. Let us denote the state activated by the segmentation grammar and the correspnding time interval as "focused state".

For example, if the segmentation ATNG has hypothesized an interval between t_b and t_c labelled VSGBA, a process with focused state: VSGBA (t_b, t_e) is generated as a KS instantiation and queued.

A traffic controller selects candidates from the ready list and allocates available resources to them.

When a process starts to run, it attempts to move on the ATNG and to arrive at a POP arc of the subnetwork whose initial state is a focused state.

If such a POP action can be completed, the interpretation process is terminated and moved into a termination list.

During the execution of a process, a PUSH to a subnetwork is seen as a generation of a child process which is terminated when the POP arc of the invoked subnetwork is reached.

Conditions are of two types. The first-type conditions are such that if they are not met when they are tested, they will never be met. The process which, attempting to verify type-1 conditions, cannot move any further on the ATNG, is aborted and put into an abort list.

The second-type conditions are those which can be met even after an unsuccessful verification. A process which cannot move any further on the ATNG because of failures in the attempt to verify second-type conditions is suspended and put into an event queue (see Brinch Hansen, 1973, for definition of event queue).

The suspended processes will be resumed when some other process will cause resumption after some conditions are changed.

In the syllabic ATNG, there are only type-2 conditions, indicated as SEGM, LABEL, and LEX; type-1 conditions are associated with POP arcs of the subnetworks relating phonetic features and acoustic cues. These conditions concern the truth of GETF LIQ, GETF NAS, GETF VB, GETF VC, and GETF VF and are respectively associated with the POP arcs of the following subnetworks invoked by the rules in Table 5.1: SILIQ, SINAS, VB, VF, VC.

A model-driven request for syllable verification corresponds to the creation of a focused state, if it has not yet been created.

This can be done on all types of PSSs and may imply the request for the extraction of acoustic cues which were not extracted before. The plausibility of an hypothesis generated in a model-driven way depends on acoustic evidences not accounted for before.

The behaviour of the syllabic control KS, in so far as concerns the operations performed after data-driven segmentation, is summarized by the following set of control (production) rules.

IF a state is activated during segmentation, and it is not yet focused, THEN create focused state;

IF a state is focused THEN try to move on its network;

IF GETF is true for a phonetic or a phonemic feature in a given time interval THEN write the feature into the blackboard;

IF POP is executed for the network of a focused state THEN terminate process;

IF type-2 conditions block a process THEN create a suspended process;

IF type-1 conditions block a process THEN abort process.

The action "try to move on a subnetwork" may be performed in a left-to-right or in a non-directional way. In a left-to-right way, the focused state becomes an active state and possible moves from this active state are attempted. The states reached from active states become active and moves from active states are attempted. When all the moves from one state have been attempted, the state becomes passive. All moves from active states are attempted and new active states are generated until a POP from the network of the focused state is attempted or no more active states remain.

In the latter case, the process is aborted. If a network is used in a non-directional way, a problem reduction representation may be useful for deriving a set of control rules for using the network. This case will be treated in detail in Section 5.5.

5.2 Extraction of Detailed Spectral Features for Sonorant Sounds

For those portions of the syllables that are sonorant, i.e., have the following phonetic features:

{VOCALIC, POSTVSON, INTERVSON, PREVSON} , the Acoustic Control
KS is requested to track formants. Formant tracking consists in
reading the samples of each spectrum of the sonorant portion of a
syllable and computing the time evolutions of the three main resonance frequencies of the vocal tract.

Short-time spectra of the speech waveform can be computed in
different ways. The method used in this approach combines FFT and
LPC spectra.

In this way, the Linear Prediction Coefficients allow good
smoothed spectra to be obtained, while the FFT recovers the spectral
peaks which are formant samples and have been missed by Linear
Prediction because the poles of the model do not match the main resonances of the vocal tract.

The approach proposed here shows how to deal with the ambiguities which may arise because of the vagueness of the speech spectra.

Early attempts at formant tracking are described in the book
by Flanagan, (1972); more recent work is reported by McCandless
(1974), Wolf and Makhoul (in Woods et al., 1976), Laface (1980).

The novelty of the algorithm which is now described consists in
the fact that a fuzzy graph, instead of the usual pattern of three
formants, is extracted from a segment of spectrogram. The arcs of
the graph are possible tracts of formants and the weight associated
with an arc expresses the possibility that the arc is a part of a
formant.

The arcs are directed concurrently with the time axis and the
procedure for representing a piece of spectrogram by a fuzzy graph
is an algorithm, because it performs a finite number of steps on a
finite number of possibilities. The algorithm contains three parts,
namely extraction of multilinked data structure from a spectrogram,
elimination of links that would generate arcs in the fuzzy graph
that cannot be part of formants, and assignment of weights.

Details on fuzzy graphs can be found in Rosenfeld (1975).

5.2.1 Extraction of a multilinked data structure from a spectrogram

Let $G(n,f)$ be the n-th short-time spectrum. Let $I(n,j)$ be the
frequency intervals of energy concentration of $G(n,f)$; $j=1,2,...,N_n$,
where N_n is the number of intervals detected in the n-th spectrum.

A binary variable LP is associated to each interval. LP = 1
means that the interval contains a local maximum of the spectrum

computed from linear prediction coefficients; LP = 0 otherwise.

Algorithm ARGEN

Step 1 - The intervals $I(n,j)$ are obtained for $n_1 \leq n \leq n_2$, where
n_1 is the time index of the first spectrum and n_2 is the
time index of the last spectrum of the spectrogram segment.

The value of LP is computed for each interval and appended
to it.

Each interval is represented by a cell in a data structure.

Each cell has five fields, containing respectively:
- the initial frequency of the interval,
- the final frequency of the interval,
- the average energy density of the interval,
- a label identifying the arc to which the interval belongs,
- LP.

Links are created between nodes corresponding to intervals
of successive spectra having partially overlapped projec-
tions along the frequency axis.

Intervals are represented by points in the time-frequency
plane and labels are assigned to points in such a way that
points belonging to the same arc have the same label, except for the graph nodes, for which two types are distinguished, according to the definition shown in Fig. 5.2.

Descriptions of type-1 nodes are stored in a data structure
T_1, descriptions of type-2 nodes are stored in a data
structure T_2, and a stack Σ_1 is introduced to store information about the intervals that have to be considered for
possible forward links.

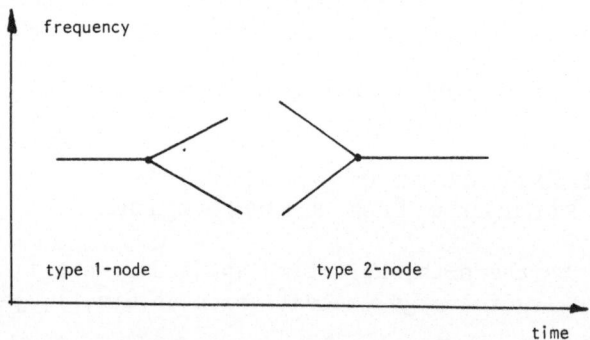

Fig. 5.2 Types of graph nodes that may appear in the extraction of
formant patterns

All the intervals that do not exhibit possible left-connections with other intervals are listed as "initial intervals"; the addresses of the corresponding cells are pushed into stack Σ_1

Step 2 - A type-1 node "generating" all the initial intervals is created in T_1.

The first element of Σ_1 is popped, i.e.:

$$\Sigma_1 \to P \qquad\qquad\qquad (5.1)$$

Notice that an arrow indicates here the extraction of an element from a list.

The first available label of a label list is assigned to P, this operation is represented as follows:

$$l(P) = \lambda_1 \qquad\qquad\qquad (5.2)$$

The interval P of the h-th spectrum may be identified by its label as follows:

$$P = \lambda_1(h) \qquad\qquad\qquad (5.3)$$

Let $\lambda_k(h)$ be the interval that has been labelled λ_k by the algorithm processing the h-th spectrum.

Step 3 - Possible "sons" of $\lambda_k(h)$ are searched, to find intervals belonging to spectra successive to the spectrum containing $\lambda_k(h)$.

These sons correspond to a possible continuation of the arc containing $\lambda_k(h)$.

The following cases are possible:

3a. $\lambda_k(h)$ has only one son;

the algorithm goes to **step 4**.

3b. $\lambda_k(h)$ has two or more sons;

the algorithm goes to **step 5**.

3c. $\lambda_k(h)$ has no sons;

the algorithm goes to **step 6**.

Step 4 - The following cases are possible:

4a. The son was not previously labelled. In this case the son is assigned the same label as "father", i.e. $l(son(\lambda_k(h))) = \lambda_k(h)$; thus the son becomes $\lambda_k(h+1)$, and the algorithm goes back to **step 3**.

4b. The son was already labelled, but not as a node.

The interval is considered a type-2 node and the data structure T_2 is updated with the definition of the new node.

A new label λ_{k+1} is derived and assigned to the line going forward from the new node to the first previously defined node.

The algorithm goes to **step 7**.

4c. The son was already labelled as a node. In this case the only operation performed is the updating of the node description.

The algorithm goes to **step 7**.

A node at which two or more lines end and from which several lines go out is defined as a type-2, node followed immediately by a type-1 node.

Step 5 - A type-1 node generating as many lines as the number of sons is defined in the data structure T_1.

All the sons generated by the node are pushed into the stack Σ_1 with reference to the node stored in T_1.

The first element of Σ_1 is popped, a new label is assigned to it, and used for updating the node description in T_1.

The algorithm goes back to **step 3**.

Step 6 - A "long linkage" to intervals far in frequency or in time is tried. The search for such linkages is controlled by rules that allow only situations of rapid shift of the formant frequencies and brief gaps in the formant lines.

Only a single long linkage is allowed and, if it is found, the algorithm goes to **step 4**.

If no linkage is found the algorithm goes to **step 7**.

Step 7 - The stack Σ_1 is now popped and two cases are possible.

7a. Σ_1 is empty. The algorithm stops.

7b. Σ_1 is not empty. The element popped from Σ_1 is assigned the next available label and the definition of the node generating such interval is updated with the label of the new path.

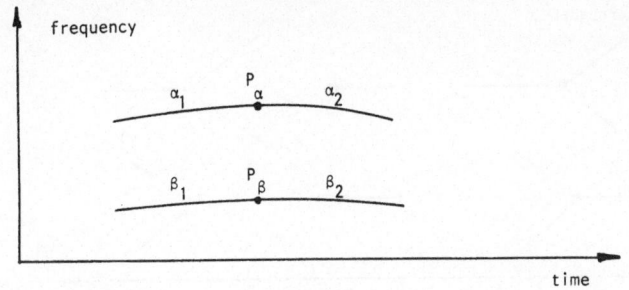

Fig. 5.3 Establishment of a long link between two formant arcs

Step 6 is controlled by a rule that may be formulated in general form, considering two arcs as in Fig. 5.3. If point P_α of an arc α is linked with point P_β of arc β, each one of the old arcs is split into two arcs, leading to four new arcs: α_1, α_2, β_1 and β_2 (see Fig. 5.3). The new link is established and the new arcs are generated if certain conditions hold. These conditions are complex and may depend on the frequency range; they involve the length of α_1, α_2, β_1 and β_2, their average amplitudes, the number of points for which LP = 1, the amplitude and slope discontinuities.

5.2.2 Deletion of unsuitable links

The graph obtained by the algorithm ARGEN contains paths that are to be eliminated because there are other paths that are much more likely to be formants than these. A formant graph simplification is performed by the following algorithm FGS.

Algorithm FGS

Step 1 - A pointer points to the first type-1 node in the data structure T_1.

Step 2 - A search is performed to establish if there is a pair of paths, not previously considered, starting from the selected generating node in T_1 and going to the same type-2 node contained in the data structure T_2.

Two cases are possible:

2a. A pair of paths is found. The algorithm goes to step 3.

2b. No pair of paths is found. The algorithm goes to step 4.

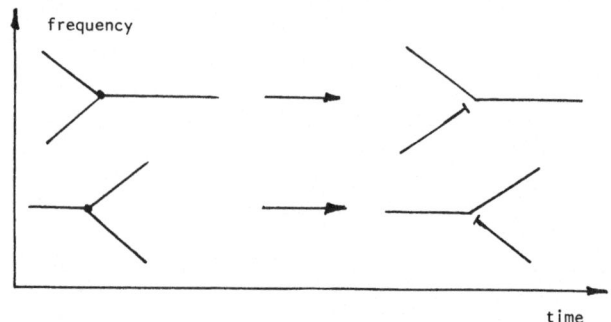

Fig. 5.4 Sample of transformational rules used for pruning the
         formant graph

Step 3 – The non-common parts of the two paths are considered for
         possible reduction to one. The decision of reduction is
         based on local transformational rules, summarized in Fig.
         5.4; the rules apply if certain conditions hold on the
         line amplitudes, number of gaps, the number of points with
         LP = 1, and the amplitude and slope discontinuities.

         If a decision of reduction is made, some cuts are performed
         in the graph in order to eliminate the redundant path.

         The algorithm goes back to **step 2**.

Step 4 – A new type-1 node is extracted from the structure T_1. If
         T_1 is empty, the algorithm stops, otherwise it goes back
         to **step 2**.

         Finally, a pruning is performed and all the arcs that do
         not belong to any path going approximately from the begin-
         ning to the end of the spectrogram are eliminated.

5.2.3 Assignment of weights to the arcs

Let Ω_1 be a segment of spectrogram that is supposed to have G
formants, namely F_1, F_2, ..., F_G. If different formant choices are
allowed, there will be different descriptions for the piece of
spectrogram Ω_1, and to each description will be assigned a membership function expressing how well the description represents a piece
of spectrogram Ω_1 Let

$$\mu(d_{11}, \Omega_1); (d_{21}, \Omega_1); \ldots; \mu(d_{i1}, \Omega_1);$$
$$\ldots; \mu(d_{E_1}, \Omega_1)$$

be such membership function and E_1 be the number of descriptions.

Let d_{i1} correspond to the selection of the paths Π_{1i} for the first formant, Π_{2i} for the second formant, Π_{3i} for the third formant. Because d_{i1} is the description of the three paths Π_1, Π_2, Π_3, the membership $\mu(d_{i1},\Omega_1)$ refers to the fact that, given Ω_1, the three paths Π_{1i}, Π_{2i}, Π_{3i} are the three formants.

The total membership is computed as follows:

$$\mu(d_{i1}, \Omega_1) = \mu(\pi_{1i}, \Omega_1) \wedge \mu(\pi_{2i}, \Omega_1) \wedge \mu(\pi_{3i}, \Omega_1) \tag{5.4}$$

Let us assume that π_1 is the concatenation of the following J_1 paths: π_{11i}, π_{12i}, π_{1J_1i}, thus:

$$\mu(\pi_{1i},\Omega_1) = \mu(\pi_{11i},\Omega_1) \wedge \mu(\pi_{12i},\pi_{11i}\Omega_1) \wedge \mu(\pi_{13i},\pi_{11i}\pi_{12i},\Omega_1) \wedge ..$$

$$.....\wedge \mu(\pi_{1J_1i},\pi_{11i}\ \pi_{12i}\ \cdots\ \pi_{1(J_1}-1)i\Omega_1) \tag{5.5}$$

A simplification is now introduced for computing (5.5):

$$\mu(\pi_{1ki},\ \pi_{11i}\ \cdots\ \pi_{1(k-1)i}\Omega_1)=\mu(\pi_{1ki}\ \ \pi_{1(k-1)i}\Omega_1) \tag{5.6}$$

This is justified by the fact that the behaviour of a formant at a certain instant depends only on the immediate past values.

The second member in (5.6) is computed, considering all the possible paths that can follow $\Pi_{1(k-1)i}$. Let us assume, for the sake of simplicity, that there are two paths, namely Π_{1ji} and Π_{1ji}; ten the following definition is used:

$$\mu(\pi_{1ki}\ \ \pi_{1(k-1)i}) = \frac{\zeta c(\pi_{1ki})+\eta g(\pi_{1ki})}{\zeta\{c(\pi_{1ki})+c(\pi_{1ji})\}+\eta\{g(\pi_{1ki})+g(\pi_{1ji})\}} \tag{5.7}$$

where ζ and η are weighting coefficients: $c(\pi_{1ki})$ is a continuity function between $\pi_{1(k-1)i}$ and π_{1ki}; $g(\pi_{1ki})$ is a function equal to the sum of the amplitudes of the points of π_{1kj} and of all points that can be reached going forward from the last point of π_{ikj}. The weight assigned to a line leading to a type-1 node is thus proportional to how close is the direction of the new line to the previous

one, how much energy can be associated with the new line, and the
paths that can be reached through it.

Minor modifications can be introduced for particular and well-
recognizable situations.

Fig. 5.5 shows an example of extraction of a fuzzy graph from
a spectrogram segment corresponding to the syllable /ma/. Figure
5.5a shows all the intervals of energy concentration extracted from
the spectrogram; each interval is represented by a vertical segment
made by repeating an alphanumeric character; the intervals marked
with an asterix are those for which LP = 1. The $\log_2$ of the energy
associated to an interval is proportional to the ASCII value of the
alphanumeric characters, the sequence of digits being placed after
the sequence of literals by change in the coefficient of proportion-
ality. (The energy represented by ϕ is the double of the energy
represented by Z. Figure 5.5b shows the links established by the
algorithm ARGEN (points of the same arc are represented by the same
label; arc labels do not have any relation with spectral energies)
after the deletion of unsuitable links performed by algorithm FGS.
Such an operation extends some arcs.

Fig. 5.5c contains the fuzzy graph extracted from the spectrogram segment. The most evident formants are made of arcs with
highest membership.

5.3 Generation of Hypotheses About Vowels

After formant tracking, the parametric graph of F_2 vs. F_1 is
obtained for each formant pattern and time is used as an additional
parameter of the points in the graph. An example of such a graph
has been given in Fig. 2.4b. Stable zones of these parametric
graphs are then looked for by the Algorithm SZDET, which is described below. Stable zones, denoted SZ, correspond to the stationary portions of the speech waveform.

Vowels, especially the non-reduced ones, are represented by at
least one stable zone in the F2-F1 graph.

5.3.1 Algorithm SZDET

Let ρ be a positive number representing the maximum distance
between two points which can be classified as belonging to the same
SZ. Let P_k be a point of the graph which is candidate to be the
first point of an SZ; let P_{k+1}, P_{k+2}, ..., P_{k+Nk} be the points
corresponding to segments successive to that represented by P_k.

The first phase of the algorithm consists in determining the number N_k of points belonging to an SZ having P_k as the starting point. This is done by the following iterative procedure.

Let S_k be a circle with center P_k and radius ρ.

IF $P_{k+1} \notin S_k$, THEN $N_k = 1$.

Otherwise, the circle S_{k+1} having center P_{k+1} and radius ρ is considered and the surface

$$I_{k+1} = S_{k+1} \cap S_k$$

is generated.

Notice that I_{k+1} contains both P_k and P_{k+1} because S_k contains P_{k+1} and S_{k+1} contains P_k, whose distance from P_{k+1} is less than ρ; moreover, all the points within I_{k+1} from P_k and P_{k+1} are less than ρ in distance.

Now,

IF $P_{k+2} \notin I_{k+1}$, THEN $N_k = 2$.

Otherwise, the surface

$$I_{k+2} = I_{k+1} \cap S_{k+2}$$

is considered. S_{k+2} is the circle with radius ρ and center P_{k+2}; I_{k+2} contains P_{k+2} which is common to I_{k+1} and S_{k+2}, and the points P_k, P_{k+1}, which both belong to S_{k+2}, having from P_{k+2} distance less than ρ (Fig. 5.6). Notice that all the points within I_{k+2} are less than ρ in distance from P_k, P_{k+1}, and P_{k+2}.

This procedure can be iterated until a point $P_{k+N_k} \notin I_{k+N_k-1}$ is found. N_k is then the number of points belonging to a possible SZ having P_k as the starting point.

It can be easily proven that I_{k+N_k-1} contains all the points P_k, P_{k+1}, ..., P_{k+N_k-1}. The number N_k is a function of K. This

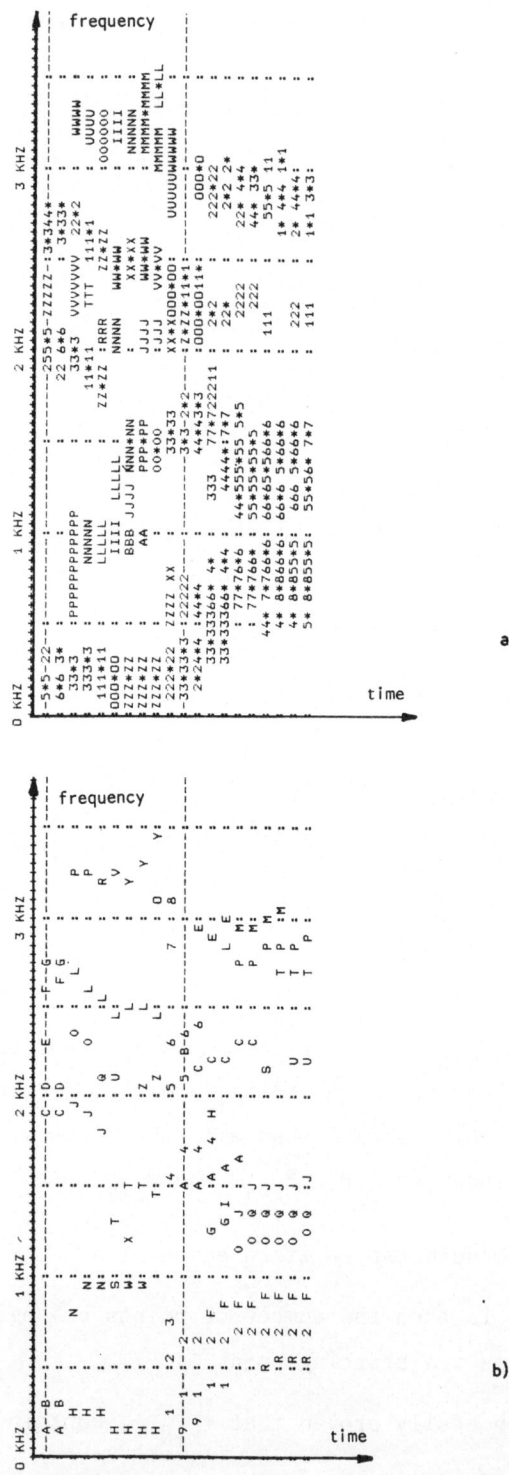

Fig. 5.5 Example of extraction of a fuzzy graph representing a
formant pattern from the spectrogram of the syllable /ma/.

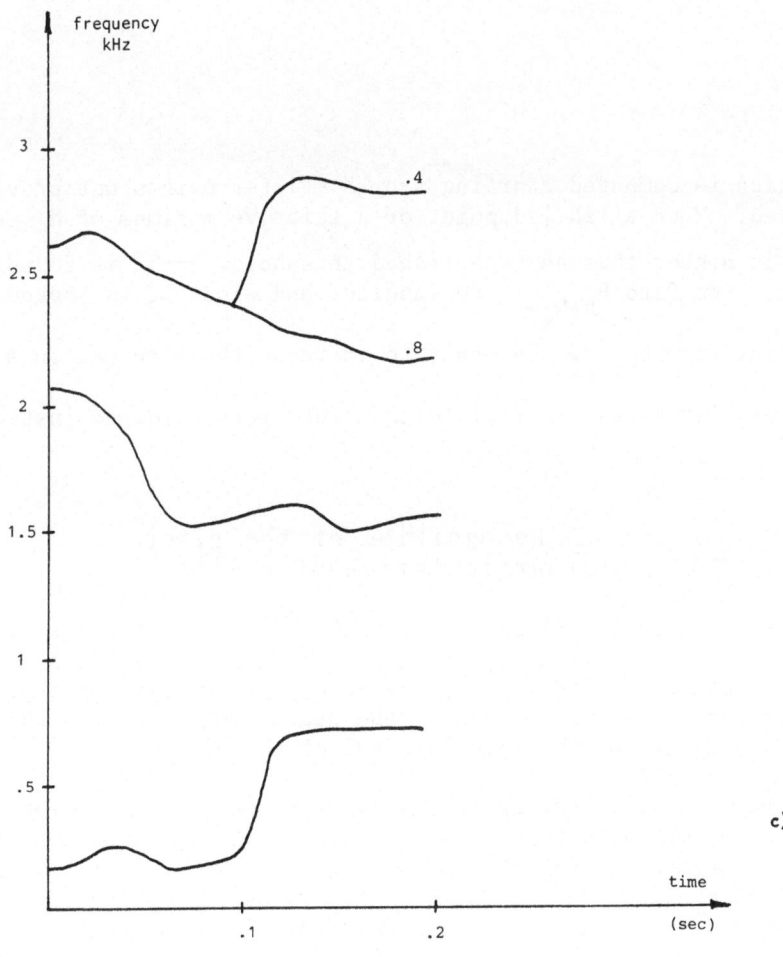

c)

Fig. 5.5 continued

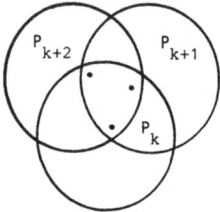

Fig. 5.6 A step of the algorithm for SZ searching

function is computed starting from K = 1 for points not previously labeled. When a labeled point or a relative maximum of N_k is found, if N is higher than an established threshold, an SZ is found, the points from P to P_{k+N_k-1} are labeled, and a new SZ is looked for, starting from P_{k+N_k}. In order to increase the computation speed, squares with sides parallel to the axes are considered instead of circles.

5.3.2 Recognition of the place of articulation of vowels

Each SZ is described by a symbol followed by four attributes: the time of its beginning, its duration, and the coordinates of the gravity center of its points. For the purpose of vowel recognition, a fifth attribute may be added: the average value of the third for-mant frequency for the time interval of the SZ.

Vowel classification is based on the detection of stable zones in speech intervals that have been previously hypothesized as VOCALIC segments. Some SZs belonging to segments that have not been previously classifed as "VOCALIC" may cause the assignment of such a feature, provided that some conditions involving relations with gross spectral features are verified. This assignment is controlled by rules that are omitted for the sake of brevity.

This may cause new items to be inserted into the queue of the vocalic hypotheses and new pseudo-syllable hypotheses to be gener-ated by the Syllabic Control KS.

Depending on the number and types of SZ detected in a segment previously labeled as VOCALIC, a dipthong may be hypothesized, re-sulting in more than one vowel hypothesis.

For each vowel hypothesis, an entry is generated in the syllabic table of the blackboard and the membership of these hypotheses in
the classes of back vowels (VB), central vowels (VC), and front vowels (VF) is computed. The phonetic features VB, VC, and VF are related by rules to linguistic values defined as fuzzy restrictions on
the values of F2. It is known that F1 is mostly related to the manner of articulation of the vowels, which mostly concerns the degree
of constriction at the place of articulation. The fuzzy sets HF2,
MF2, LF2 are defined over F2. These fuzzy sets are characterized by
vectors of break points.

For example, the vector of break points V(F2) of the fuzzy sets
defined over F2 is:

$$V(F2) \quad = \quad [1000, 1400, 1700, 2500] \qquad\qquad (5.8)$$

and is used as follows.

HF2 has membership 1 for frequencies higher than 1700 Hz; the
membership decreases linearly from 1 to 0, reaching the value 0 at
F2 = 1400 Hz. MF2 has membership 1 for 1400 ≤ F2 ≤ 1700; from these
two values the membership decreases linearly, reaching the value
zero at F2 = 1000 Hz and F2 = 2500 Hz. LF2 has membership 1 for
F2 ≤ 1400 Hz and decreases linearly, reaching the value zero at
F2 = 1700 Hz. These membership functions are shown in Fig. 5.7.

A vector of break-points V(F1) is introduced to define fuzzy
sets over F1, in order to detect SZs which do not belong to vowels.
It is defined as follows:

$$V (F1) \quad = \quad [400, 600, 1000] \qquad\qquad (5.9)$$

The rules for hypothesizing the place of articulations of the
vowels are the following:

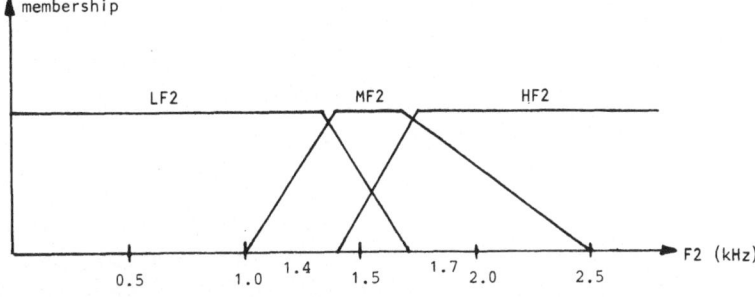

Fig. 5.7 Membership functions for the fuzzy restrictions LF2, MF2,
          HF2

$$
\begin{aligned}
VF &= LF1 \cdot HF2 \\
VC &= HF1 \cdot MF2 \\
VB &= LF1 \cdot LF2
\end{aligned}
\qquad (5.10)
$$

The fuzziness of the rules for the definition of the place of articulation accounts also for little differences in the formant loci due to speaker-dependent effects.

The rules (5.10) give the place of articulation of a vowel (the location of the point of greatest constriction of the vocal tract), provided that the hypotheses VOCALIC has previously been generated for a given segment.

5.3.3 Hypothesis generation and problem solving

The generation of a hypothesis HPV about the place of articulation of a vowel can be seen as the solution of a problem named HPV, given the pattern p. Following the theory of problem solving (see Nilsson, 1971, for details), the solution of HPV given p can be represented as follows:

$$
HPV/p \qquad (5.11)
$$

Because a direct solution of (5.11) is too complex, the problem is split into a collection of subproblems, each one of which has already been solved. Let:

- V be the solution of the problem consisting in the detection of the feature VOCALIC in p;

- FP be the extraction of a formant pattern from p;

- SZ be the extraction of a stable zone from FP.

The problem (5.11) can be represented as follows:

$$
\begin{aligned}
&X1 &&: \quad HPV/p, SZ, FP, V \\
&X2 &&: \quad SZ/p, FP, V \\
&X3 &&: \quad FP/p, V \\
&X4 &&: \quad V/p
\end{aligned}
\qquad (5.12)
$$

The problem (5.11) is solved if *all* the problems (5.12) are solved. This can be represented by an AND graph as in Fig. 5.8 a).

An equivalent representation of problem reduction is shown in Fig. 5.8 b), where problems are ordered in a hierarchy, which is useful for deciding the execution and scheduling of the processes that have to attempt to solve the problems. The representations are useful for deriving semantic rules of evidence composition.

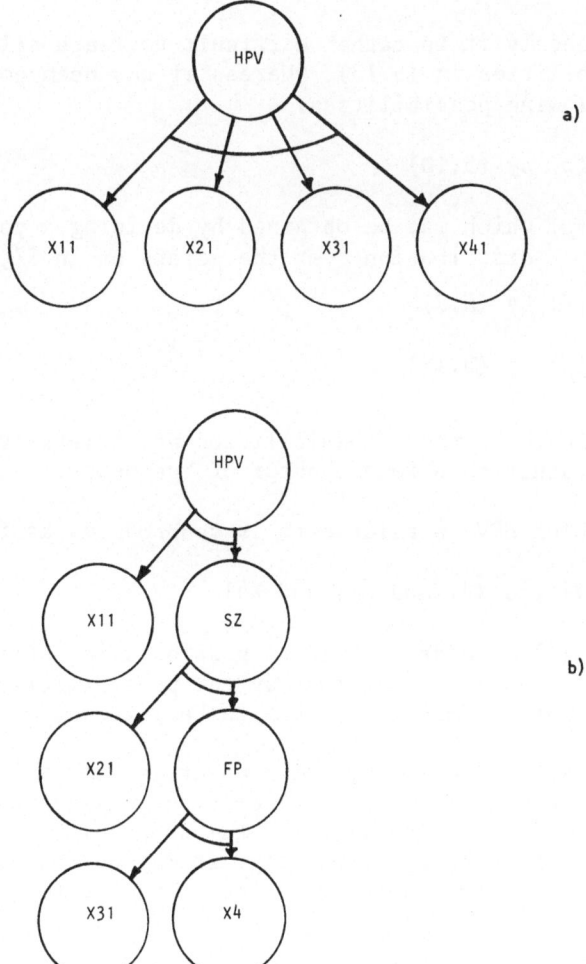

Fig. 5.8 AND-graphs of the problem HPV

Because of the various sources of vagueness involved in the solution of these problems, HPV is seldom solved with full evidence; rather it is often solved with a degree of worthiness. If this degree were expressed by a probability value, the probability of HPV/p would be expressed as follows:

$$p(HPV/p) = p(HPV/p, SZ, FP, V) \cdot p(SZ/p, FP, V) \cdot p(FP/p, V) \cdot p(V, p)$$

$$(5.13)$$

Unfortunately it is rather difficult to learn all the conditional probabilities in (5.13), whereas it has been possible to compute the following possibilities:

- π_{X11} (p) by (5.10)

- π_{X21} (p) which may be obtained by defining a possibility distribution over the points of an SZ,

- π_{X31} (p) by (5.7)

- π_{X41} (p) by (3.18)

Notice that all these possibilities are related to a concept of evidence rather than to a concept of frequency.

The problem HPV is related to is subproblems as follows:

HPV = X11 and X21 and X31 and X41   (5.14)

From (5.14) a relation between possibilities similar to (5.13) can be derived. Notice that the way the possibilities of the subproblems are defined may cause them to be *interactive*.

If this is not the case, the connective <and> is used, corresponding to the min operator in a relation between possibilities. Such an assumption is plausible for the subproblems of HPV. Furthermore, for the sake of simplicity, it may be assumed that $\pi_{X21}(p)$ and $\pi_{X31}(p)$ are both equal to 1, leading to:

$$\pi_{HPV} (p) = \pi_{X11} (p) \wedge \pi_{X41} (p)$$

The places of articulation of the vowels are used as context-constraints for the recognition of nasals and stop sounds.

If phoneme recognition has to be performed, a vocalic SZ is named by one or more vowel symbols and a membership function is assigned to each label. This operation uses a definition of vowel loci as fuzzy sets in the F1-F2 plane.

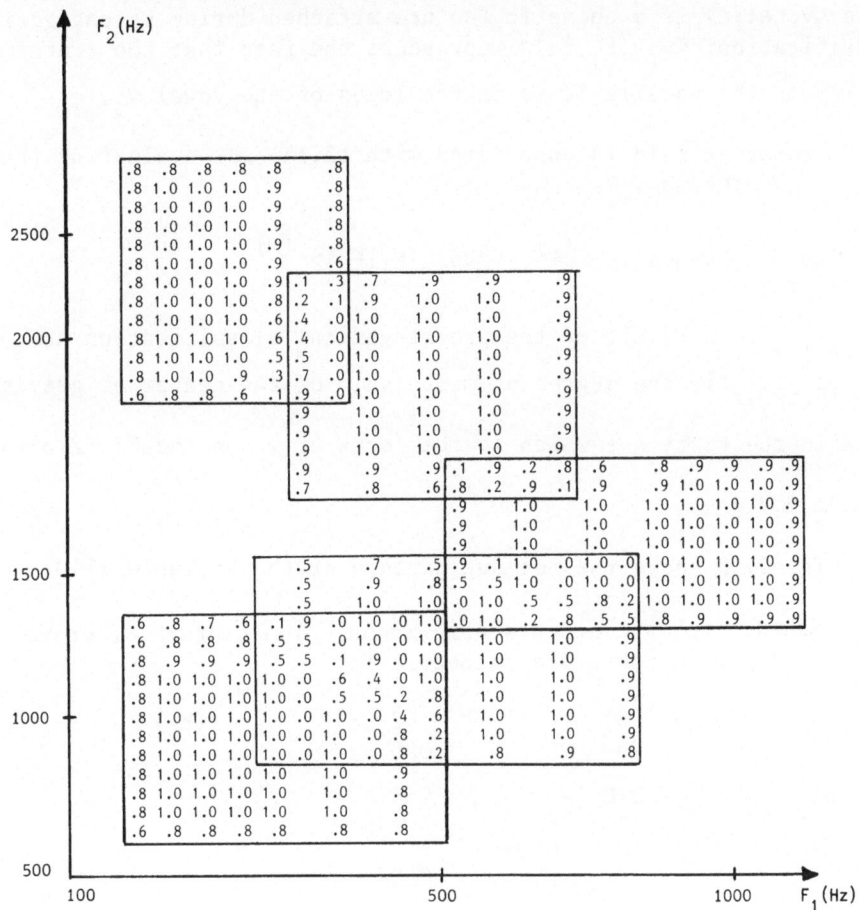

Fig. 5.9 Vowel loci as fuzzy sets in the F1-F2 plane

Fig. 5.9 shows an example of such loci. Each SZ has the coordinates of its center of gravity as attributes. The coordinates identify a point in the F1-F2 plane that may belong to one or more vowel loci with certain membership functions.

Vowel loci, defined as fuzzy sets, have an assignment of membership functions that can be made by a subjective adjustment of probability densities estimated after a long period of learning.

The resulting assignment is speaker dependent.

With the same assumptions as for the place of articulation, and using membership instead of possibilities for the sake of simplicity, a vocalic hypothesis v_i is defined by the following rule:

$$\langle v_i \rangle :: = \langle vocalic \rangle \langle and \rangle \langle v_i (F_1, F_2) \rangle \qquad (5.15)$$

where <vocalic> is a phonetic feature attached during precategorical classification; $<v_i \ (F_1,F_2)>$ represents the fact that the center of gravity of the vocalic SZ is in the locus of the vowel v_i.

A semantic rule is associated with (5.15) for evaluating the vocalic hypothesis v_i:

$$\mu_{<v_i>} = \mu_{<vocalic>} \ (p) \ \wedge \ \mu_{<v_i \ (F_1,F_2)>} \ (p)$$

$\mu_{<vocalic>}$ is a result of the precategorical classification and $\mu_{<v_i \ (F_1,F_2)>}$ is the degree of membership of the center of gravity of SZ in the fuzzy set which is the locus of v_i in the F1-F2 plane.

Example 5.3.1

Fig. 5.10 shows the formant pattern of the syllable /ilu/.

Dashed circles contain stable zones. The centers of gravity of the two stable zones are respectively:

	F1	F2	(Hz)
SZ_1	292	2068	
SZ_2	310	955	

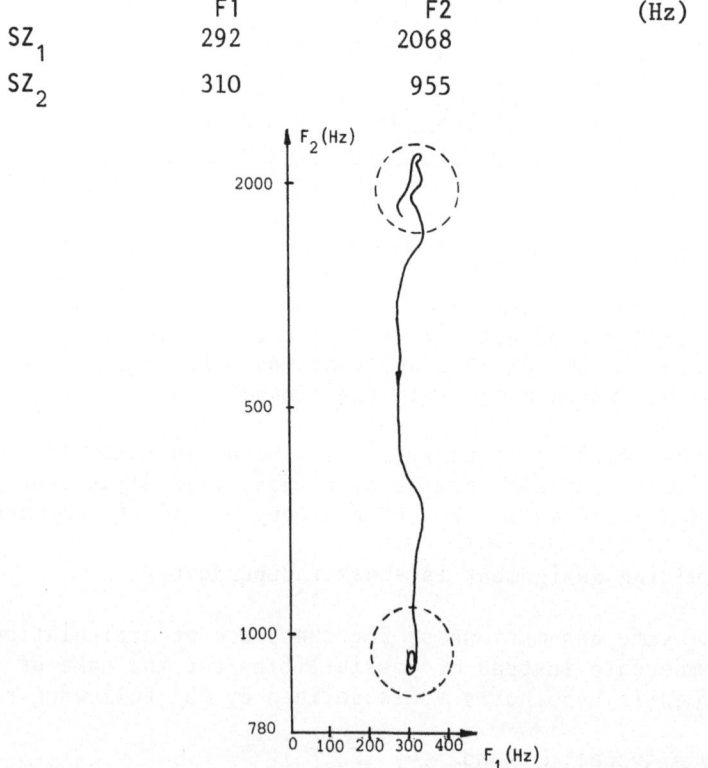

Fig. 5.10 Formant pattern of the syllable /ilu/

TABLE 5.2

Percentage of missed vowels   :   0.6%

Percentage of vowels recognized with highest
    membership   :   98%

Percentage of vowels recognized within two
    choices   :   99%

Such centers of gravity belong respectively to the fuzzy sets
of <i> and <u>, as defined in Fig. 5.9, with membership 1.

The experimental results relating to 200 vowel samples are
reported in Table 5.2 and confirm the effectiveness of the method.

Vowel loci are among the most speaker-dependent cues. Several
techniques have been proposed for their normalization. It would be
too long to recall all of them here. The reader interested in the
details of this problem may refer to Stevens and House (1955), Fant
(1960 and 1973), Kasuya, Suzuki and Kido (1968), Wakita (1972),
Isñizaki (1977, 1978), Nakajima and Suzuki (1978), Itahaski and
Yokoyama (1979); also very interesting is the work by Ishinazi,
Nakajima and Ohmura (1977) on the segmentation of vowels in diphthongs.

5.4 Use of Formants for the Recognition of Liquids and Nasals

After experiments on liquid and nasal intervocalic consonants,
it has been found that if the algorithm used for vowels finds consistent paths with no discontinuities joining the formant lines of
the two vowels, it generally tracks the right formants.

This result, together with the constraint that the first formant is allowed to have only a downward shift in its frequency interval, made it possible to track correctly the formants for all the
liquid and for some of the nasal consonants. This correctness was
confirmed in many cases by synthesis experiments.

For some nasal consonants, particularly in the context of two
front vowels, formant tracking appears to be very difficult (especially for the second formant of /m/) even if FFT and up to 16-pole
LPC spectra are used together. This is due to the complex interaction of the anti-resonances of the vocal tract and the resonances
of the pharyngeal and nasal tract.

Such difficulties have been successfully avoided by controlling

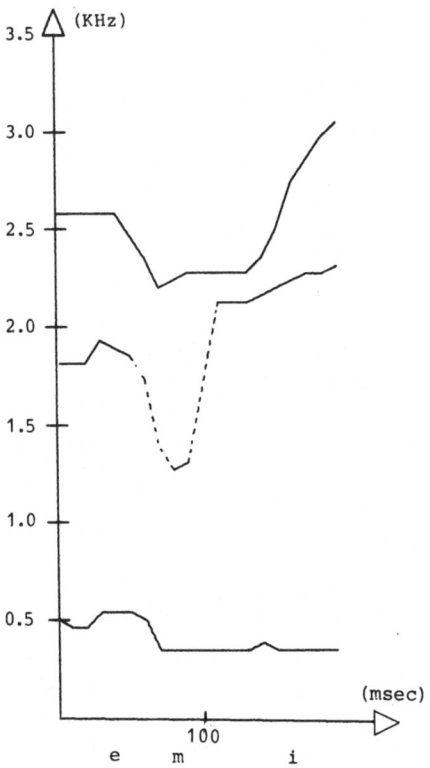

Fig. 5.11 Example of formant tracking on the syllable /emi/

formant tracking with rules for those cases for which there is not a single, consistent path in the consonant region that joins, with no discontinuities, the formant lines of the vowels.

The special rule inferred after experiments is briefly introduced in the following. After first formant tracking, the most consistent tract of energy concentration in the consonant segment and in the frequency range from 2 kHz to 2.8 kHz is assigned to the third formant. Consistency is based on stationarity in time, spectral energy, and number of LPC spectra peaks along the candidate tract. Then the first consistent tract' of energy concentration below this third formant is assigned to the second formant of the consonant. Such a rule is consistent with synthesis experiments. Figure 5.11 shows an example of formant tracking of a PSS /emi/; dashed lines represent the connections established by the rule just described.

5.4.1 Liquid-nasal classification

The liquid-nasal classification is a process of generation of hypotheses about the consonant sounds classified as sonorants with high membership function value in the precategorical classification. The decision within the sonorant consonant class is the answer to the composite question:

$$Q_L \stackrel{\wedge}{=} B_L$$

The answer to Q_L is obtained by a relation on the answers to the component questions Q_{Li} ($i = 1,2,\ldots,K_L$); the answers to the component questions Q_{Li} are fuzzy linguistic variables, defined over the continuous intervals of the previously defined and the following parameters.

- F_i (nT): the n-th sample of the i-th formant frequency (T is a sampling period = 10 ms);

- A_i (nt): the n-th sample of energy associated with the i-th formant;

- D_{12} = {A_1 (n*T) - Max [A_2 (n*T), A_3 (n*T)]} where n*T is the time interval for which A_2(nT) has the absolute minimum in the syllable.

- DF1 = Max (F_1 (nT) - F_1 (nT - T)) at the onset of a vowel;

- DR1 = Max (A_1 (nT) - A_1 (nT - T)) at the onset of a vowel;

- DF2 = F2 (of the SZ of the following vowel) - F2 (n'T), where n'T is the time for which:

$$\underset{n}{\text{Max}} \; (A_1 \; (n \; T) - A_1 \; (n \; T-T)))$$

Among the just introduced parameters, D_{12} is one of the most useful for the distinction liquid/nasal. Fig. 5.12 shows an example in which the formant frequencies and the formant amplitudes of the liquid /l/ and the nasal /n/ are plotted in the context of a back and a front vowel. A_1 (nT) is a continuous line; A_2 (nT) is a dashed line and A_3 (nT) is a dotted line. It is easy to see that D_{12} is high for the nasal and low for the liquid.

The answers to Q_L are represented in an algebraic form which defines the two phonetic features PLIQ and PNAS. These rules apply to prevocalic as well as to intervocalic sonorant consonants (case of SILIQ and SINAS in the ATNG of Section 5.1).

The hypotheses about the place of articulation of the vowel following the consonant act as context constraints in the rules and are represented by the symbols b (back), c (central), and f (front) inside brackets.

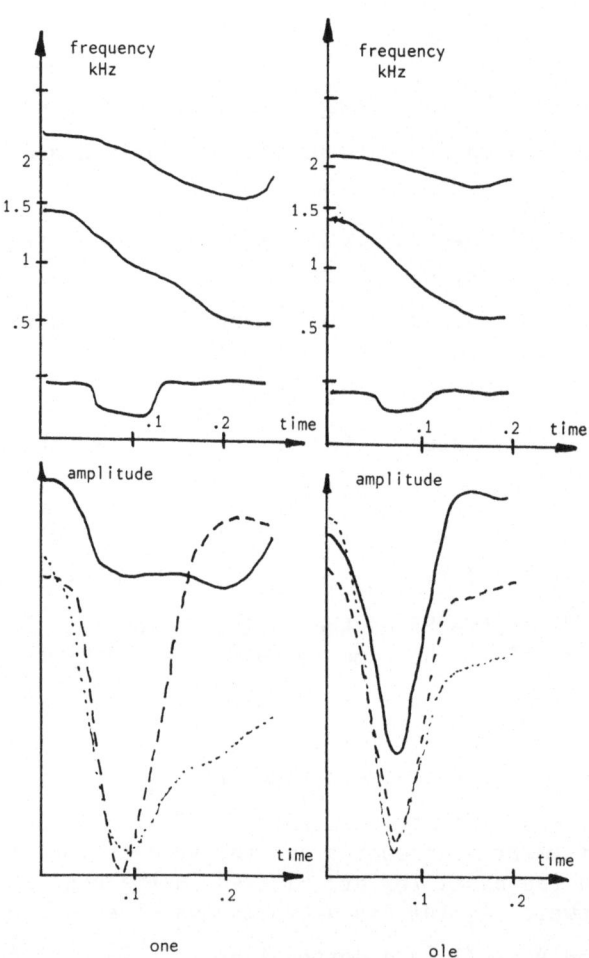

Fig. 5.12 Formant frequencies and amplitudes of the syllables /eno/, /one/, /elo/, /ole/ extracted from continuous speech. Amplitudes of F1 are solid lines, amplitudes of F2 are dashed lines, amplitudes of F3 are dotted lines.

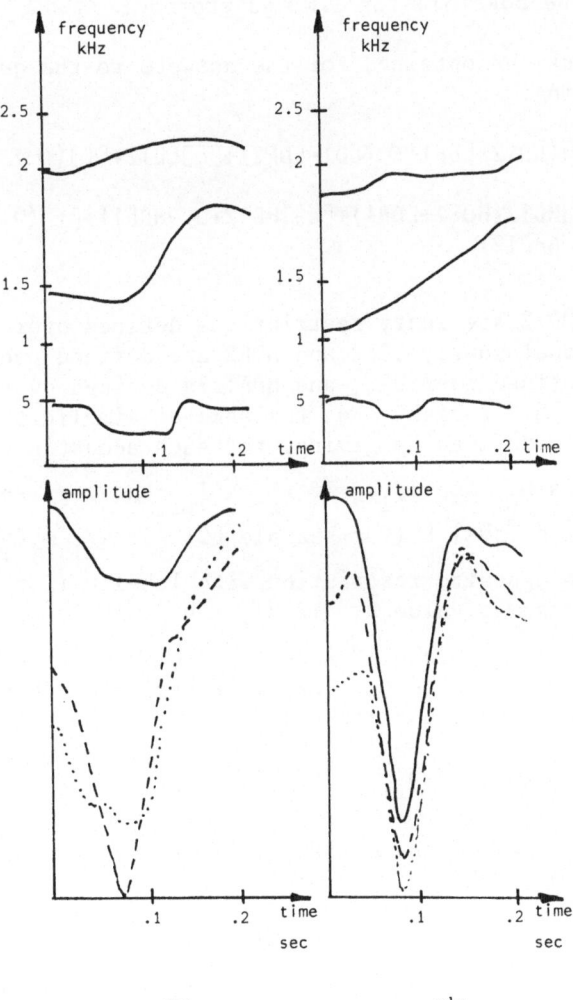

eno   elo

Fig. 5.12 continued

The fact that the relation derived shows a context dependency in the distinction between liquids and nasals means that context dependency is relative to the particular acoustic parameters used.

Nevertheless, the experiments carried out for deriving rules in general cases make it appear unlikely that the distinction liquid/nasal is feasible with a context-independent procedure even if this could be possible for limited protocols.

The expressions obtained for the answers to the question Q_L are the following:

$$PLIQ = [b](LD12 \cdot LDF1 + 0.6LD1 \cdot LDF2) + [c]LD12 + [f](0.8\ LD12 + 0.9HDR1)$$

$$PNAS = [b]HD12(HDF1 + LDR1) + [c](HD12 + 0.9HDF1) + [f](0.8(HD12 + HDF1) + 0.6HD12)$$

$$(5.16)$$

LD12 and HD12 are fuzzy restrictions defined over the acoustic variable D12; analogously LDF2 and HDF2 are defined over DF2, HD23 and LD23 are defined over D23, and HDF1 is defined over DF1; L stands for low, H for high. For the sake of simplicity these fuzzy restrictions are defined by giving, for each acoustic variable x_{Li}, two break points h_{Li} and l_{Li}. If $x_{Li} \leq l_{Li}$ the corresponding fuzzy restrictions with label L (for example LD12 if $x_{Li} = D_{12}$) assumes value 1 and the opposite restriction with label H (for example HD12 if $x_{Li} = D_{12}$) assumes value 0. If $x_{Li} \geq h_{Li}$, the corresponding fuzzy restriction with label H assumes value 1 and the restriction with label L assumes value 0. The two restrictions have a membership which is a linear function of:

$$x_{Li} \text{ for } l_{Li} \leq x_{Li} \leq h_{Li}\ .$$

Table 5.3 contains the break points of the fuzzy restrictions used for PLIQ and PNAS.

TABLE 5.3

i	x_{Li}	h_{Li}	l_{Li}	Dimension
1	D_{12}	10	5	dB
2	DR1	5	1	dB
3	DF1	200	0	Hz/centisec
4	DF2	200	50	Hz/centisec

The semantic rules associated with (5.16) allow one to compute the membership values $\mu_{PLIQ}(p)$ and $\mu_{PNAS}(p)$ for a pattern p.

Coming back to the network of Section 5.1, the expression for PLIQ can be represented by a subnetwork which is invoked by moving on the PUSH SILIQ arc. The actions associated with the arcs of this subnetwork have to compute the evidence of the hypothesis "liquid" in order to set the Condition GETF LIQ. The computation of this evidence may be seen as the solution of the problem "LIQ is in p", which may be decomposed into the two subproblems "(liquid is in p) /(sonorant is in p)" and "(sonorant is in p)"; assuming no ambiguities in formant tracking, analogous conclusions may be obtained for "nasal", leading to the following problem representation:

LIQ is in p = (sonorant is in p) <and> (liquid is in p)/
(sonorant in p)

NAS is in p = (sonorant is in p) <and> (liquid is in p)/
(sonorant is in p)

It may be assumed that the connective <and> applies to non-interactive statements; representing the right-side possibilities by memberships, for the sake of brevity, one gets:

$$\text{Poss } \{LIQ \text{ is in } p\} = \mu_S(p) \wedge \mu_{PLIQ}(p)$$

$$\text{Poss } \{NAS \text{ is in } p\} = \mu_S(p) \wedge \mu_{PNAS}(p)$$

(5.17)

where:

$$\mu_{PLIQ} = \text{Poss } \{(\text{liquid is in } p)/(\text{sonorant is in } p)\}$$

$$\mu_{PNAS} = \text{Poss } \{(\text{nasal is in } p)/(\text{sonorant is in } p)\}$$

5.4.2 Applications to the classification of liquids

The classification of liquids mainly concerns "1" and "r" because the liquid "gl" (as in the Spanish word Sevilla) is easily distinguished by a syntax-controlled procedure that recognizes typical evolutions in the F1-F2 graph. The acoustic parameters involved in this classification are dur (dip (S)), D_{23}, dur (cons), i.e., the duration of the consonant tract, F_{1c} and F_{2c} that are, respectively, the first and the second formant frequencies corresponding to the minimum of the second formant energy.

The question Q_{LQ} for the classification of liquids has compo-

TABLE 5.4

i	x_{Li}	h_{Li}	l_{Li}	DIMENSION
1	Dur (dip(S))	40	20	msec
2	D_{23}	2	0	dB
3	Dur (conson.)	50	30	msec
4	F_{1c}	560	400	Hz
5	F_{2c}	1300	800	Hz

nent questions and fuzzy set definitions according to Table 5.4; the expressions derived are the following:

$$\langle l \rangle = h_{LQ1} + m_{LQ1}(l_{LQ2} + h_{LQ3} + m_{LQ3}(l_{LQ4} h_{LQ5}))$$

$$\langle r \rangle = l_{LQ1} + m_{LQ1}(l_{LQ3} + h_{LQ4}) \quad .$$

$$(5.18)$$

The rules (5.18) apply under the hypothesis that the consonant is liquid. m is a fuzzy set "medium" defined between low and high.

It may happen that, if the evidence of a feature such as LIQ is not very high, both the features $\langle l \rangle$ and $\langle r \rangle$ obtain high membership. This is because the rules have been inferred with the attempt of making evident the differences between $\langle l \rangle$ and $\langle r \rangle$ only when the sounds are clearly liquid. This suggests considering LIQ and $\langle l \rangle$ or $\langle r \rangle$ as having interactive possibilities. Thus the total possibilities of the phonemic hypotheses /L/ and /R/ can be expressed by the following products:

Poss {/L/ is in p} = Poss {LIQ is in p}·Poss {$\langle l \rangle$ is in p}

Poss {/R/ is in p} = Poss {LIQ is in p}·Poss {$\langle r \rangle$ is in p}

Simpler, but less effective rules can be used for the recognition of liquid-sonorant sounds. They are based only on gross acoustic cues and are given in the sequel.

LIQUID ::=((deep AND short) OR (NOT deep AND long))dip of S

NASAL ::=(deep AND long)dip of S

$\langle r \rangle$ ::=(high mAD1) AND (high MAD1)

$\langle l \rangle$ ::=(low mAD1) AND (low MAD1) .

The logical connectives have been indicated in capital letters here.

AD1 is the time-derivative of the first formant amplitude; mAD1 is its minimum value in the consonantal time-interval; MAD1 .is the maximum value of AD1 in the consonantal time-interval.

5.5 Detailed Recognition of Nasal Sounds

5.5.1 Introductory acoustical and perceptual considerations

Accurate recognition of nasal consonants may improve the performance of a Speech Understanding System because of the remarkable frequency of occurrence of such sounds in many languages (about 11% for English, Italian, and Polish). Moreover, an investigation of their acoustic patterns in VCV utterances extracted from continuous speech may make it possible to infer rules for describing a variety of coarticulation effects exhibited in different contexts by many speakers.

The investigation reported in Subsection 5.5.2 (De Mori et al., 1979) is limited to the characterization of the bilabial /m/ and the alveolar /n/ in every possible VCV context. The velar nasal /ŋ/ was not considered because it does not appear in such contexts. Rules for other contexts will be presented later.

The rules inferred after experiments reflect a theoretical basis (see Öhman, 1966, and Liberman, 1970) whereby the generation of VCV sequences involves the encoding of three discrete phonemes to reflect the continual changes in vocal tract shape during the utterance of the whole syllable, exhibiting various degrees of the influence of the vowels on the consonant. Several perception experiments (Flanagan, 1972, Malécot, 1956, Nakata, 1959, Kanamori and Kido, 1975) indicate that vowel transitions interact with nasal resonances. This aspect may be confirmed on the basis of an articulatory model of production and of recent results of spectrogram measurements (Gillman, 1974).

It has long been known (Fujimura, 1962 and Dukiewicz, 1967) that nasal spectra exhibit more zones of energy concentration than vowel spectra. Recently, experimental results by Gillman (1974) for English, in accordance with Dukiewicz (1967) for Polish, and the acoustic patterns of Italian nasals analyzed for inferring the rules presented here, have shown that nasal resonances do not vary very much for a single speaker.

Thus, the distinction between /m/ and /n/ as well as the coar-

ticulation effects due to the adjacent vowels depend principally on
the anti-resonance produced in the vocal tract because the mouth is
closed during the pronunciation of nasals.

The anti-resonances introduce a variety of deformations in the
spectra of nasal and pharynx resonances, leading to complex spectral
patterns. The experimental work that forms the basis of this Section
shows, in accordance with Takeuchi et al. (1975) that, in spite of
the complex spectral configurations, it is still possible to track
formants in the nasal segments between two vowels. Plots of formant
frequencies versus time sometimes show rapid variations, due to the
attempt to join the vowel formants with traces of the nasal resonances not completely masked by antiresonances. Furthermore, in the
frequency range of the second formant, the energy is split into more
than the normal three resonance traces, each one of which turns out
to be very weak.

The effectiveness of a description of nasal sounds based only
on formant patterns is also supported by synthesis experiments (see,
for example, Nakata 1959).

In order to complete the introductory considerations of this
section, it is worth mentioning some studies on the importance of
formant transitions in VN (vowel-nasal) and NV syllables.

It has been proven that in general the VC formant transitions
provide better cues for consonant identification than the CV ones.
For nasal consonants, in particular, it has been found (Brady et al.,
1961) that the velar opening is assimilated by vowels more when a
nasal consonant follows the vowel than vice versa. These aspects are
still present in VNV syllables, but in some cases it appears that the
unconstricted parts of the vocal tract assimilate, in the consonant
part, the shape of initial and final vowels. Moreover, it is possible that the tongue makes a distorted vowel gesture while the articulatory system is executing the nasal consonant.

All the above motivations support the assumption that the rules
for nasal classification should depend, generally, on both vowels of
a VNV syllable.

5.5.2 Inference of the recognition rules

5.5.2.1 **Speech material**. The nasals /m/ and /n/ spoken in the intervocalic position were extracted from sentences pronounced by four
male speakers. The sentences were constructed in such a way that all
the possible coarticulations of nasals with five Italian vowels (/i/,
/e/, /a/, /o/ and /u/) and their combinations were obtained. A total
number of 50 utterances for each speaker was analyzed. In the present study particular attention was paid to nasals coarticulated with

frontal vowels because this case turned out to be the most difficult.

The results obtained showed that the place of articulation is the main feature of the vowels affecting the formant pattern of the nasals. This justifies the circumstance that no phonetic distinctions among the vowels more detailed than those implicit in their orthographic representation were made for the purpose of these experiments.

5.5.2.2 **Parameters of the atomic questions.** The classification of nasals is obtained as an answer to the composite question Q_N. This answer is obtained by a relation on the answers to the component questions Q_{Ni} $(i=1,2,\ldots,K_N)$; the answers to the component questions Q_{Ni} are fuzzy linguistic variables h_{Ni}, 1_{Ni} defined by Table 5.5 on the basis of the acoustic parameters previously defined and the following ones:

- F_{21}: the value of $F_2(nT)$ at the time corresponding to the beginning of the downward shift of the first formant frequency of the nasal consonant;

- F_{22}: the value of $F_2(nT)$ at the start of the rise in the first formant frequency of the nasal consonant.

5.5.2.3 **The recognition rules.** The recognition rules for the classification of nasals were inferred subjectively, the aims being to avoid a wrong answer with a degree of worthiness equal to one and to try to minimize the number of times the wrong answer assumes a degree of worthiness higher than the right answer.

The vocalic contexts are represented by places of articulation and the break points at the end of the intervals where $(\mu_{1_{Ni}} = 1) \wedge (\mu_{h_{Ni}} = 0)$, or $(\mu_{h_{Ni}} = 1) \wedge (\mu_{1_{Ni}} = 0)$ are reported in Table 5.5 with column heading H_{Ni} and L_{Ni} respectively. In Table 5.5 the type of parameters on which the questions are asked is reported under column heading P_{Ni}.

An analytic representation of the inferred rules is the following:

$$\langle m \rangle = [ff] \; (1_{N1}+1_{N2})+[fc] \, (1_{N3}+1_{N4}+1_{N5})+[fb] \, (1_{N6}+1_{N7}+1_{N8})$$

$$+ \; [cf] \; (1_{N9}+1_{N10})+[cc] \, (1_{N11}+1_{N12})+[cb] \, (1_{N13}+1_{N14}) \qquad (5.19)$$

$$1 \; [bf] \; (1_{N15}+1_{N16}+1_{N17})+[bc] \, (1_{N18}+1_{N19}+1_{N20})+[bb] \, (1_{N21}+1_{N22})$$

$$\langle n \rangle = [ff] \ (h_{N1}+h_{N2}) + [fc](h_{N3}+h_{N4}+h_{N5})+[fb](h_{N6}+h_{N7}+h_{N8})$$
$$+ \ [cf] \ (h_{N9}+h_{N10})+[cc](h_{N11}+h_{N12})+[cb](h_{N13}+h_{N14}) \qquad (5.20)$$
$$+ \ [bf] \ (h_{N15}+h_{N16}+h_{N17})+[bc](h_{N18}+h_{N19}+h_{N20})+[bb](h_{N21}+h_{N22})$$

Generally, for a vocalic segment, no more than two hypotheses are generated, resulting in no more than four hypotheses about the places of articulation of the vowels between which the nasal is pronounced. Only the context for which both the preceding and the following vowel have high membership (e.g. > 0.7) are considered plausible contexts for the nasal, because the final membership is found to be high for at least one of the two nasal sounds /m/ and /n/.

As the recognition system performs very well on the vowels, more than 80% of the cases have only one vocalic context hypothesized with membership equal to 1.

In a general case, the membership of the vocalic context is ANDed with the content in the parentheses following it and the general membership is obtained using min-max operators. By abbreviating (Poss $\langle m \rangle$ is in p) by $\mu_{\langle m \rangle}$ and Poss ($\langle n \rangle$ is in p) by $\mu_{\langle n \rangle}$ one gets:

$$\mu_{\langle m \rangle} = (\mu_{f-f} \wedge (\mu_{1_{N1}} \vee \mu_{1_{N2}})) \vee (\mu_{f-c} \wedge (\mu_{1_{N3}} \vee \mu_{1_{N4}} \vee \mu_{1_{N5}})) \vee$$

$$\vee (\mu_{f-b} \wedge (\mu_{1_{N6}} \vee \mu_{1_{N7}} \vee \mu_{1_{N8}})) \vee (\mu_{c-f} \wedge (\mu_{1_{N9}} \vee \mu_{1_{N10}})) \vee$$

$$\vee (\mu_{c-c} \wedge (\mu_{1_{N11}} \vee \mu_{1_{N12}})) \vee (\mu_{c-b} \wedge (\mu_{1_{N13}} \vee \mu_{1_{N14}})) \vee$$

$$\vee (\mu_{b-f} \wedge (\mu_{1_{N15}} \vee \mu_{1_{N16}} \vee \mu_{1_{N17}})) \vee \qquad (5.21)$$

$$\vee (\mu_{b-c} \wedge (\mu_{1_{N18}} \vee \mu_{1_{N19}} \vee \mu_{1_{N20}})) \vee$$

$$\vee (\mu_{b-b} \wedge (\mu_{1_{N21}} \vee \mu_{1_{N22}}))$$

As for other cases b is for "back", c is for "central", and f is for "front".

TABLE 5.5

Definition of the component questions QN for different types
of coarticulation

Context	Question	P_{Ni}	L_{Ni}	H_{Ni}	Dimension
ff	Q_{N1}	F_{21}	1880	2320	Hz
	Q_{N2}	D_{12}	0	15	dB
fc	Q_{N3}	F_{21}	1750	2150	Hz
	Q_{N4}	F_{22}	1500	1700	Hz
	Q_{N5}	D_{12}	6	17	dB
fb	Q_{N6}	F_{21}	1400	2250	Hz
	Q_{N7}	F_{22}	900	1400	Hz
	Q_{N8}	D_{12}	9	23.5	dB
cf	Q_{N9}	F_{21}	1500	1700	Hz
	Q_{N10}	D_{12}	8	17	dB
cc	Q_{N11}	F_{21}	1300	1500	Hz
	Q_{N12}	F_{22}	1300	1500	Hz
cb	Q_{N13}	F_{21}	1300	1480	Hz
	Q_{N14}	F_{22}	960	1100	Hz
bf	Q_{N15}	F_{22}	940	1200	Hz
	Q_{N16}	F_{22}	900	1100	Hz
	Q_{N17}	D_{12}	6	20.5	dB
bc	Q_{N18}	F_{21}	900	1000	Hz
	Q_{N19}	F_{22}	980	1400	Hz
	Q_{N20}	D_{12}	10	19	dB
bb	Q_{N21}	F_{21}	980	1050	Hz
	Q_{N22}	F_{22}	900	1000	Hz

$$\mu_{<n>} = (\mu_{f-f} \wedge (\mu_{h_{N1}} \vee \mu_{h_{N2}})) \vee (\mu_{f-c} \wedge (\mu_{h_{N3}} \vee \mu_{h_{N4}} \vee \mu_{h_{N5}})) \vee$$

$$\vee (\mu_{f-b} \wedge (\mu_{h_{N6}} \vee \mu_{h_{N7}} \vee \mu_{h_{N8}})) \vee (\mu_{c-f} \wedge (\mu_{h_{N9}} \vee \mu_{h_{N10}})) \vee$$

$$\vee (\mu_{c-c} \wedge (\mu_{h_{N11}} \vee \mu_{h_{N12}})) \vee (\mu_{c-b} \wedge (\mu_{h_{N13}} \vee \mu_{h_{N14}})) \vee \quad (5.22)$$

$$\vee (\mu_{b-f} \wedge (\mu_{h_{N15}} \vee \mu_{h_{N16}} \vee \mu_{h_{N17}})) \vee$$

$$\vee (\mu_{b-c} \wedge (\mu_{h_{N18}} \vee \mu_{h_{N19}} \vee \mu_{h_{N20}})) \vee (\mu_{b-b} \wedge (\mu_{h_{N21}} \vee \mu_{h_{N22}}))$$

Notice that some linguistic variables, such as l_{N1} and l_{N3} are defined over the same parameter (F_{21}), giving different memberships for the same frequency because they are different variables related to different contexts. Relations (5.21) and (5.22) invariably give a unique membership for every nasal hypothesis <m> or <n>.

μ_{f-f} is the membership of the hypothesis that the vocalic context is front-front, and is computed as follows:

$$\mu_{f-f} = \mu_{f_p} \wedge \mu_{f_f}$$

where μ_{f_p} is the membership of the hypothesis that the preceding vowel is front and μ_{f_f} is the hypothesis that the following vowel is front; the memberships of the other contexts are defined analogously.

In terms of phonetic features, <m> can be seen as a conditional detection of the phonetic features:

(alveolar/nasal-sonorant-consonant);

analogously <n> can be seen as a conditional detection of:

(bilabial/nasal-sonorant-consonant).

The evidence of the phoneme hypotheses /m/ and /n/ can be computed from the following representations:

$$/m/ \;:\; \begin{bmatrix} consonant \\ sonorant \\ nasal \\ alveolar \end{bmatrix} \qquad /n/ \;:\; \begin{bmatrix} consonant \\ sonorant \\ nasal \\ bilabial \end{bmatrix}$$

as follows:

/m/ : (alveolar/nasal-sonorant-consonant) and (nasal/sonorant-
      consonant) <and> (sonorant-consonant),

/n/ : (bilabial/nasal-sonorant-consonant) and (nasal/sonorant-
      consonant) <and> (sonorant-consonant).

As for liquid sounds, the possibilities (5.19) and (5.20) are interactive with Poss (NAS in in p), and the global possibility of phonemes /m/ and /n/ is obtained by multiplying (5.19) and (5.20), respectively, by Poss (NAS is in p).

Example 5.5.1

Let us consider as an example, the pseudo-syllable /ano/ whose plot of formant frequencies versus time is reported in Fig. 5.13a.

The vocalic context is central-back [cb]; assume that the vocalic context is hypothesized with membership equal to 1; thus, the recognition rules (5.19) and (5.20) are reduced to:

$$<m> = [cb] \ (l_{N13} + l_{N14})$$

$$<n> = [cb] \ (h_{N13} + h_{N14})$$

The fuzzy variables l_{N13} and h_{N14} are defined as piecewise linear restrictions with break-points given in Table 5.5.

Notice that:

$$\mu_{l_{N13}} = 1 - \mu_{h_{N13}}$$

$$\mu_{l_{N14}} = 1 - \mu_{h_{N14}}$$

From Fig. 5.13 one gets:

$$F_{21} = 1560 \ Hz$$

$$F_{22} = 1160 \ Hz$$

$$\mu_{h_{N13}} = \mu_{h_{N14}} = 1$$

and

$$\mu_{l_{N13}} = 1 - \mu_{h_{N13}} = \mu_{l_{N14}} = 1 - \mu_{h_{N14}} = 0,$$

$$\mu_{<m>} = \mu_{1_{N13}} \vee \mu_{1_{N14}} = \text{Max } (0,0) = 0,$$

$$\mu_{<n>} = \mu_{h_{N13}} \vee \mu_{h_{N14}} = \text{Max } (1,1) = 1.$$

Thus, if the consonant is nasal, it is certainly /n/.

Example 5.5.2

Let us consider the pseudo-syllable /ine/ whose plot of formant frequencies versus time is shown in Fig. 5.14a.

Assume, as for the previous example, that the vocalic context front-front [ff] is recognized with membership equal to 1; the recognition rules (5.19) and (5.20) are reduced to:

$$<m> = [ff] \; (1_{N1} + 1_{N2})$$

$$<n> = [ff] \; (h_{N1} + h_{N2})$$

Measuring D_{12} on Fig. 5.14 one gets:

F_{21}	=	2300 Hz
$A_1 (n*T)$	=	-10 dB
$A_2 (n*T)$	=	-24 dB
$A_3 (n*T)$	=	-16 dB
D_{12}	=	6 dB
$\mu_{h_{N1}}$	=	0.9
$\mu_{1_{N1}}$	=	$1 - \mu_{h_{N1}} = 0.1$
$\mu_{h_{N2}}$	=	0.3
$\mu_{1_{N2}}$	=	$1 - \mu_{h_{N2}} = 0.7$
$\mu_{<m>}$	=	$\mu_{1_{N1}} \vee \mu_{1_{N2}} = 0.7$
$\mu_{<n>}$	=	$\mu_{h_{N1}} \vee \mu_{h_{N2}} = 0.9$

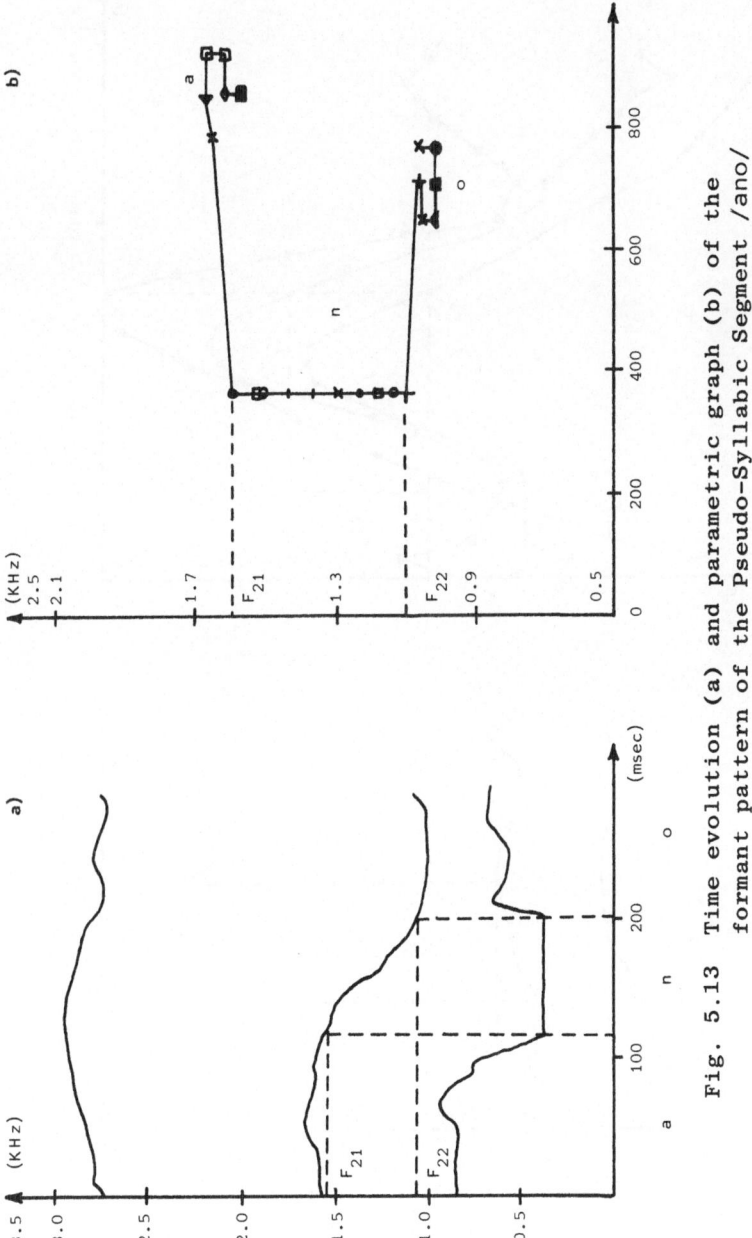

Fig. 5.13 Time evolution (a) and parametric graph (b) of the
           formant pattern of the Pseudo-Syllabic Segment /ano/

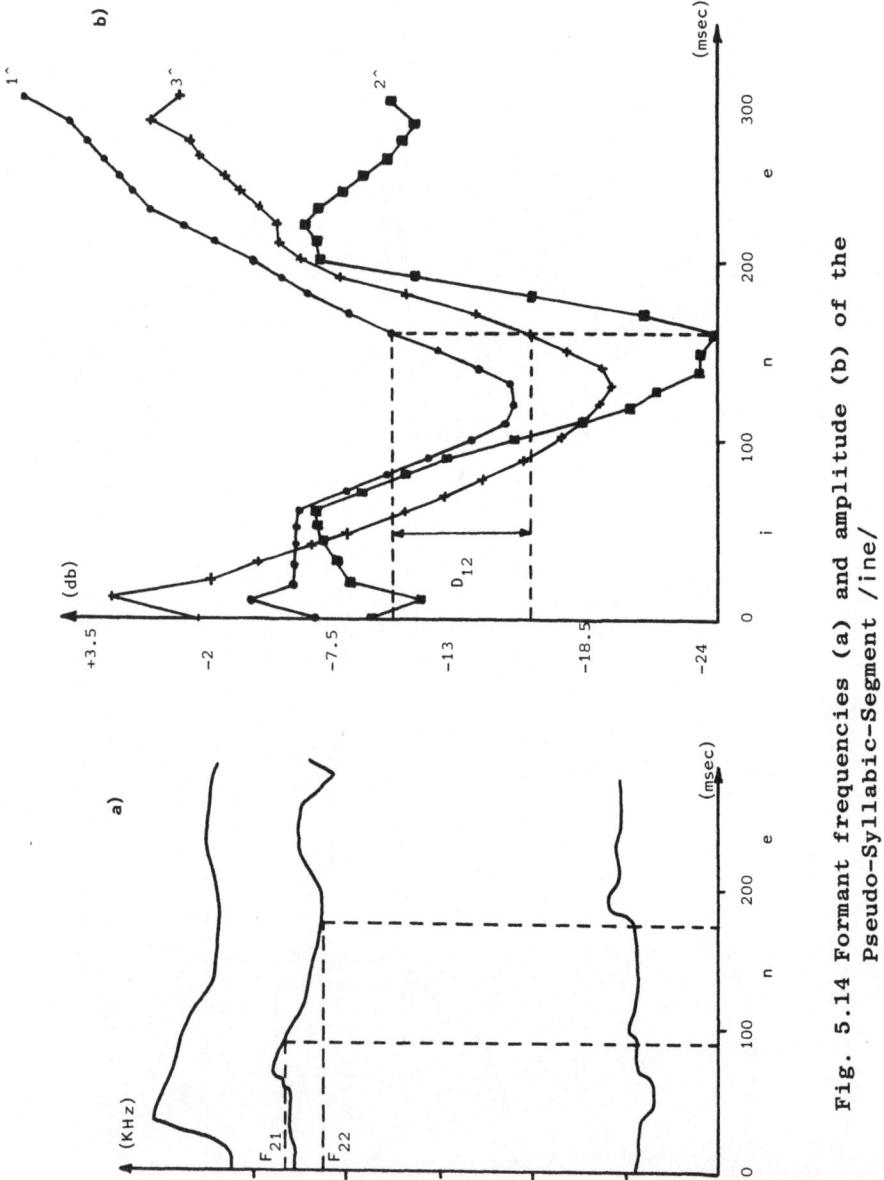

Fig. 5.14 Formant frequencies (a) and amplitude (b) of the
Pseudo-Syllabic-Segment /ine/

The membership of the nasal consonant in the set of <m> is 0.7, while the membership in the set of <n> is 0.9; thus, if the evidence of the feature NAS is not very high, both hypotheses may be considered by the lexical component of the SUS, even though the right hypothesis is the most plausible.

5.5.3 Experimental results

An approach to the recognition of two classes of nasals, bilabial /m/ and alveolar /n/, has been presented. The classification of nasal sounds spoken in the intervocalic position and extracted from continuous speech is performed under the control of a knowledge source.

A set of rules in the form of fuzzy relations for generating hypotheses about intervocalic nasal consonants was inferred and an endeavour made to minimize the number of times that the wrong answer assumes a degree of worthiness higher than the right one.

It should be emphasized that the description of the vocalic context is based only on articulatory representation of vowels, e.g. on their places of articulation, determined very simply from the values assumed by the second formant on the stationary portion of the vocalic segment. There is no need to represent the vocalic context in terms of phonemes. Moreover, the chance of a wrong coarticulation rule due to errors of vowel recognition at the phonemic level is reduced.

Table 5.6 shows the (quite satisfactory) results of the recognition of nasals obtained for 200 utterances by 4 male speakers.

The results show an overall error rate of 6%

However, there are still some problems to be solved, especially in the case of coarticulation with front vowels, for which the points F_{21} and F_{22} are sometimes rather difficult to determine with sufficient accuracy.

None of the errors listed in Table 5.6 corresponds to the assignment of unit degree of worthiness to the wrong hypothesis, but only to the fact that the wrong answer obtained a degree of worthiness higher than the right one.

As for the liquid consonants, the possibilities of the phonemes are conjunctions of interactive events and are expressed by the product of the possibilities of these events as follows:

TABLE 5.6

Four speakers' detailed distribution of the error rates
for each coarticulation instance

Coarticulation instance	Error rate
front-front	10%
front-central	9%
front-back	16%
central-front	9%
central-central	-
central-back	9%
back-front	-
back-central	-
back-back	-
average	6%

$$\text{Poss } \{/M/ \text{ is in } p\} = \frac{\text{Poss } \{NAS \text{ is in } p\}}{\text{Poss } \{<m> \text{ is in } p\}}$$

$$\text{Poss } \{/N/ \text{ is in } p\} = \frac{\text{Poss } \{NAS \text{ is in } p\}}{\text{Poss } \{<n> \text{ is in } p\}} \qquad (5.23)$$

where $\text{Poss } \{<n> \text{ is in } p\} = \mu_{<n>}$

$\text{Poss } \{<m> \text{ is in } p\} = \mu_{<m>}$

The rules for the recognition of nasal sounds have been inferred
for intervocalic consonants; nevertheless it is expected that very
similar rules hold for either postvocalic or prevocalic consonants
in clusters.

Recognizing clusters and delimiting the postvocalic or prevocalic portion of a sonorant consonant is not difficult when there is
at least a nonsonorant consonant in the cluster. Recognizing clusters of all sonorant consonants and extracting the proper parameters
to be used in rules of the type introduced in this chapter is a difficult task and the problem is still open for research.

Segmentation techniques based on a total spectral derivative may be useful for this purpose; an example of the application of these segmentation techniques has been given by Liénard et al. (1974).

5.5.4 On the extension of the rules to other contexts

In the Italian Language, if the liquid and nasal sounds do not appear in a VCV context, they may appear only in a consonant in either a prevocalic or postvocalic context.

Characterizing the coarticulation effects for sonorant consonants in contexts which are not VCV is easier. Simple rules, similar to the ones presented in this chapter, can be inferred for liquids and nasals in a VCC context, such as:

V (LIQ) C and V (NAS) C.

The acoustic cues are the amplitudes of the first two formants in the consonant segment and the maximum time derivative of. F1. The rules in these contexts tend to be less context dependent because often the formants reach some steady-state values corresponding to the consonant loci.

A syntactic procedure, applied to the LPC spectral shape on the consonant loci, seems to be very promising in extracting acoustic cues when the formants reach the consonant loci.

Similar considerations apply to the detailed recognition of nasal sounds. In contexts which are not VCV and, particularly in V (NAS) (PLOSIVE) contexts, the nasals tend to be lengthened in the Italian Language, allowing the system to extract quasi-stationary consonant patterns and to apply context-independent rules to them.

Nevertheless, rules on F21 for the V (NAS) C contexts and on F22 for the C (NAS) V contexts turn out to be very useful.

These rules are similar to the corresponding ones for the V (NAS) V context. Break-points of the fuzzy sets may depend on the context. Their details are omitted here because these values, although they seem to be speaker independent, certainly depend on the preprocessing and formant tracking methods.

Putting together context-dependent and context-independent rules may appear to be an unnecessary redundancy. Although this still has to be proven, there are no doubts that such a redundancy increases the recognition efficiency. Small but important advantages can be achieved by using rules more complex than those introduced here and inferred by the algorithm that will be described in Chapter 9.

5.5.5 On the evaluation of
binary features

As has been pointed out at the beginning of this book, fuzzy set theory has been used for computing evidences of hypotheses, linguistic evidences have been introduced for performing Approximate Reasoning, but decision is based on the statistical distributions of probability densities over the evidence values. Based on these distributions, the classical Receiving Operating Characteristics (ROC) have been obtained. The ROC theory can be found in the book by Green and Sweets (1966).

Applying that theory to our case, each generator of a type of hypothesis (for example the hypothesis "sonorant") can be seen as a receiver which receives evidences (for example μ_{SON} and μ_{NONS}) and decides whether a feature has to be hypothesized or not (for example, whether GETF SON should be true or false). The effectiveness of the method used for measuring the evidences can be seen on the ROC diagram which is the plot of $Pr(H(F)/F)$ versus $Pr(H(F)/\overline{F})$.

F is a feature; H(F) is the hypothesis of the feature F. $Pr(H(F)/F)$ is the probability of hypothesizing F when F is present at the input; $Pr(H(F)/\overline{F})$ is the probability of hypothesizing F when F is not present at the input. Fig. 5.15 shows the ROC curves for the following hypotheses:

    1:    sonorant,
    2:    nonsonorant,
    3:    liquid,
    4:    nasal.

The fact that $Pr(H(F)/F)$ reaches values very close to one for small values of $Pr(H(F)/\overline{F})$ confirms that the algorithm for mapping acoustic cues into phonetic features is very good.

The control statement CS1 introduced in Chapter 4 roughly corresponds to accepting 99% of the right hypotheses by setting the corresponding GETF X to true. The dashed line in Fig. 5.15 corresponds to a probability of 0.99. Its intersections with the ROC curves give the false-alarm probabilities of the generators of phonetic hypotheses. Figure 5.15 shows that these false-alarm probabilities lie between 0.1 and 0.3, depending on the feature.

5.6 Structure of the
Procedural Knowledge

Fig. 5.16 shows a problem representation suitable for deriving

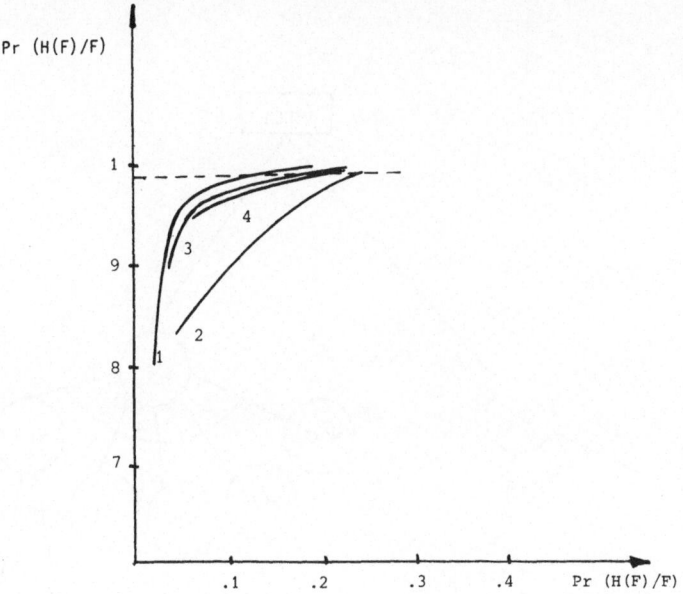

Fig. 5.15 ROC for phonetic features: 1: sonorant
                                      2: nonsonorant
                                      3: liquid
                                      4: nasal

the procedural rules used by the syllabic controller for generating
more detailed hypotheses about a syllable having the structure:

    V VSGBA V.

    The controller already knows which one of the problems
INTERVSON or NONS has been solved. Let us assume that INTERVSON
has been solved, meaning that there is enough evidence in a time
interval for an intervocalic segment made of one or more sonorant
consonants.

    The controller, acting as a problem solver, will attempt to
solve the problem SISON, i.e., it will apply the rules relating the
phonetic feature "single-intervocalic-sonorant" to acoustic cues.

    If enough evidence is found for SISON, the problem solver will
attempt to find evidences for the features LIQ (liquid) and NAS

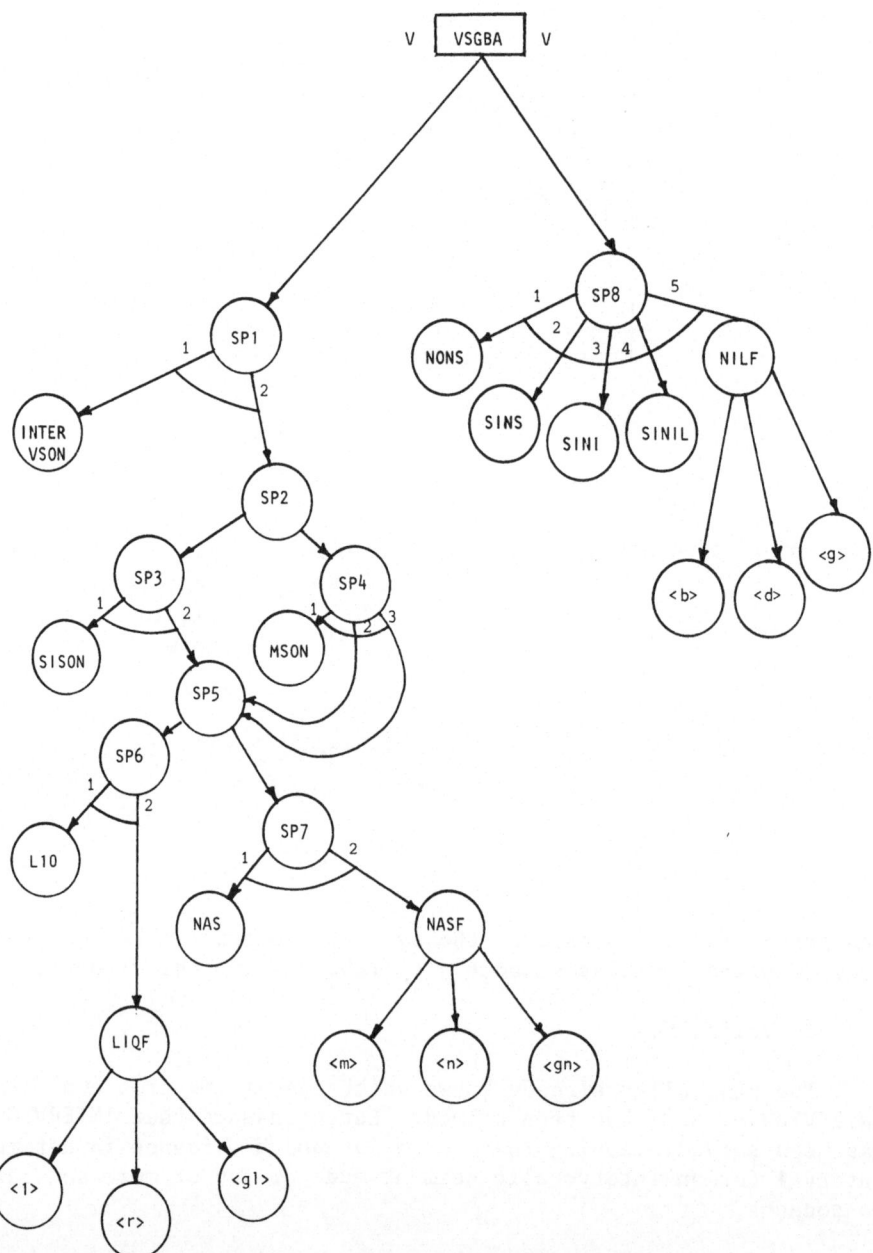

Fig. 5.16 Problem space representation for interpreting a conson-
        ant interval

(nasal). If no acceptable evidence is found for SISON, the corresponding problem is considered *unsolved* and the particular task of the problem solver is suspended until a model-driven request is advanced for a cluster of sonorant consonants. Let us assume that enough evidence has been found for LIQ; the corresponding problem is solved and the generation of hypotheses about the liquid phonemes /l/, /r/ or /gl/ (an Italian phoneme) is attempted by applying specific context-dependent rules for these hypotheses in an intervocalic context.

The evidence conditions such as GETF LIQ are evaluated on the basis of the evidence computed as an action associated with the POP arc of the subnetwork relating phonetic or phonemic features to acoustic cues.

This is done by making SILIQ a subnetwork in which the rules (5.16) are applied. Then, by applying the semantic rules (5.17), Poss {LIQ is in p} and Poss {NAS is in p} are computed and, based on the statistics of these possibilities, the truth of GETF LIQ is obtained by an action associated with the POP of the subnetwork SILIQ. If the POP is performed, the hypothesis LIQUID, together with the time bounds of the interval in which it holds and its possibility value are written into the blackboard. Otherwise, the problem is considered unsolved and the arc labelled PUSH SILIQ cannot be followed.

The knowledge representation in the problem space is equivalent to the representation by ATNG except for the conditions and actions which are related to context dependencies and cannot be embedded into a PRR that represents a context-free grammar (Hall, 1973). The representation in the problem space is useful for applying the semantic rules with which the evidences have to be composed and for designing a "non-directional" parser. This parser does not follow the ATNG from left to right; rather, it allows the focus of attention of the system to apply sequentially to non-contiguous portions of the ATNG. The sequence of application of these portions may depend only upon the a priori knowledge about the complexity of the problems. In this case, subproblems are ordered a priori in such a way that when, for example, an AND node is encountered, its subproblems are solved in a specific order. When a problem turns out to be unsolved, the process which has been created for solving the other subproblems of the AND node is completed, making the problem corresponding to the AND node unsolved.

Some of the subproblems in Fig. 5.16 are labelled by integers, indicating the order in which they have to be solved.

For the other problems in Fig. 5.16 ordering is not necessary and their solution may be obtained by following the ATNG from left-to-right. Problem orderings are affected by the context dependencies in the ATNG.

Semantic rules can be applied according to the sequence in which the problems have been solved.

The Problem Reduction Representation is used for deriving a set of procedural (production) rules which describe how to use the ATNG when moves on it are not strictly left-to-right. Each terminal problem in Fig. 5.16 corresponds to a PUSH arc of the ATNG. Its solution requires the application of relations between phonetic features and acoustic cues on a specific time-interval and the computation of the evidence of the phonetic features. Then a statistical criterion is applied to feature evidences and the truth of the corresponding GETF is computed. If the GETF of a problem turns out to be true, the problem is solved, otherwise it is not.

The procedural rules used for expanding the data structures of Fig. 5.1 are given in the following.

Comments inside parentheses describe their effects on the ATNG.

Let p be a problem. It is active if it has been activated by a focused state, or by other problems, provided that the associated conditions are met.

1) *If* p active *and* p has no successors *then* first son of p becomes active.

2) *If* p active *and* terminal *then* try to solve it by performing a PUSH to the corresponding subnetwork.

3) *If* p terminal *and* the corresponding GETF is true *then* p is solved.

4) *If* p solved *and* has successors *then* first successor becomes active (a move on an arc has been performed).

5) *If* p solved *and* has no successors *and* father is AND node *then* father is solved (a POP action).

6) *If* p solved *and* father is OR node *then* father is solved (a move from an active state is possible).

7) *If* p unsolved *and* father is AND node *then* father is unsolved (a path in the ATNG is interrupted).

8) *If* p is OR node *and* all sons are unsolved *then* p is unsolved (a state ceases to be active).

9) *If* the root problem is solved *then* the corresponding process is terminated.

10) *If* the root problem is unsolved *then* the corresponding process is aborted.

11) *If* p is solved *and* corresponds to a phonemic or a phonetic feature *then* a feature hypothesis is written into the blackboard.

A successor of a subproblem is one of its brothers following it in the order specified on the AND/OR graph.

Notice that a PUSH action started by the above rule 2 corresponds to the use of a subnetwork in a left-to-right manner for which the procedural rules need not be specified. The above rules mostly represent the non-directional use of the ATNG.

These rules can be applied simultaneously to different PSS-hypotheses generated by the segmentation parser. Each application of this set of rules to a PSS-hypothesis is performed by a concurrent process corresponding to an *instantiation* of the syllabic Knowledge Source. Execution of concurrent processes requires special system architectures capable of providing suitable synchronization primitives for the access to shared variables. Brinch-Hansen (1973) proposes solutions to these problems.

5.7 References

The Augmented Transition Network Grammars were first introduced by W. A. Woods (1970).

Interesting perceptual and articulatory investigations on the nasal sounds can be found in the Journal of the Acoustical Society of America (J.A.S.A.), the Journal of Phonetics, the IEEE Transactions on Acoustics, Speech and Signal Processing, the Transactions of the Acoustical Society of Japan, the Proceedings of the ARSO Conference (USSR), the Quarterly Progress Reports of the Massachussets Institute of Technology, the Haskins Laboratories (New Haven, Conn.) and the Royal Institute of Technology (Stockholm).

Also important for this subject are the report by Hughes and Hemdal (1965), the papers by Stevens (1973), Broad and Soup (1975), Bondarko (1969), and the books by Cole (1980), Jakobson et al. (1963), Lea (1980), Niimi (1979).

6. RULES FOR CHARACTERIZING

THE NONSONORANT SOUNDS

6.1 Introduction

The nonsonorant sounds are fricative and stop consonants whose spectra often do not have marked resonances or broad peaks. *Fricative* consonants are produced by an incoherent noise excitation of the vocal tract. The points of the constrictions where the noise is generated are the place of articulation of the consonant. Radiation of fricatives normally occurs from the mouth.

The vocal cord source can operate in conjunction with the noise source, making the fricative a voiced (or lax) one; if only the noise source is used, the fricative is unvoiced (or tense).

Both voiced and unvoiced fricatives are *continuant* sounds. Pairs of fricatives corresponding to the same place of articulation are called *cognates*.

Creation of *stop* consonants depends upon vocal tract dynamics. To produce these sounds, a closure is formed at some point of the vocal tract; this causes the pressure to build up. The occlusion is suddenly opened and the pressure released by an abrupt motion of the articulators. The explosion and aspiration of air help to characterize the stops. The position of the occlusion is the place of articulation. The stop can be produced with or without voicing. In the latter case, the glottal excitation can be used to form the pressure. Both voiced and unvoiced stops are *interrupted* sounds and pair of stops corresponding to the same place of articulation are called *cognates*.

Many languages have speech sounds which are sequences of a

228

stop and a fricative like /tʃ/ of "church". These sounds are called affricates and may be voiced or unvoiced.

Speech originates as neural impulses resulting in a continuous signal transmitted throught the air to the listener's ear. Here it is converted to displacements of points along the basilar membrane. These movements, which are related to the frequency components of the sound wave, finally result in neural impulses to the VIIIth cranial nerve.

An utterance could be described physically at any point along the speech chain. Traditionally, however, the articulatory description of an utterance has formed the basis for phonetic description.

We shall attempt to follow the articulatory description, trying to establish relations between it and the acoustic cues which can be extracted from spectrograms. In order to achieve this task for the nonsonorant sounds, we shall introduce some items of acoustic and articulatory phonetics; such items may involve also the characterization of other sounds.

Systems for classifying articulatory positions generally propose a number of acoustic cues or parameters for specifying the different articulations. The primary phonetic parameters are places of articulation (locations of constrictions along the vocal tract) and manner of articulation. Secondary parameters such as laryngeal action, air release, tongue and lip shape provide a finer classification.

The following are the principal *places of articulation*:

bilabial	:	involving both lips,
labio-dental	:	involving one lip and the teeth,
lingua-dental	:	involving the teeth and the tongue tip,
alveolar	:	involving the alveolar ridge,
alveolo-palatal	:	behind the alveolar ridge,
palatal	:	along the hard palate,
velar	:	along the soft palate,
uvular	:	near the uvula,
pharyngeal	:	along the back wall of the pharynx,
glottal	:	at the vocal folds.

The following are the most important *laryngeal actions*:

voiceless	:	the vocal folds are drawn apart,

whispered	:	a partial closure of the vocal folds gener- ates turbulence,
breathy	:	simultaneous oscillation and turbulence at the vocal folds,
voiced	:	quasi-periodic oscillation of the vocal folds.

The following are the most important *manners* of articulation:

nasal	:	air flow through the nose and oral closure,
stop	:	occlusion and release,
fricative	:	incoherent noise excitation,
sonorant	:	the constriction is sufficient to produce non-laminar flow through the vocal tract but not sufficient to produce turbulence,
trill	:	oscillations of the tongue tip or the uvula,
vocalic	:	laminar air flow through the pharynx and the oral cavity,
nasalized	:	air flow through the nose for a non-nasal sound,
rounded	:	lips are rounded and extended,
palatalized	:	an additional constriction is placed between the blade of the tongue and the palate,
lateralized	:	a central contact between the tongue tip and the palate,
retroflexed	:	the tongue tip is curled back,
velarized	:	the tongue is humped toward the velum,
pharyngealized	:	constriction in the pharynx,
aspirated	:	turbulent release of a plosive.

The articulatory features just introduced define a phonetic space.

Sounds having different images in the phonetic space can be represented by phonetic symbols, or *phones*. In principle, a phonetic description of an utterance is a sequence of phonetic symbols, each of which can be related to some aspect of the physical articulatory event without any consideration of the particular language to which the utterance belongs.

Phonology, on the other hand, describes an utterance using a symbolic alphabet of *phonemes* based on the structure of a particular language.

Phonemes are phonological units, but, unlike letters of the alphabet, they are systematically related in the sense that each phoneme may be viewed as a set of phonetic features and these relations may be affected by the context.

This is the approach proposed in this book.

Relations between phonemes and the articulatory features mentioned above are usually redundant; the redundancy increases when sequences of features are used in a sentence for which only the meaning is essential.

While the latter type of redundancy can be used at the lexical or higher levels, the former type is used at the syllabic level for finding reliable subsets of features which are sufficient for characterizing each phoneme.

Efforts have been made in this research to characterize small sets of phonemes (phoneme classes) in terms of phonetic features and to relate them to acoustic forms and parameters available from the spectrograms. Attention has been given to characterizing those phonetic features which can be detected with context-independent rules. The relations between phonetic and acoustic features should be language independent.

To distinguish among the phonemes of the same class, direct relations between phonemes and acoustic cues have been established.

All the imprecisions involved in feature extraction and in the inference of the rules have been incorporated into fuzzy algorithms.

The phonetic and phonemic knowledge described in this chapter is incorporated into the Syllabic Augmented Transition Network Grammar of Table 6.1.

The networks containing the rules for the recognition of nonsonorant sounds are described in the following, using a LISP-like notation with some comments.

Q9 is called from VSGBA described in Chapter 5.

TABLE 6.1

```
(Q9        (PUSH SINI/T
           (TO Q10)))
 (Q10      (PUSH SINIL/T
           (TO Q11)))
```

```
(Q11      (POP/(SEGM))
          (PUSH SINILF / (LABEL)
          (TO QP7)))
(QP7      (POP))
(UN       (PUSH SINS/T
          (SETR TIMER NIL)
          (TO Q18))
          (PUSH NSCL/T
          (SETR TIMER NIL)
          (TO QP13)))
(Q18      (PUSH SINI /T
          (TO Q19))
          (PUSH SINA /T
          (TO QP13))
          (PUSH SINC T/
          (TO QP13)))
(Q19      (PUSH SINIT /T
          (TO Q20)))
(Q20      (POP/(SEGM))
          (PUSH SINITF / (LABEL)
          (TO QP13)))
(QP13     (POP))
(SINC     (PUSH SINCONT/T              nonsonorant continuant
          (TO Q21)))
(Q21      (POP/(SEGM))
          (PUSH SINCT / (LABEL)
          (TO Q22))
          (PUSH SINCL / (LABEL)
          (TO Q23)))
(Q22      (PUSH SINCTF /T
          (TO QP14)))
(Q23      (PUSH SINCLF /T
          (TO QP14)))
```

```
(QP14        (POP))
(SINA        (PUSH SINAFF/T        nonsonorant affricate
             (TO Q24)))
(Q24         (POP/(SEGM))
             (PUSH SINAT / (LABEL)
             (TO Q25))
             (PUSH SINAL / (LABEL))
             (TO Q26)))
(Q25         (PUSH SINATF /T
             (TO QP15)))
(Q26         (PUSH SINALF /T
             (TO QP15)))
(QP15        (POP))
(SINITF      (PUSH /P//T           nonsonorant interrupted
                                   tense
             (TO QP16))
             (PUSH /K//T
             (TO QP16))
             (PUSH /T//T
             (TO QP16)))
(QP16        (POP))
(SINILF      (PUSH /B//T           nonsonorant interrupted
                                   lax
             (TO QP17))
             (PUSH /G//T
             (TO QP17))
             (PUSH /D//T
             (TO QP17)))
(QP17        (POP))
(SINCTF      (PUSH /S//T           nonsonorant continuant
                                   tense
             (TO QP18))
             (PUSH /F//T
             (TO QP18))
             (PUSH /ʃ//T
```

```
                    (TO QP18)))
    (QP18           (POP))
    (SINCLF         (PUSH /Z//T              nonsonorant continuant
                                             lax
                    (TO QP19))
                    (PUSH /V//T
                    (TO QP19)))
    (QP19           (POP))
    (SINATF         (PUSH /TZ//T             nonsonorant affricate
                                             tense
                    (TO QP20))
                    (PUSH /tʃ//T
                    (TO QP20)))
    (QP20           (POP))
    (SINALF         (PUSH /DZ//T
                    (TO QP21))               nonsonorant affricate
                                             lax
                    (PUSH /dʒ//T
                    (TO QP21)))
    (QP21           (POP))
```

When the state Q9 is reached, if the phonetic feature "nonson-orant" is sufficiently evident, GETF SINS is true, the condition is met, and the subnetwork SINI (Single Intervocalic Nonsonorant Interrupted) is called into operation. If the rules of SINI, whose details will be given in Sec. 6.2, are applied and the corresponding feature is detected with enough evidence, the control goes to the state Q10 of the subnetwork VSGBA. Here GETF SINI will be true (Intervocalic Nonsonorant Interrupted is enough evident), and the rules of the subnetwork SINIL (Single Intervocalic Nonsonorant Interrupted Lax) will be applied.

After the application of these rules, whose details will also be given in Sect. 6.2, the state Q11 of VSGBA is reached. Now, if the Syllabic Control KS is performing segmentation, the variable SEGM is true and the variable LABEL is false; if also GETF SINIL is true, a POP action is performed from the state QP7 of VSGBA. If LABEL is true and GETF SINIL is true, the subnetwork SINILF is called into operation and the rules for the recognition of the voiced plosives /b/, /d/, /g/ are applied. In a similar way one can interpret the remaining subnetworks of the Syllabic ATNG.

As has been pointed out in Chapter 4 above, a special parser uses the ATNG for segmenting continuous speech into Pseudo Syllabic Segments (PSS). Then the ATNG is used for refining the hypotheses inside each PSS. The ATNG is ambiguous because many states are followed by more than one PUSH arc (for example UN is followed by arcs labelled PUSH SINS and PUSH NSCL).

An equivalent way of using evidences to condition the movements of the parser through the network consists in applying the conditions of the POP arcs of the subnetworks containing relations between phonetic or phonemic features and acoustic cues rather than on the arcs following the destination state. For example, starting from UN, a PUSH SINS is performed, the destination state Q18 is reached and all the movements from this state are conditioned by the truth of GETF SINS, which means that the feature "Single Intervocalic Non Sonorant" has been detected with enough evidence. The same effect is obtained if the condition GETF BSINS is removed from the arcs leaving the state Q18 and associated with POP arc of SINS which will assume the following structure:

$$(SINS \quad (PUSH \; RSINS/T \quad (EVALUATE \; (MU(SINS))) \; (TO \; QPSINS))$$
$$(QPSINS \quad (POP \quad /(GETF \; BSINS))). \tag{6.1}$$

When the network SINS is invoked, the arc labelled PUSH RSINS is followed and the rules relating the phonetic feature SINS with acoustic cues are applied; furthermore, the action (EVALUATE(MU (SINS))) is performed. This action consists in the application of the semantic rules corresponding to the syntactic rules of RSINS. After the application of the semantic rules, the possibility

Poss {SINS is in p}

is obtained as a membership value MU(SINS) and the state QPSINS is reached. From this state the truth of GETF SINS is obtained by verifying if MU(SINS), or a function of it, falls into a specific interval; only if the condition is verified, the POP action is performed and the state Q18 is reached.

The condition LEX appears in the subnetwork NSCL, whose details are omitted here for the sake of brevity.

Sections 6.2 and 6.3 will describe the syntactic and semantic rules used for generating detailed phonetic and phonemic hypotheses for single intervocalic non-sonorant sounds.

Figure 6.1 shows the Problem-Reduction-Representation of all the nonsonorant segments. This representation is used by the ATNG parser.

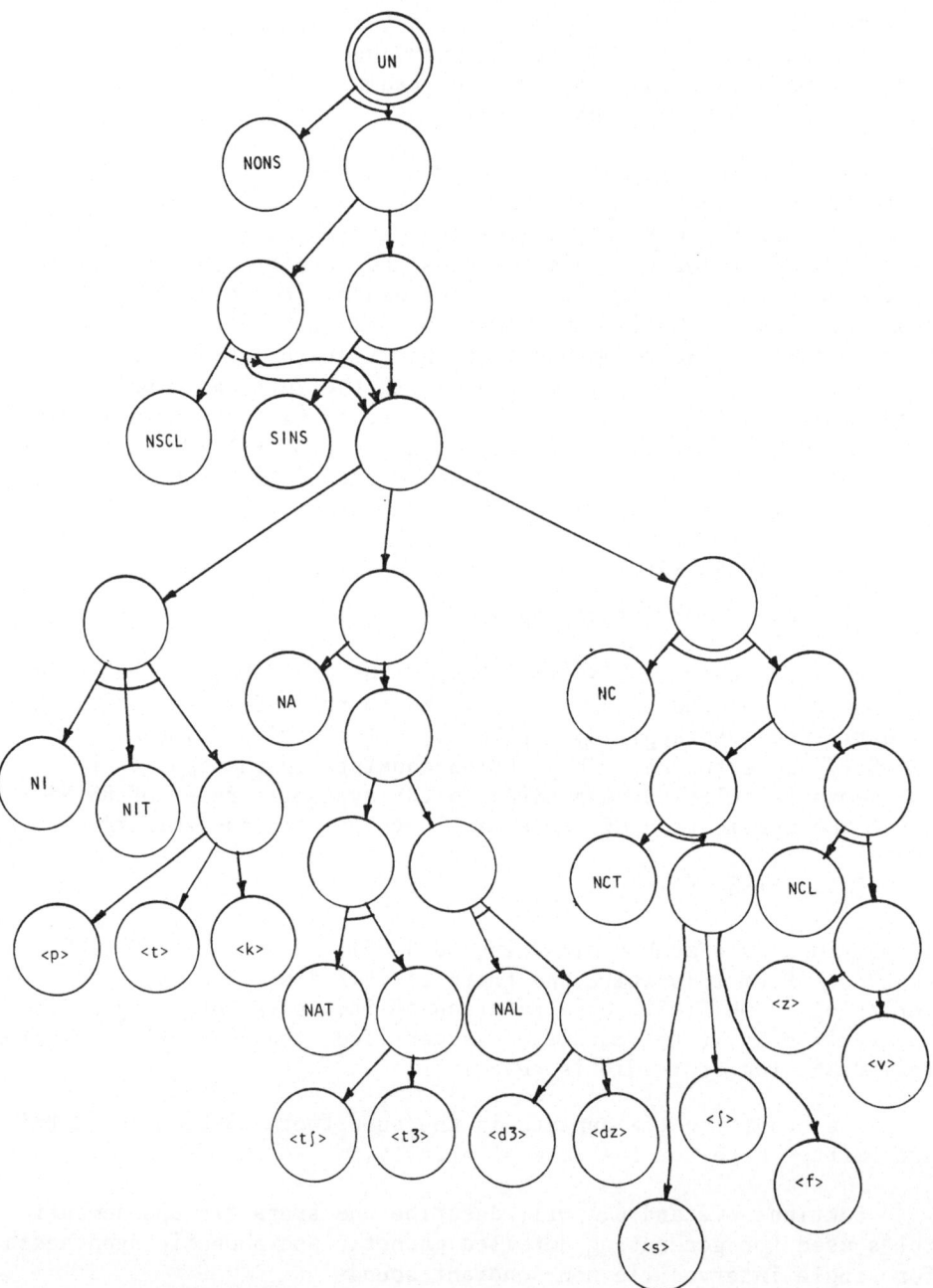

Fig. 6.1 PRR representation for the segments labelled UN during
        segmentation

6.2 Recognition of the Phonetic
Features of Nonsonorant Sounds

Once the feature "nonsonorant" has been found in the acoustic patterns with enough evidence, the evidences of the features "continuant", "interrupted", and "affricate" are looked for, using subnetworks containing rules for relating phonetic features to acoustic cues.

In order to detect possible clusters of nonsonorant consonants and to perform a preliminary crude classification of each sound, the following acoustic cues, represented schematically in Fig. 6.2, are looked for in the descriptions of gross spectral features:

HFE   : high frequency energy characterized by a peak in S and a simultaneous long dip in R (characteristic of the continuant consonants except /v/);

DDTE : deep dip in the total energy S (characteristic of some of the plosives and, sometimes, /v/);

DRFDS : dip in R following a dip in S (characteristic of all affricate sounds and some of the plosives).

Using these binary variables and other fuzzy restrictions, defined over certain acoustic parameters, a set of rules for generating detailed phonetic hypotheses of nonsonorant sounds have been obtained. For the sake of brevity, only the rules for the case of single intervocalic nonsonorant consonants will be given in the following.

The acoustic parameters are extracted from the time evolutions of the gross spectral features introduced in Chapter 3 plus the "buzz bar energy", that is, the average energy in the 100-400 Hz band in the dip of S.

A summary of the acoustic parameters which have been used for this purpose is given in the following:

SMIN-SSIL is the minimum value of S with respect to the value assumed by S on the silences;

RATIO is equal to $\frac{\text{SMIN-SSIL} + \alpha}{\text{MINIMUM R} + \beta}$, where α and β are consonants;

MINIMUM R is the minimum value of R;

DIP IN R - DIP IN S is the value of the differences between the instant when R is minimum and the instant when S is minimum in the same consonant;

DURCONS is the duration of the dip in the signal amplitude between two vowels;

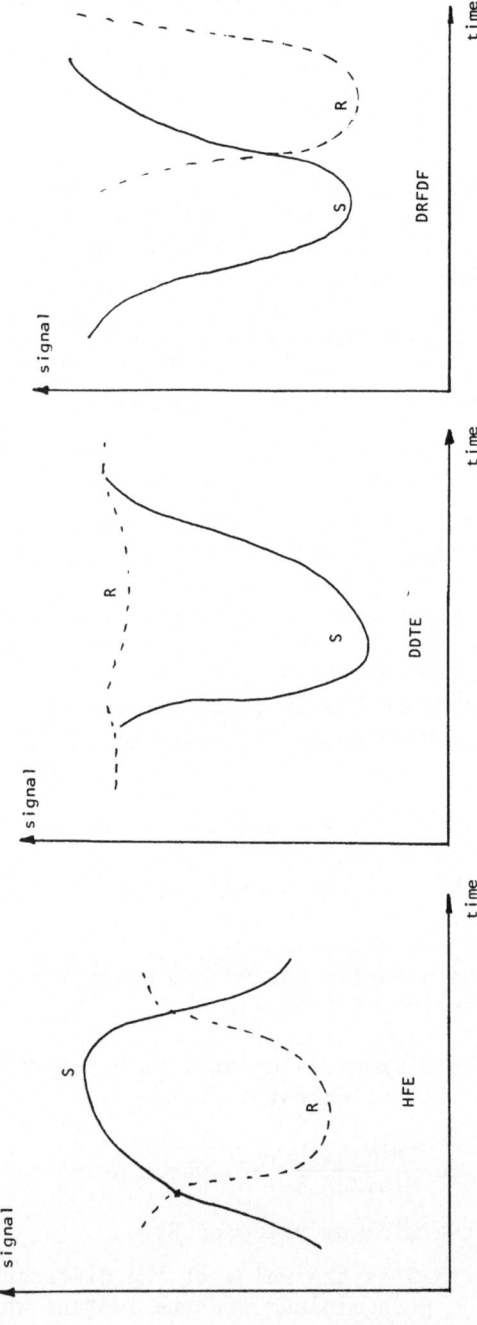

Fig. 6.2 Definition of the acoustic cues HFE, DDTE, and DRFDS

DUR DIP R is the duration of the dip in R;

BUZZ BAR ENERGY is the average energy in the 100-400 band in the consonant.

The linguistic variables of the terminal alphabet of the rules contain fuzzy restrictions defined over the just introduced acoustic parameters. The fuzzy linguistic variables are indicated by a single capital letter, followed by an index, according to the general scheme shown in Fig. 6.3.

Here the fuzzy linguistic variables L_j, A_j, B_j, ..H_j are defined as fuzzy sets over the axis corresponding to the acoustic parameter x_j. These fuzzy sets are defined in practice by giving a vector of break-points $\{x_{j0}, x_{j1}, x_{j2}, \ldots, x_{jNi}\}$. The soft bounds of these fuzzy sets are segments connecting two points having as abscissae two successive break-points and a projection on the vertical axis in the $[0,1]$ interval.

Table 6.2 contains the acoustic parameters used, their dimensions, and the break-points characterizing the fuzzy sets defined over them, and labelled, following the general scheme of Fig. 6.2.

Notice that the fuzzy sets L_j (low) and H_j (high) are defined,

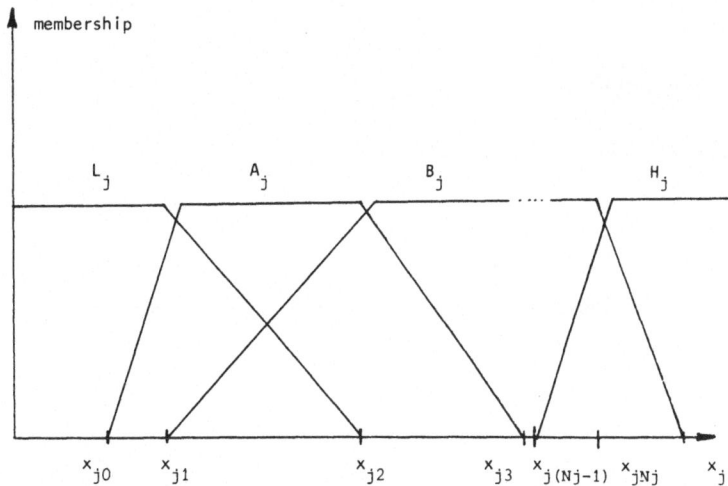

Fig. 6.3 Linguistic variables defined over the acoustic parameters of Table 6.2

TABLE 6.2

j	Parameter	Dimensions	Break-points
1	SMIN-SSIL	c.u.	$x_{11}=0.5$, $x_{12}=3$, $x_{13}=4$, $x_{14}=$ 10, $x_{15}=18$, $x_{16}=30$, $x_{17}=42$, $x_{18}=45$
2	RATIO	c.u.	$x_{21}=1$, $x_{22}=0$, $x_{23}=0.3$, $x_{24}=$ 0.5, $x_{25}=1$, $x_{26}=1.5$, $x_{27}=2$, $x_{28}=2.5$, $x_{29}=7$, $x_{2\ 10}=9$, $x_{2\ 11}=12$
3	MINIMUM R	c.u.	$x_{31}=14$, $x_{32}=12$, $x_{33}=7$, $x_{34}=$ 0.3, $x_{35}=1$, $x_{36}=0$, $x_{37}=1.5$
4	DIP IN R - DIP IN S	msec	$x_{41}=100$, $x_{42}=30$, $x_{43}=0$, x_{44} =10, $x_{45}=60$, $x_{46}=100$
5	DURCONS	msec	$x_{51}=50$, $x_{52}=70$, $x_{53}=100$, x_{54} = 140, $x_{55}=180$, $x_{56}=220$
6	DUR DIP R	msec	$x_{61}=20$, $x_{62}=40$, $x_{63}=60$, x_{64} =80
7	BUZZ BAR ENERGY	c.u.	$x_{71}=11$, $x_{72}=10$, $x_{73}=7$

"c.u." means "conventional unit".

with different break-points, for every j; the number of fuzzy sets A_j, B_j, ... located between L_j and H_j depends on the particular acoustic parameter. All the fuzzy sets between L_j and H_j are labelled from left to right in alphabet order.

The symbols of the nonterminal alphabet satisfy the relations written below in algebraic form (comments on their possible intuitive meaning are written in lower case letters):

```
SINI    = DDTE.(VVLTE.NHR2 + NHTE.DSPDR.LR2 + 0.6.NHR2.NHTE)
          + HFE.(RTHR1 + NVLR1.RTLTE.DSPDR.RTLDC)
          + DRFDS.L6.DSNPDR

SINA    = DDTE.(VVLTE.HR2.DSPDR + 0.6.HR2.RTLTE)
          + DRFDS.(NATA + NALA)
          + HFE.ADRDS.D1

SINC    = 0.6.DDTE.(H1 + L4 + A4.A6 + LDURC.B4)
          + HFE.(H4 + DRPDS)

SINIT   = NBZ.BRT
          + 0.9.(NBZ + BRT)
          + 0.8.ULB.(HTG + AVTG + LTG.LDURC)
          + 0.8.HLFE.HDURC.(HTG + LTG)
          + 0.7.RLLFE.(HTG + AVTG + UBR)
          + 0.6.RLLFE.VLTG
          + 0.6.HLFE.HTG.B5
          + 0.6.HLF.UBR.C1

SINIL   = CBZ.BRL
          + 0.9.(CBZ + BRL)
          + 0.8.ULB.UBR
          + 0.7.ULB.LTG.VLDURC
          + 0.7.RLLFE.RLTG
          + 0.7.HLFE.AVTG
          + 0.6.HLFE(HTG.VLDURC + LTG.LDURC)
          + 0.6.HLF.UBR.RTHTE

BRT     = LR.B3                 burst of tense sounds

VLTG    = C3.LL2

RLTG    = RLR2.C3

AVTG    = AVR2.C3

BRL     = D3.(B2 + F2)          burst of lax sounds
          + E3.(RLR2 + AVR2)
          + HR1.(RLR2 + AVR2)

RLLFE   = A7.(D1 + E1 + HTE)    rather low low frequency energy

HLFE    = HLF.(L1 + C1 + RTHTE)

ULB     = B7.(VLTE + C1 + D1)

CBZ     = B7.(E1 + HTE)         evident buzz-bar
          + 0.9.VLTE1.H7

HTG     = RHR2.C3

UBR     = D3.(C2 + MR2)

LTG     = LR2.C3

NBZ     = VLTE.LFFE + NLTE.VLFFE    no buzz-bar

NATA    = DRFDS.(H4 + E1 + H5 + H6)
```

NALA   = DRFDS.(ADRDS.(LTE + L6))

VVLTE   = L1 +A1

VLTE   = VVLTE + B1   very low total energy

LTE   = VLTE + C1   low total energy

RTLTE   = LTE + D1 + E1

NHTE   = RTLTE + F1

HTE   = F1 + G1 + H1

RTHTE   = D1 + E1 + HTE

NLTE   = RTHTE + C1

VLTE1   = A1 + B1

LR2   = L2 + RLR2

NHR2   = LR2 + MR2 + F2 + G2 + I2 + J2

HR2   = H2 + K2

RLR2   = A2 + B2 + C2

AVR2   = MR2 + F2 + G2

RHR2   = HR2 + I2 + J2

MR2   = D2 + E2

VLR1   = L3

VHR1   = H3

NVLR1   = A3 + B3 + C3 + D3 + E3 + F3 + VHR1

RTHR1   = D3 + E3 + F3 + VHR1

LR1   = VLR1 + A3

HR1   = F3 + VHR1

DSPDRV   = E4 + H4   dip of S precedes dip of R

DSPDR   = DSPDRV + ADRDS

ADRDS   = D4 + E4

DSNPDR   = DRPDS + C4

DRPDS   = L4 + A4 + B4

VLDURC   = L5 + A5

LDURC   = VLDURC + B5   low consonant duration

AVDC   = C5 + D5

RTLDC   = LDURC + AVDC

MLLDC   = A5 + B5 + AVDC

```
MLHDC   = D5 + E5

RTHDC   = HDURC + AVDC

HDURC   = MLHDC + H5

VLFFE   = L7                        very low first formant energy

LFFE    = VLFFE + A7

HLF     = H7

LL2     = very L2

SINCL   = HFE.((A4 + D4) + (L4.LR2.
          (A6 + B6)) + (L4.L2.AVDC.
          H6) + C4.A6) + DDTE.C4

SINCT   = HFE.(DSPDRV + (LR2.L4.
          .(E1 + H5 + L6 + A5.H6)) +
          + (C4(B2 + L6 + H6))) + DDTE.
          .(DRPDS + B4)

SINAT   = NATA + HFE.D4.MLHDC

SINAL   = NALA + DDTE + HFE.MLLDC
```

As for the sonorant sounds, the possibilities obtained with the semantic rules associated to these syntactic rules are conditional. Thus, for example, the semantic rule associated with the definition of SINI gives the possibility μ_{SINI} that an acoustic pattern contains the feature "interrupted", provided that it contains the feature "nonsonorant". As the feature and the parameters used for computing the two possibilities are interactive, the logical conjunction cannot correspond to the "min" operator in the semantic rules; for this reason the arithmetic product has been chosen for such rules:

$$\text{Poss } \{<\text{interrupted}> \text{ is in } p\} = \mu_N \, (p) \cdot \mu_{SINI} \, (p) \qquad (6.2)$$

Similar considerations hold for the other phonetic features.

The binary variables such as (GETF BSINI), assume values "true" or "false" according to an algorithm similar to the one introduced in Chapter 4, Section 3. In the case of ternary features, such as "interrupted/continuant/affricate", the control strategy uses the following probabilities:

p (H/x), where

$$x = \mu_H - \text{Max } \{\mu_{H^*}\},$$
$$\forall \; H^* \neq H$$

H = {interrupted, continuant, affricate}.

Using this approach, the following control rules have been obtained:

(GETF BSINI) = TRUE iff:

$$\mu_{INTERRUPTED} - Max\ \{\mu_{CONTINUANT}, \mu_{AFFRICATE}\} > -0.5 \qquad (6.3)$$

(GETF BSINIL) = TRUE iff:

$$\mu_{LAX} - \mu_{TENSE} > -0.5 \qquad (6.4)$$

(GETF BSINIT) = TRUE iff:

$$\mu_{TENSE} - \mu_{LAX} > -0.1 \qquad (6.5)$$

Using these types of rules, the probability of missing the right hypothesis is less than 0.01; the average number of competing hypotheses among the possible nine candidates in the set F defined in the following is 2.4, and the probability of having the right hypothesis with the highest membership is 0.86.

$$F = \begin{cases} VOCALIC,\ SONORANT\text{-}LIQUID,\ SONORANT\text{-}NASAL,\ NONSONORANT\text{-} \\ INTERRUPTED\text{-}TENSE,\ NONSONORANT\text{-}INTERRUPTED\text{-}LAX, \\ NONSONORANT\text{-}CONTINUANT\text{-}TENSE,\ NONSONORANT\text{-}CONTINUANT\text{-} \\ LAX,\ NONSONORANT\text{-}AFFRICATE\text{-}TENSE,\ NONSONORANT\text{-}AFFRICATE \\ \text{-}LAX \end{cases}$$

The test set contained about 300 samples of all sounds extracted from sentences spoken by 4 men and 1 woman.

The rules described in this section have been inferred by an expert phonetician after inspection of the data of the training set.

A fully automatic learning algorithm proposed by De Mori and Saitta (1980) has been applied and simpler rules with better performances have been obtained. As the algorithm is very complex, it will be described in Chapter 9.

Here it is worth noticing that fuzzy relations can be established subjectively by an expert and used for reaching a preliminary evaluation of the importance of the acoustic cues and their combinations. Then, for the most important of them, the learning algorithm, which requires too much computation if there are too many features, can be applied.

Example 6.2.1

Fig 6.4 shows the time evolution of S and R for the syllable

/agu/ extracted from continuous speech. From the acoustic pattern, the following parameter values have been obtained:

SMIN-SSIL	=	1.3 c.u.
RATIO	=	0.66 c.u.
MINIMUM R	=	0.1 c.u.
DIP IN R - DIP IN S	=	20 ms
DURCONS	=	90 ms
DUR DIP R	=	0 ms
BUZZ BAR ENERGY	=	-5 c.u.

Based on these values, the following non-zero membership values have been found (the symbol μ is represented by MU in this example for the sake of simplicity):

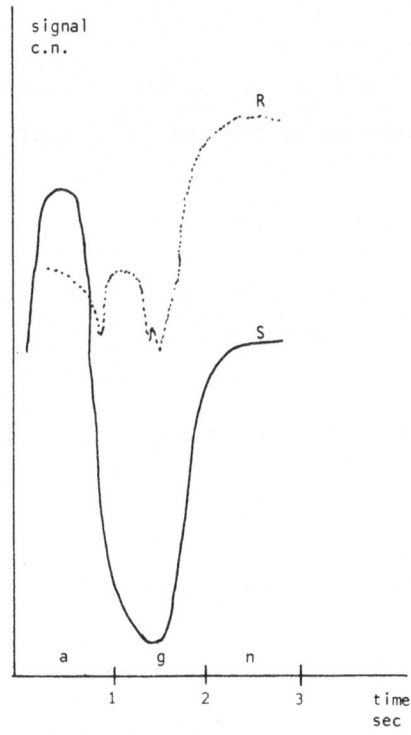

Fig. 6.4 Time evolutions of some gross spectral features of the syllable /agu/

MU (C1) = 0.6   MU (D1) = 1   MU (E1) = 0.4

MU (C2) = 0.7   MU (D2) = 1   MU (E2) = 0.3

MU (E3) = 0.9   MU (F3) = 1   MU (H3) = 0.1

MU (C4) = 0.8   MU (D4) = 1   MU (E4) = 0.2

MU (A5) = 0.7   MU (B5) = 1   MU (C5) = 0.7

MU (L6) = 1   MU (B7) = 0.6   MU (H7) = 1

Furthermore, the feature DDTE has been recognized.

Now the following acoustic cues, corresponding to the nonterminal symbols ANDed with DDTE, have obtained non-zero memberships:

MU (NHTE) = 1

MU (RTLTE) = 1

MU (LR2)   = 0.7

MU (NHR2) = 1

MU (DSPDR) = 1

MU (LDRUC) = 1

Based on the above data, the following memberships for the
phonetic features are obtained:

$$\mu_{SINI} = 0 \wedge 1 \vee 1 \wedge 1 \wedge 0.7 \vee 0.6 \wedge 1 \wedge 1 \ = 0.7$$

$$\mu_{SINA} = 0 \wedge 0 \wedge 1 \vee 0.6 \wedge 0 \wedge 1 \ = 0$$

$$\mu_{SINC} = 0.6 \wedge (0 \vee 0 \vee 0 \wedge 0 \vee 1 \wedge 0) \ = 0.$$

The rules for the phonetic features "sonorant" and "nonsonorant" have been given the following results:

$$\mu_S = 0.7 \ ; \quad \mu_N = 0.88 \ ; \quad thus$$

Poss {<interrupted> is in p} = 0.88 x 0.7 = 0.61.

Applying now the rules for the classification tense/lax, the
following non-zero memberships have been obtained:

MU (BRL) = 1

MU (HLFE) = 1

MU (ULB) = 0.6

MU (CBZ) = 0.4

It is easy to verify that μ_{SINIT} = 0 and μ_{SINIL} = 0.9 because of the term 0.9·BRL in the rule.

Thus the burst is the acoustic cue which determines the final decision; furthermore:

Poss {<interrupted-lax> is in p}

      = Poss {interrupted is in p}·

μ_{SINIL} = 0.61 x 0.9 = 0.55.

Poss {<interrupted-tense> is in p} = 0.

6.3 Bottom-Up Generation of Phonemic Hypotheses of Plosive Sounds

Following the syllabic ATNG, each hypothesized syllable can be further analyzed, in order to generate all the detailed hypotheses which are compatible with the data and involve subnetworks whose activation is conditioned by the label PHT.

Again, the generation of hypotheses is controlled by the parser, which acts as a problem solver on a Problem Reduction Representation (PRR). There is one PRR of this kind for each structural type of hypothesizable syllable. While the possible time sequences of features for the Pseudo Syllabic Segments are those allowed by the Syllabic ATNG, PRR indicates an ordering of the subproblems based on their ease of solution.

6.3.1 Review of research concerning the plosive consonants

This sub-section gives an historical overview of some of the diverse forms of plosive research, to show the basis for the choice of the parameters used in this approach.

The psychological studies on speech perception are of two types: those that use synthetically generated speech and those that use real speech. The synthetic speech experiments have discovered the fundamental characteristics of plosives, while the real speech experiments have tended to clarify the interaction of these characteristics and the variability of their realizations in real speech.

In synthetic speech experiments (Cooper et al., 1955) it was found that certain stop burst spectra were characteristic of the three stop types. In later experiments they found that the formant

transitions from the plosive consonant to the following vowel produced a successful synthetic voiced plosive. The first formant transition appeared to contribute to the voicing of the stop while the
second formant transition provided a basis for distinguishing between the stop types. Further experiments led to development of the
now famous "locus theory" which postulates that the second formant
transitions should point to a frequency locus no matter what the
following vowel is. This appeared to be particularly characteristic
of /d/, not quite so reliable for /b/, while for /g/ there were two
loci, a high one if the following vowel was a front vowel and a low
one if the following vowel was a back vowel.

Attempts to locate third formant loci were also made but these
were not conclusive, although it was obvious that the third formant
transitions could be an important secondary plosive cue.

In real speech experiments Öhman (1966) studied spectra of VCV
coarticulations for Swedish, using all possible combinations in which
the consonant was a plosive.

He deduced that each VCV coarticulation was a "basic dipthongal
gesture with an independent stop consonant gesture superimposed on
its transitional portion". By the end of the 1960's there was no
significant agreement on what provided the best perceptual cue for
classification within the class of plosives. Each researcher had excellent experimental material to support his view of whether burst or
transitions was more significant. This dichotomy is still unresolved.

Turning to the mechanism of perception, we are faced with the
problem of how the signal received by the ear is processed and ultimately recognized by the brain. This is of particular importance in
the case of plosives, as they are characterized by several features.
Research with adaptation-followed-by-testing paradigms indicates that
there exist feature detectors at some level, i.e., detectors that in
some as yet unknown way respond to events such as "burst" and "transition". Furthermore, feature detectors can be tuned. Such tuning
would account for the manner in which slight differences in articulation, as say in the production of /t/ in Italian and /t/ in English,
can be perceived after sufficient training. The mode of integration
of the information from the feature detectors is not known with any
certainty. A possible explanation is given in a model presented by
Oden (1978). He proposes that each feature is fuzzily identified as
belonging to one of several classes. After this, there is prototype
matching in which fuzzy identification rules for each phoneme are
applied to the results output by the feature detectors. Finally the
phoneme is classified as the phoneme whose composite feature detection pattern is most closely matched by the composite feature detection pattern of the test phoneme. Other interesting research on the
use of fuzzy set theory for modeling speech perception is reported by
Massaro and Oden (1978).

It has been shown that the plosives are characterized by several features; transitions, bursts, and timing. It seems that for
each plosive there exists several sufficient but not necessary combinations of appropriate features. The articulation of plosives is
influenced by the sounds produced in conjunction with them.

The perception mechanism consists of feature detectors, the
information from which is integrated to produce identification under the control of rules that may be nondeterministic and have to
account for the different evidences with which features have been
detected.

The work of Halle et al. (1957) is a standard work on plosive
recognition. The hierarchy of their recognition system has been
adopted in almost all later recognition schemes. When considering
ways to locate a plosive in continuous speech they point out that
the production of a plosive requires several centiseconds during
which there is no energy except for a possible voicing component.
Transitions and bursts may or may not be perceived, depending on
the context. As regards the voiced/unvoiced decision, Halle et al.
point out that the essential distinction between /b/, /d/, /g/ and
/p/, /t/, /k/ in English is that in "the production of the latter
more pressure is built up behind the closure than in the production
of the former". Thus they prefer that the distinction be called a
tense/lax decision. Having found that suitably trained subjects
could learn to identify accurately individual plosives when the
burst portion alone was presented to them, Halle et al. concluded
that the burst was a sufficient cue and proceeded to develop a
burst recognition scheme. This scheme consisted of first measuring
the intensity in the 700-10,000 Hz and 2,700-10,000 Hz ranges.
The sounds, with significant energies in the higher frequency
range, were /t/, /d/ and /k/, /g/ before front vowels. These
sounds were labelled acute. The grave class consisted of /p/, /b/
and /k/, /g/ before back vowels. This grave class was then subdivided into two subclasses by measurement of the frequency position
of the highest spectral peak, and the acute class was subdivided on
the basis of further energy band measurements. Up to 85% correct
recognition was obtained by this method.

Halle et al. also attempted to develop rules for plosive recognition using the transitions. An initial problem here arises
when one wishes to specify the transitions accurately in position
and time. Nevertheless gradient rules were found which, while they
were consistent with the Haskins locus theory (Cooper et al., 1955),
were considerably more complex than had been suggested by the
Haskins research.

Recent interesting work was carried out by Weinstein et al.
(1974) at the Lincoln Laboratories, MIT. This work is notable because of its effort to provide very accurate acoustic-phonetic data.

To train the system, male and female voices were used and the re-
cordings were done in a terminal room atmosphere. One hundred and
eleven sentences were used. These sentences in general concerned
command of a speech data base. No attempt was made to achieve
phonetic balance. Careful initial segmentation and classification
were carried out and during this process plosives were located as
a conjunction of a silence with or without voice bar, a plosive
burst (if present), and aspiration (if present). The duration of
the silence following the burst is critical to plosive detection.
A silence duration exceeding 70 ms means that the sound is tagged
as a fricative. A voiced/unvoiced decision was made according to
the output of the pitch detector. However only a small number of
the /b/, /d/, /g/ sounds were tagged as voiced, although no /p/,
/t/, /k/ sounds were ever so tagged. Recognition according to
place of articulation was done by "finding the frequency location
and the relative strength of the major concentrations of energy in
the burst spectrum". When an extremely low main energy concentration frequency was found it was concluded that the burst location is incorrect. An algorithm to detect post-vocalic /k/ and
/g/ from formant transitions into the pre-burst silence was also
developed. It was felt that recognition rates would have been improved if the rules had been made speaker dependent.

Another project in which quite a lot of attention was paid to
plosive recognition is described by Woods et al. (1976). Acoustic-
phonetic rules were initially developed by noting the performance
of speech researchers in parameter reading sessions. Energy band
parameters were used for initial classification into broad phonetic classes. Thus voiced plosives were tagged by a characteristic
dip in the overall energy. The voiced/unvoiced decision was made
by a voice onset time (VOT) measurement. For plosives occurring
before vowels, a two-pole frequency approximation to the peak for
the 20 ms analysis window centered on the burst was used for classification according to place of articulation. An auxiliary clas-
sificatory measurement was the change in the third formant just
before the silence. To improve classification it was later decided that plosive allophone algorithms should be developed, partic-
larly for cases of plosives followed by /r/. Through each stage
of the recognition, probabilistic scoring was employed.

Recently Datta et al. (1980) derived a system for plosive
recognition from examination of a data set of 600 plosive-vowel
combinations spoken by three male speakers of Telogu. The plosive
sounds of Telagu differ from those of English in that they are of
four different articulatory place types - labial, dental, alveolar,
and velar - and they have no associated aspiration. The three
cues used for classification were the first and second formant
transitions and the duration of these transitions. The transition
measurements were done by hand, by extrapolating the formants back

to the point of release of the burst and then measuring the difference between these points and the formant positions during the
steady state.

Tanaka (1978) introduced a dynamic programming approach for
CV syllables, based on a model of the vocal tract.

Blumstein and Stevens (1979) have obtained good results using
context-independent property detectors on the burst spectra. Property detectors can be implemented using syntactic methods. Adding
context-independent rules improves the recognition accuracy, which
is not very high if only the burst spectra are used as acoustic
cues.

6.3.2 Recognition of plosive sounds

An algorithm is presented in this subsection for the bottom-
up generation of phonemic hypotheses for plosive consonants in VCV
or CV contexts.

Let $H_p(p)$ be a hypothesis about a plosive sound in the acoustic pattern p. $H_p(p)$ takes values on the following set P of phoneme labels for plosive sounds:

$$P = P_T \vee P_L \tag{6.6}$$

where:

$$P_T = \{p, t, k\}$$

is the set of labels for the tense plosive sounds and:

$$P_L = \{b, d, g\}$$

is the set of labels for the lax plosive sounds.

Hypotheses in P are assigned by a fuzzy algorithm, which is
executed whenever a consonant is hypothesized in an interval $[t_i,
t_j]$ of the acoustic pattern and the possibility that the consonant
may be nonsonorant and interrupted is high enough.

The fuzzy algorithm for the recognition of plosive consonants
consists of two parts, one for each subset P_T and P_L appearing in
(6.6).

Again, the execution of one or two parts of this algorithm depends on the possibilities of the phonetic hypotheses "tense" and "lax" previously assigned to consonantal intervals which the hypotheses "nonsonorant" and "interrupted" have generated.

The fuzzy algorithm for generating plosive hypotheses is based on a set of fuzzy composite questions of the type:

$$QH_P = \text{"is } H_P \text{ in } p \ (t_i, t_j)\text{"} \qquad (6.7)$$

where H_P takes values in P defined by (6.6).

The universe of discourse U of Q_{H_P} is the set of all possible acoustic patterns, the body B is a structured linguistic variable having a label belonging to P_L or P_T and the answer set A of Q_{H_P} is a set of linguistic a posteriori possibilities, expressing the evidence of H_P in p, based on the evaluation of the possibilities:

$$\text{Poss } \{H_P \text{ is in } p \ (t_i, t_j)\} \qquad (6.8)$$

$$\forall \ H_P \in P = P_T \cup P_L$$

The linguistic probabilities are obtained following the approach outlined in Chapter 4.

Each structured linguistic variable representing a phonemic hypothesis is a triple $\tilde{H} = (H_P, U, R(H))$, where U is the previously defined universe of acoustic patterns, H_P is a label in $P = P_L \cup P_T$, and R(H) is a fuzzy restriction of U associated with H_P; R(H) defines the *meaning* of H_P.

Each plosive phoneme becomes a fuzzy linguistic variable defining a set PL on which H takes values:

$$PL = \{/b/, \ /d/, \ /g/, \ /p/, \ /t/, \ /k/\} \qquad (6.9)$$

Each element of PL is related by fuzzy rules to previously hypothesized phonetic or acoustic cues, which are themselves fuzzy linguistic variables, because they correspond to hypotheses generated with vagueness.

Moreover, the rules are fuzzy to represent the imprecision of knowledge. Fuzzy rules contain a syntactic portion, which establishes non-numerical relations between labels, and a semantic portion, which contains expressions for computing the evidence of a hypothesis, given the evidences of the components and the degrees of worthiness of the rules which relate the hypothesis and its components.

The plosive sounds are perceived by their manner and place of articulation.

Automatic detection of the place of articulation from the spectra is very difficult.

It is believed (Stevens and Blumstein, 1978) that cues for place of articulation which are more likely to remain invariant under the context are to be found in the spectrum at the onset of the consonant, in an aperiodic portion of the speech signal where the vibration of the vocal chords has not yet started.

Much knowledge about the spectral dynamics of the plosive sounds comes from perception experiments. Furthermore, little is known about the importance and the evidence of the spectral cues that can be extracted from plosive consonants in continuous speech. There is evidence, at least for the Italian Language, that burst spectra are often undetectable in continuous speech, especially for lax plosives in VCV contexts. In these cases formant transitions are the only cues for recognition.

For these reasons, all of the acoustic cues which can be effectively extracted from spectral data have been considered and the search for their best combinations into analytic rules has been performed by a learning algorithm for fuzzy relations. While the rules will be introduced in this Section, the details of the algorithms have been published in De Mori and Giordano (1980a).

For the recognition of each plosive consonant, two structural variables have been introduced; they can be considered as nonterminal symbols of a fuzzy grammar. The structural variables are of the type XFOR and XSP. X takes values on P as defined by (6.6), generating six pairs of structural variables. XFOR is related to acoustic cues extracted from the formants; such features are formant pseudo-loci and slopes of formant transitions.

XSP is related to acoustic cues such as burst spectra and voicing characteristics.

According to the syllabic grammar, features depend on the type of pseudosyllabic segment which has been identified by the segmentation algorithm. In this chapter VCV syllables are considered for lax plosives and CV syllables for tense plosives. Of course these syllables appear in continuous speech.

The syntactic category XFOR is, in turn, expressed in terms of the following variables, which are themselves syntactic categories belonging to the nonterminal alphabet V_N:

XFL : formant pseudo-loci;

XFS : formant slopes.

Formant pseudo-loci are defined as the last formant samples which are detected at the end of the formant transitions from the vowel preceding the plosive or as the first formant samples which are detected at the beginning of the transition before the vowel following the plosive. To describe the former case, the syntactic category XFLP is introduced, while the latter case is described by syntactic categories labelled XFLF $(X \in P = P_L \cup P_T)$.

Similar considerations hold for formant slopes, for which the symbols XFSP and XFSF are used.

The syntactic category XSP is expressed in terms of the features of the burst spectra (XS) and other features (XZ) related to voicing characteristics.

The features XZ, XFSP, XFLP appear only in the rules of the lax plosives.

The relations between phoneme hypotheses and features are of the type:

$$X \xrightarrow{\alpha_x} XFOR$$

$$X \xrightarrow{\beta_x} XSP \qquad\qquad (6.10)$$

$$X \xrightarrow{\gamma_x} XFOR \cdot XSP$$

The α_x, β_x and γ_x are real numbers belonging to the unit interval and represent the degrees of worthiness of the rules.

For the sake of simplicity, (6.10) can be rewritten in an algebraic form as follows:

$$X = \alpha_x \ XFOR + \beta_x \ XSP + \gamma_x \ XFOR \cdot XSP \quad (\forall \ X \in PL)$$

From this rule, using the usual *min* and *max* operators, the following semantic rule is obtained:

Poss {X is in p} = $\alpha_x \wedge$ Poss {XFOR is in p} $\vee$

$\beta_x \wedge$ Poss {XSP is in p} $\vee$

$\gamma_x \wedge$ Poss {XFOR is in p}$\wedge${XSP is in p}

The syntactic categories in (6.10) are further related to fuzzy linguistic variables belonging to the terminal alphabet V_T. The variables in V_T describing the formant pseudo-loci are now introduced.

Let FiY be the pseudo-locus of the i-th formant (i = 2,3). Y is a variable taking values in the set of positions {P: preceding, F: following}.

Learning has been performed on a set of about 200 examples taken from male and female speakers pronouncing sentences in a laboratory environment through a noise-cancelling microphone. The data were processed by pitch-synchronous FFT and calculation of LP coefficients, from which smoothed spectra were obtained.

Then formants were extracted, using the algorithm described in the previous chapter and pseudo-loci were obtained automatically. From the collected data, some broad "fields of existence" of the plosive sounds in different contexts were identified and represented on the diagrams of Figs. 6.5a and 6.5b.

It is important to notice that these fields have been drawn on the basis of some experimental data and on the *expectation* of the designer, who used all of his phonetic knowledge and experience in analyzing the plosive sounds and modeling their production.

Based on these fields, a set of intervals for the second and the third formant were identified and represented by labels on Fig. 6.5a and 6.5b.

The index 9 was used for the intervals of F3 and the index 10 for F2. Each one of the symbols shown in the figures has been used to label a fuzzy restriction defined over the axis of the formant frequencies.

The fuzzy restrictions defined over F3P are shown in Fig. 6.6.

The "fuzzification" of the intervals delimiting the fields of existence in Fig. 6.5a has been introduced to represent the imprecision of the subdivision made by the designer.

The possible contexts considered here for the plosives are so numerous that it would have been impractical to obtain statistics of the pseudo-loci. Rather, the designer has used some observations and his *knowledge* in defining the fields of existence of the pseudo-loci, and has expressed his imprecisions by introducing fuzzy restrictions (soft bounded intervals) instead of binary restrictions (hard bounded intervals).

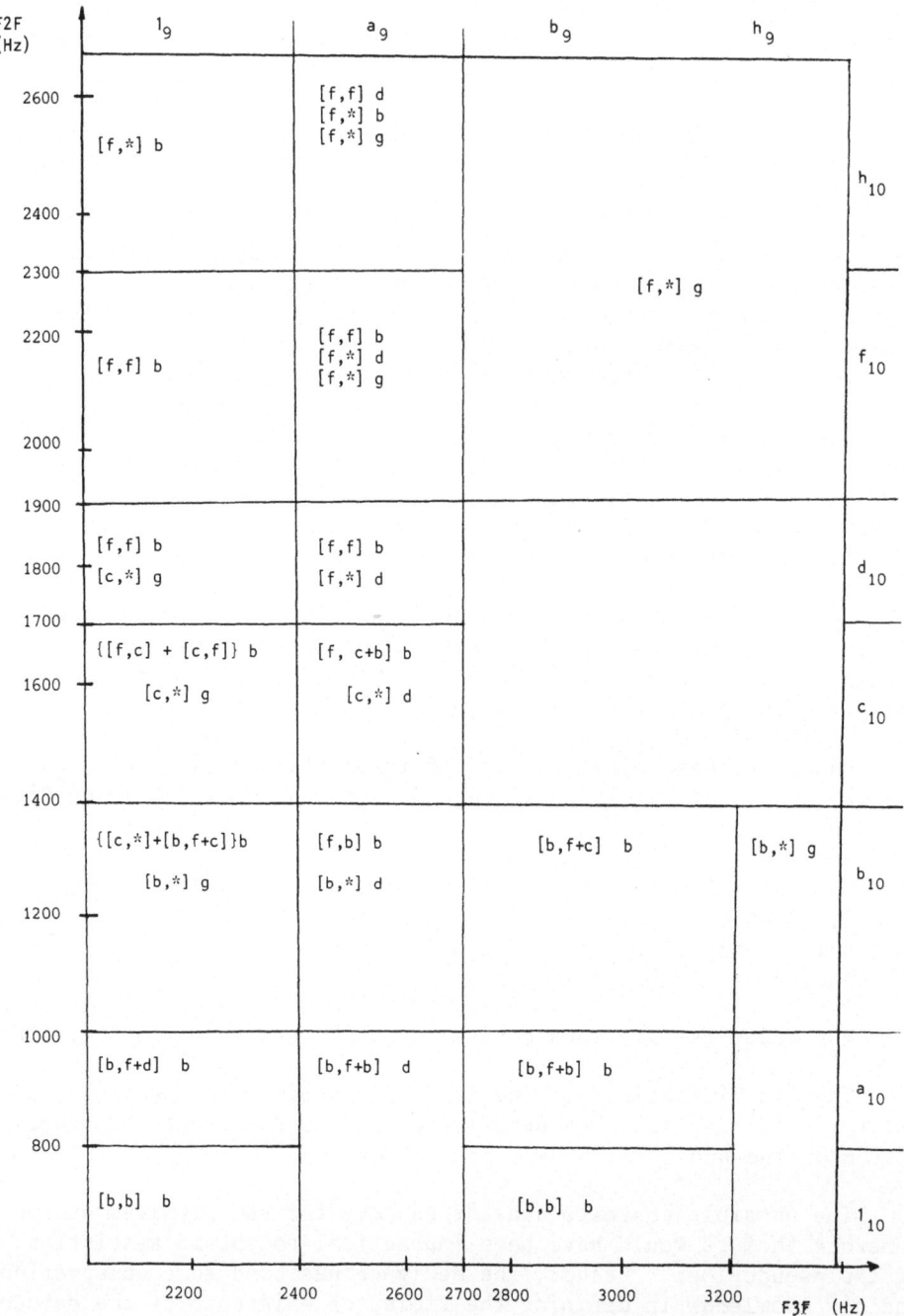

Fig. 6.5 Field of existence of the pseudo-loci of the plosive
          sounds

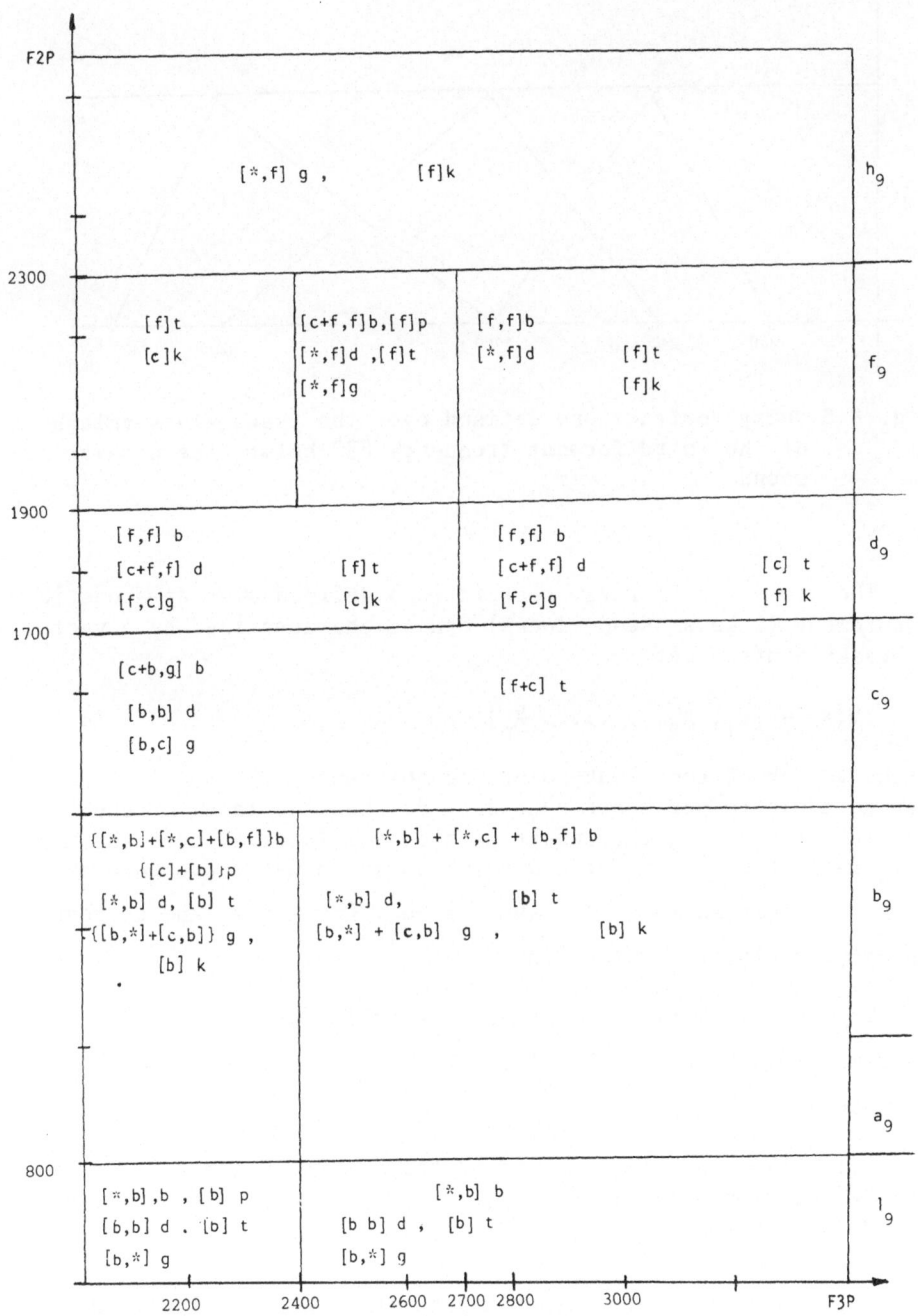

Fig. 6.5 continued

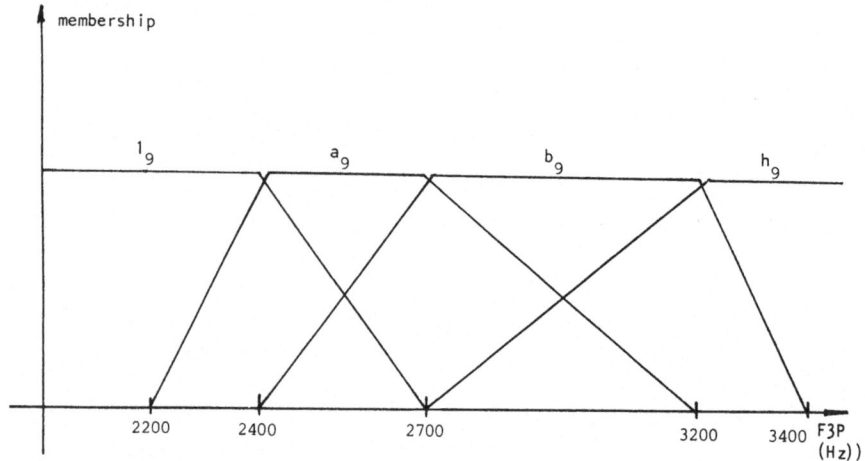

Fig. 6.6 Fuzzy restrictions defined over the pseudo-loci values
of the third formant frequency F3P before the plosive
sound

The whole set of fuzzy restrictions defined over an acoustic
measurement (like a pseudo-locus) can be characterized by a vector
of break-points. Let

$$V(X) = [X_1, X_2, \ldots\ldots, X_n] \qquad (6.11)$$

be the vector of the break-points of the fuzzy relations defined
over an acoustic parameter such as F3P. Let the label of the fuzzy
restriction covering the lowest part of the interval over the x-axis
be l_j (j is the suffix corresponding to an acoustic cue such as
F3P). The membership of l_j takes value 1 for $x \leq x_2$ and decreases
linearly, taking the value 0 on x_3.

The fuzzy restriction immediately following l_j has label a_j;
its membership assumes value 1 for $x_2 \leq x_1 \leq x_3$ and decreases linearly to 0 from x_2 to x_1 and from x_3 to x_4.

Going from left to right along the x-axis, fuzzy restrictions
have been assigned labels b_j, c_j,... in alphabetic order, excluding
l_j and h_j.

The fuzzy set whose membership function takes value 1 for $x \leq$
x_{n-1} is labelled h_j; its membership decreases linearly from x_{n-1}

to x_{n-2} and takes value 0 for $x \leq x_{n-2}$.

Fig. 6.6 shows an example where $j = 9$, $X = F3P$, $n = 5$; the vector of break-points is: V (F3P) = [2200, 2400, 2700, 3200, 3400].

The same fuzzy sets are used for F3P. For F2P and F2F the fuzzy restrictions $\{l_{10}, a_{10}, b_{10}, c_{10}, d_{10}, l_{10}, h_{10}\}$ are defined with the following vector of break-points:

V (F2P) = [700, 800, 1000, 1400, 1700, 1900, 2300, 2400].

The formant slope before the consonant is described by the difference between the frequency of the second formant when the formant amplitude reaches the absolute maximum on a vowel and the frequency of the pseudo-locus; the opposite of such a difference describes the formant slope after the consonant.

The fuzzy restrictions of the formant slopes have labels A (ascendent), H (horizontal), D (descendent) and the following vector of break-points.

V (XFSF) = [-220, -160, +160, +220]

The same vector defines the fuzzy sets D (descendent), H (horizontal), A (ascendent) for XFSP.

The feature XZ is related to parameters such as the maximum increase of the derivative of the first formant amplitude measured in conventional units, the voice onset time, and the consonant duration. The details of this relation are omitted for the sake of brevity.

XS describes the spectral characteristics of the burst for the plosive X. The burst is detected by finding a short peak in the total energy with low ratio between low and high frequency energies before the beginning of the next vowel.

The acoustic parameters of the burst spectra and the corresponding restrictions are defined in the following, and the fields of the plosives for these parameters are shown in Fig. 6.7.

1) Ratio R between low (200-900 Hz) and high (5-10 kHz) frequency energies measured in conventional units.

The labels for the fuzzy restrictions are:

$$\{l_{14}, a_{14}, b_{14}, c_{14}, d_{14}, e_{14}, f_{14}, h_{14}\}$$

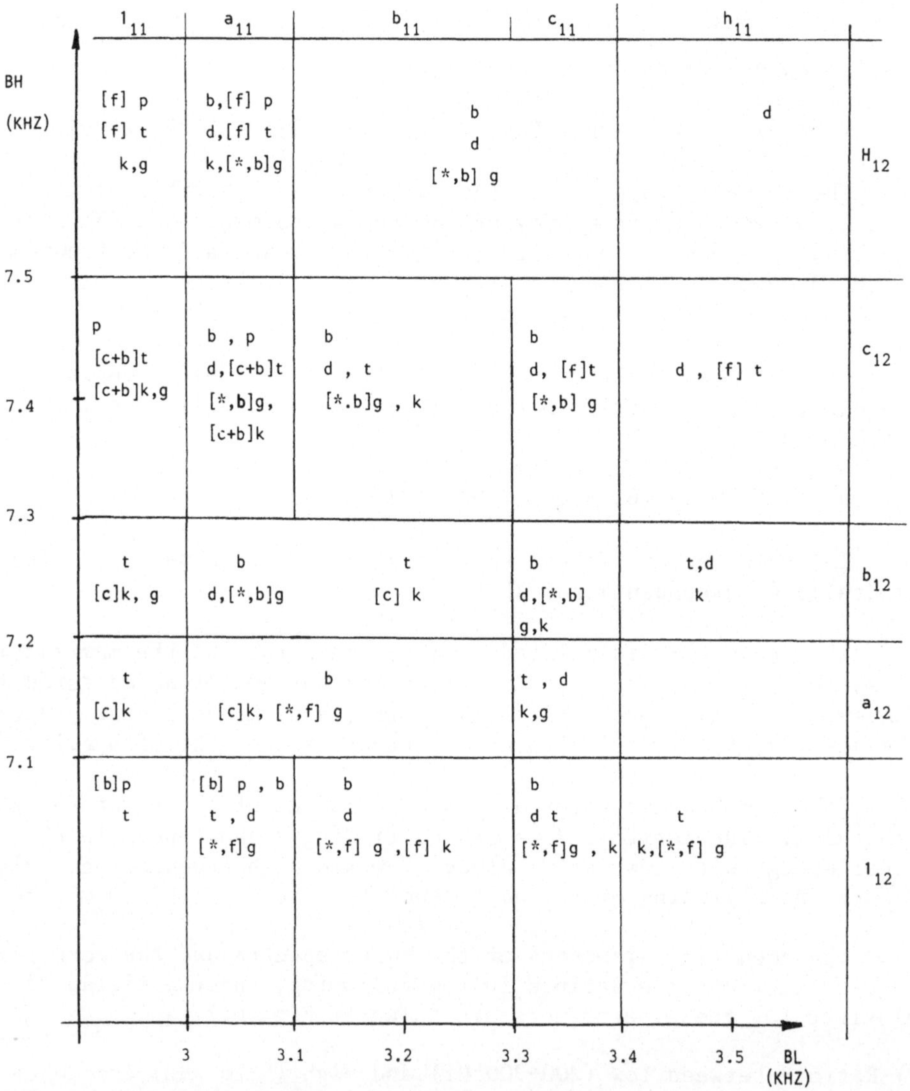

1_{11} a_{11} b_{11} c_{11} h_{11}

BH

(KHZ)

	1_{11}	a_{11}	b_{11}	c_{11}	h_{11}	
	[f] p [f] t k,g	b,[f] p d,[f] t k,[*,b]g	b d [*,b] g		d	H_{12}
7.5						
7.4	p [c+b]t [c+b]k,g	b , p d,[c+b]t [*,b]g, [c+b]k	b d , t [*.b]g , k	b d, [f]t [*,b] g	d , [f] t	C_{12}
7.3						
7.2	t [c]k, g	b d,[*,b]g	t [c] k	b d,[*,b] g,k	t,d k	b_{12}
7.1	t [c]k	b [c]k, [*,f] g		t , d k,g		a_{12}
	[b]p t	[b] p , b t , d [*,f]g	b d [*,f] g ,[f] k	b d t [*,f]g , k	t k,[*,f] g	1_{12}

3   3.1   3.2   3.3   3.4   3.5   BL
(KHZ)

Fig. 6.7 Fields of existence of the parameters of the burst spectra

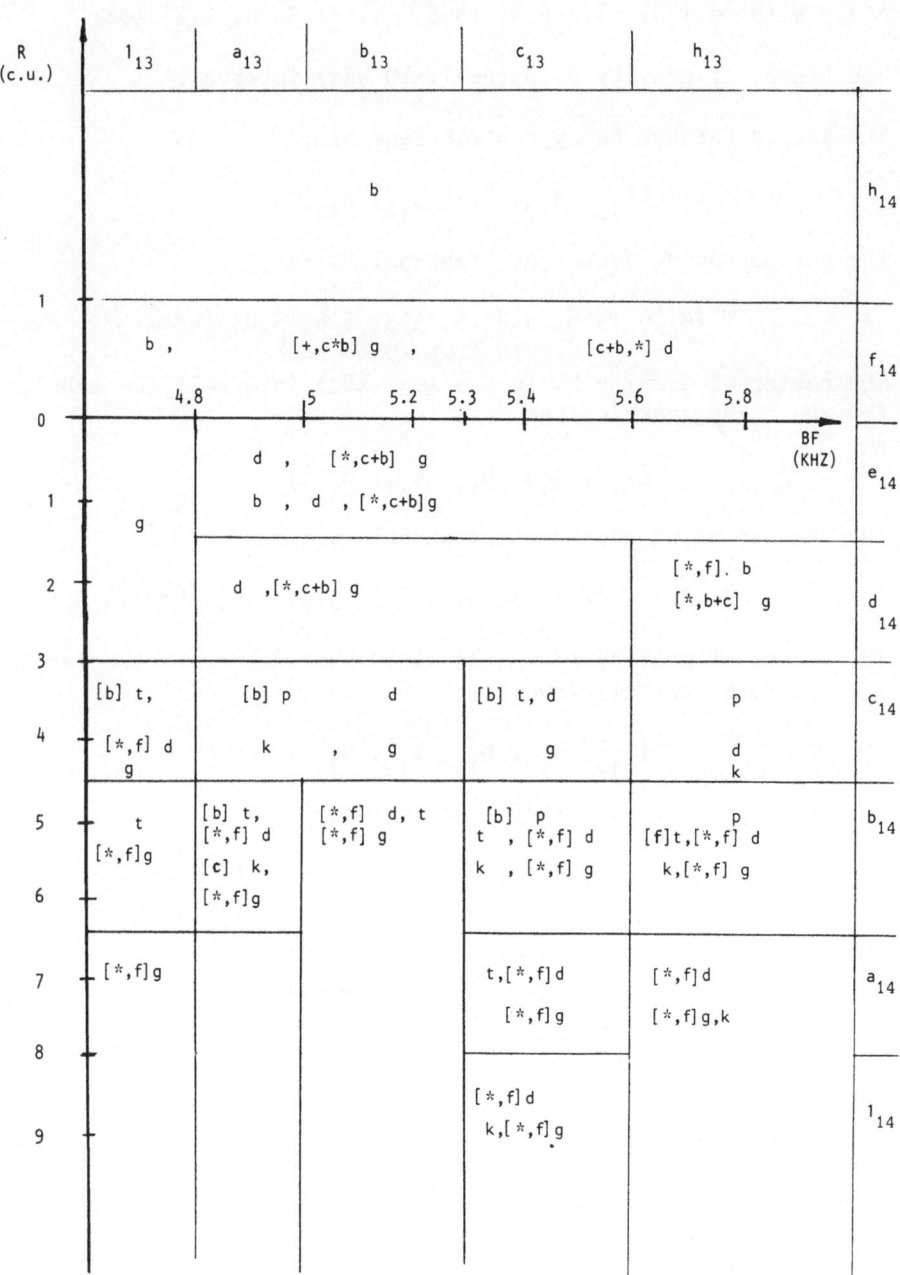

Fig. 6.7 continued

with the corresponding vector of break-points:

$$V (R) = [-9, -8, -6.4, -4.5, -3, -1.3, 0, 1.2] \text{ (dB)}.$$

2) The center of gravity B in the (1-10 kHz) interval.

The labels for the fuzzy restrictions are:

$$\{l_{13}, a_{13}, b_{13}, c_{13}, h_{13}\};$$

the corresponding vector of break-points is

$$V (B) = [4.6, 4.8, 5, 5.3, 5.6, 5.8] \text{ (kHz)}.$$

3) The center of gravity BL in the (1-5 kHz) interval; the labels for the fuzzy restrictions are:

$$\{l_{11}, a_{11}, b_{11}, c_{11}, h_{11}\};$$

the corresponding vector of break-points is:

$$V (BL) = [2.8, 3, 3.1, 3.3, 3.4, 3.6] \text{ (kHz)}.$$

4) The center of gravity BH in the (5-10 kHz) interval; the labels for the fuzzy restrictions are:

$$\{l_{12}, a_{12}, b_{12}, c_{12}, h_{12}\};$$

the corresponding vector of break-points is:

$$V (BH) = [7, 7.1, 7.2, 7.3, 7.5, 7.6] \text{ (kHz)}.$$

6.4 Rules for the Recognition of Plosive Sounds

6.4.1 Rules for formant loci, formant slopes and burst spectra

The detailed rules for the recognition of plosive sounds in vocalic contexts will be introduced in this section. The vocalic context is expressed by the place of articulation and is indicated inside brackets. For the tense plosives only the following vowel acts as a contextual constraint, while for the lax plosives in VCV utterances the places of articulation of both the preceding and the following vowels appear in the rules.

As for the cases considered in Chapter 5, b is for back, c is

for central and f is for front, and a contextual constraint of the
type [b,f] means that the vowel preceding the plosive has to have a
"back" place of articulation and the vowel following the plosive
has to have "front" place of articulation in order to make the
rules following the brackets applicable to a pattern. In a general
case, the vocalic context will be hypothesized with imprecision and
its degree of vagueness will be combined by a semantic rule with
the imprecision with which the acoustic pattern exhibits the features appearing in the rules affected by this context. This is due
to the fact that places of articulation are defined as fuzzy sets
on a two-dimensional space having the first two formants as dimensions. The vocalic hypotheses are generated and evaluated after
the detection of the zones of stability of the formants, the computation of the centers of gravity of these zones, and the evaluation of their grades of membership in the fuzzy sets of the places
of articulation. The symbol + represents the logical disjunction
and the * means "every type of vowel". The symbol ⌐ is for negat-
ion.

The rules are presented in an algebraic notation for the sake
of simplicity.

$$BFLP = [f,*] \, (l_9 + a_9) \, h_{10} + [f,f] \, (l_9 + a_9)(l_{10} + d_{10}) +$$
$$+ [f,c] \, (l_9 + a_9) \, c_{10} + [c,f] \, l_9 \, c_{10} + [f,b] \, a_9 \, c_{10} +$$
$$+ [b,f + c] \, (l_9 + b_9) \, b_{10} + [c,*] \, l_9 \, b_{10} + [f,b] \, a_9 \, b_{10} +$$
$$+ [b,f + b] \, (l_9 + b_9) \, a_{10} + [b,b] \, (l_9 + b_9) \, l_{10}$$

$$DFLP = [f,f] \, a_9 \, h_{10} + [f,*] \, a_9 \, (l_{10} + d_{10}) + [c,*] \, a_9 \, c_{10} +$$
$$+ [b,*] \, a_9 \, b_{10} + [b,f + b] \, a_9 \, (a_{10} + l_{10})$$

$$GFLP = [f,*] \, (a_9 + b_9 + h_9)(h_{10} + l_{10}) + [c,*] \, l_9 \, (d_{10} + c_{10}) +$$
$$+ [b,*] \, (l_9 + h_9) \, b_{10}$$

$$BFLF = [c,f] \, a_9 \, l_{10} + [f,f] \, (a_9 + b_9 + h_9)(l_{10} + d_{10}) +$$
$$+ [c + b,f] \, c_{10} + ([*,b] + [*,c] + [b,f])(a_{10} + b_{10}) +$$
$$+ [*,b] \, l_{10}$$

$$DFLF = [*,f] \, f_{10} \, (a_9 + b_9 + h_9) + [c + f,f] \, d_{10} + (⌐[b,b]) \, c_{10} +$$
$$+ [*,b] \, (a_{10} + b_{10}) + [b,b] \, l_{10}$$

$$\text{GFLF} = [*,f]\ (h_{10} + a_9\ f_{10}) + [f,c]\ d_{10} + [b,c]\ c_{10} +$$
$$+ ([b,*] + [c,b])(a_{10} + b_{10}) + [b,*]\ l_{10}$$

$$\text{PFL} = [f]\ a_9\ f_{10} + ([c] + [b])\ l_9\ (a_{10} + b_{10}) + [b]\ l_9 . l_{10}$$

$$\text{TFL} = [f]\ f_{10}\ (a_9 + b_9 + h_9) + [f]\ (l_9 + a_9)\ d_{10} + [c]\ (b_9 + h_9)$$
$$d_{10} + + [f + c]\ c_{10} + [b]\ (a_{10} + b_{10} + l_{10})$$

$$\text{KFL} = [f]\ (h_{10} + (d_{10} + f_{10})(b_9 + h_9)) + [c]\ (l_9 + a_9)\ d_{10} +$$
$$[b]\ (a_{10} + b_{10})$$

$$\text{BFSP} = [f + c,*]\ D + [b,*]\ (H + D)$$

$$\text{BFSF} = [f + c]\ A + [*,b]\ (A + H)$$

$$\text{DFSP} = [b,*]\ (A + H) + [c,*]\ H + [f,*]\ (D + H)$$

$$\text{DFSF} = [*,b]\ D + [*,c]\ H + [*,f]\ (A + H)$$

$$\text{GFSP} = [f,*]\ A + [b,*]\ (H + D) + [c,f + c]\ A + [c,b]\ (H + D)$$

$$\text{GFSF} = [*,f]\ (H + D) + [*,c]\ D + [*,b]\ (D + H)$$

$$\text{PFS} = [f + c]\ A + [b]\ (A + H)$$

$$\text{TFS} = [b]\ D + [c]\ H + [f]\ (A + H)$$

$$\text{GFS} = [b]\ D + [c]\ H + [f]\ (A + H)$$

The rules for the burst spectra are the following:

$$\text{BS} = h_{14} + 0.4\ f_{14} + (0.6\ e_{14}(a_{13} + b_{13} + c_{13} + h_{13}) + [*,f]d_{14}h_{13})$$
$$((a_{11} + b_{11} + c_{11})(h_{12} + c_{12} + b_{12} + l_{12}) + a_{12}(a_{11} + b_{11}))$$

$$\text{DS} = [c + b,*]\ f_{14} + ((a_{13} + b_{13} + c_{13} + h_{13})(0.6\ e_{14} + d_{14} + c_{14} +$$
$$+ [*,f]\ b_{14}) + [*,f]\ (a_{14}\ h_{13} + c_{13}l_{14}))((a_{11} + b_{11} + c_{11} +$$
$$h_{11})(h_{12} + c_{12} + b_{12}) + (a_{11} + b_{11} + c_{11})(a_{12} + l_{12}))$$

$$GS = 0.4 \ [c+b,*] \ f_{14} + (0.6 \ e_{14}(l_{13} + [*,c+b] \ (a_{13} + b_{13} + c_{13} + h_{13})) +$$

$$+ \ [*,c+b] \ d_{14}(a_{13} + b_{13} + c_{13} + h_{13}) + c_{14}(l_{13} + a_{13} + b_{13} + c_{13}) +$$

$$+ \ [*,f] \ (b_{14} + (l_{13} + c_{13} + h_{13}) \ a_{14} + c_{13}l_{14}))$$

$$((l_{11} + [*,b] \ (a_{11} + b_{11} + c_{11}))(h_{12} + c_{12} + b_{12}) +$$

$$+ \ ([*,f] \ (a_{11} + b_{11}) + c_{11}) \ a_{12} + [*,f] \ l_{12}(a_{11} + b_{11} + c_{11} + h_{11}))$$

$$PSP = (([b] \ (a_{13} + b_{13}) + h_{13}) \ c_{14} + ([b] \ c_{13} + h_{13}) \ b_{14})$$

$$((l_{11} + a_{11})(h_{12} \ f + c_{12}) + l_{12} \ [b] \ (l_{11} + a_{11}))$$

$$TSP = ([b] \ (l_{13} + c_{13}) c_{14} + (l_{13} + [b] \ a_{13} + b_{13} + c_{13} + [f] \ h_{13}) b_{14} +$$

$$+ \ c_{13}a_{13})([f] \ (l_{11} + a_{11}) \ h_{12} + ([c+b] \ (l_{11} + a_{11}) + b_{11} + [f](c_{11} + h_{11})$$

$$c_{12} + b_{12}(l_{11} + a_{11} + b_{11} + c_{11}) \ a_{12} + (l_{11} + a_{11} + c_{11} + h_{11}) l_{12})$$

$$KSP = ((a_{13} + b_{13} + h_{13}) c_{14} + ([c] \ a_{13} + c_{13} + h_{13}) b_{14} + h_{13}a_{14} + c_{13}l_{14})$$

$$((l_{11} + a_{11}) h_{12} + ([c+b] \ (l_{11} + a_{11}) + b_{11}) c_{12} + ([c] \ (l_{11} + a_{11} + b_{11}) +$$

$$+ \ c_{11} + h_{11}) b_{12} + ([c](l_{11} + a_{11} + b_{11}) + c_{11}) a_{12} +$$

$$+ \ ([f] \ b_{11} + c_{11} + b_{11}) \ l_{12})$$

A single rule is very seldom useful for hypothesis generation if used alone. Rather, rules have to be combined in order to make the evidence of the right hypothesis emerge over the competing ones.

As has been mentioned in Section 6.3, expressions for the spectral characteristics and formant features have been inferred using the algorithm which will be introduced in Chapter 9. The details of the learning procedure followed for inferring these rules are reported in a book chapter by De Mori, Laface and Saitta, 1981. A summary of their inference will be given in Chapter 9.

6.4.2 Rules for spectral characteristics of plosives

$$BSP = 0.6 \ BZ + 0.5BS + BZ \cdot BS$$

DSP = 1DZ + 0.7 DS

GSP = 0.8 GZ + 0.2 GS + GZ · GS

6.4.3 Rules for formant features

BFOR = 0.95 BFLF · BFLP + 0.7 BFSP · BFSF

DFOR = 0.95 DFLF · DFLP + 0.8 DFSP · DFSF

GFOR = 0.95 GFLF · GFLP + 0.7 GFSP · GFSF

PFOR = 0.8 PFL + 0.95 PFL · PFS

TFOR = 0.8 TFL + 0.95 TFL · TFS

KFOR = 0.8 KFL + 0.95 KFL · KFS

6.4.4 Rules for phonemic hypotheses

<b> = 0.8 BFOR + 0.93 BSP

<d> = 0.63 DFOR + 0.9 DSP + 0.95 DFOR · DSP

<g> = 0.9 GFOR + 0.9 GSP + 1 GFOR · GSP

<p> = 0.7 PFOR + 0.6 PSP + 1 PFOR · PSP

<t> = 0.7 TFOR + 0.6 TSP + 1 TFOR · TSP

<k> = 0.7 KFOR + 0.6 KSP + 1 KFOR · KSP

Example 6.4.1

Fig. 6.8 shows the time evolutions of the formants of the syllable /agu/.

From the spectra of this syllable, the following values are extracted:

F2P = 1520 Hz

F3P = 2145 Hz

F2F = 1072 Hz

F3F = 2300 Hz

As MIN R = 1.39, the burst is not significant for the detailed classification of voiced plosives, giving a similar low contribution (~0.5 for the three sounds). As the vocalic context is recognized correctly with membership 1, BFLP and DFLP have a membership of less than 0.5 while GFLP has membership 1. For the loci of the formants after the burst, BFLF, DFLF, GFLF all reach membership 1.

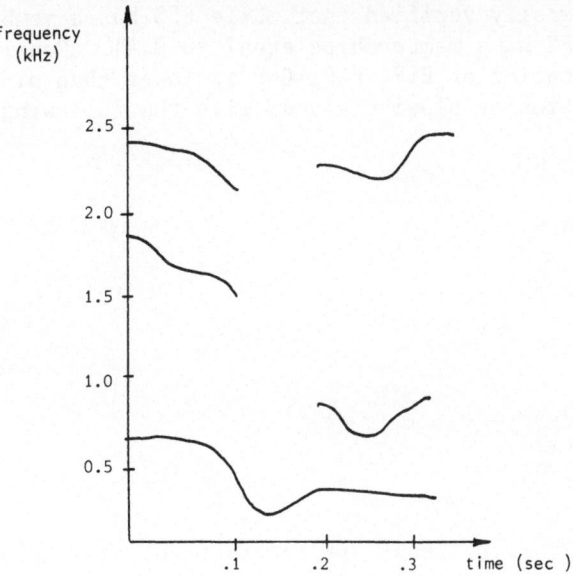

Fig. 6.8 Time evolutions of the formants of the syllable /agu/

Thus the memberships of BFOR and DFOR are less than 0.5 and the membership of GFOR is 0.95.

6.4.5 Composition of evidences

As for the nasal sounds, the rules introduced in this section apply only to segments for which the hypothesis "nonsonorant-interrupted-lax" (NIL) or "nonsonorant-interrupted-tense" (NIT) has been previously generated. Thus, the rules introduced in this section solve the subproblem of attaching a phonemic label to a segment, provided that other subproblems consisting in assigning phonetic hypotheses have already been solved.

For example, the fact that for the PSS /agu/ a degree of worthiness $\mu_{<g>}$ = 0.95 has been found means that a possibility of having /g/, given the fact that the segment was NIL, has been obtained. As we are interested in the unconditional possibility of having the "phoneme g" in the segment, we have to combine $\mu_{<g>}$ with the membership of the hypothesis NIL.

It can be easily verified that while BFS has a membership equal to 0, DFS and GFS have memberships equal to 0.41. Considering now that the contribution of BSP, DSP, GSP is lower than 0.5, the rules for the recognition of plosive sounds give the following results:

$$\mu_{BFOR} \quad < \quad 0.5$$

$$\mu_{DFOR} \quad < \quad 0.5$$

$$\mu_{GFOR} \quad = \quad 0.95$$

$$\mu_{<b>} \quad < \quad 0.5$$

$$\mu_{<d>} \quad \leq \quad 0.5$$

$$\mu_{<g>} \quad = \quad 0.95$$

As the two possibilities are interactive, their product is taken to obtain the unconditional total possibility. Let

Poss {/g/ is in p} be such a possibility, then one gets:

Poss {/g/ is in p} =Poss {NIL is in p}· $\mu_{<g>}$.

<g> can be seen as <palatal/consonant-nonsonorant-interrupted-lax>.

The rules relating this feature to acoustic cues are context dependent. Similar considerations hold for the other plosive sounds.

In the case of the example 6.4.1, a value Poss {NIL is in p} of 0.55 was found, leading to the following unconditional possibilities for the voiced plosive hypotheses:

Poss {/b/ is in p} $\leq 0.5 \times 0.55 = 0.275$

Poss {/d/ is in p} $\leq 0.5 \times 0.55 = 0.275$

Poss {/g/ is in p} $\leq 0.95 \times 0.55 = 0.52$

From these data and from a priori knowledge about the statistical distributions of global possibilities of competing hypotheses, a linguistic value of the a posteriori probability can be obtained.

A linguistic value is more suitable than a precise value because its vagueness reflects all the imprecisions introduced in computing the a posteriori probabilities. The use of possibility theory for combining evidences is motivated by the need to combine hypotheses using imprecise rules, whose symbols are labels for soft-bounded sets. A set of this type is softly bounded because the only thing that is known is that its kernel, where the membership takes value 1, is not common to other sets.

It can be verified that the rules proposed here can be derived from the following phonetic representations:

$$
p = \begin{bmatrix} \text{consonant} \\ \text{nonsonorant} \\ \text{interrupted} \\ \text{tense} \\ \text{labial} \end{bmatrix}
\qquad
t = \begin{bmatrix} \text{consonant} \\ \text{nonsonorant} \\ \text{interrupted} \\ \text{tense} \\ \text{alveolar} \end{bmatrix}
$$

$$
k = \begin{bmatrix} \text{consonant} \\ \text{nonsonorant} \\ \text{interrupted} \\ \text{tense} \\ \text{palatal} \end{bmatrix}
\qquad
b = \begin{bmatrix} \text{consonant} \\ \text{nonsonorant} \\ \text{interrupted} \\ \text{lax} \\ \text{labial} \end{bmatrix}
$$

$$
d = \begin{bmatrix} \text{consonant} \\ \text{nonsonorant} \\ \text{interrupted} \\ \text{lax} \\ \text{alveolar} \end{bmatrix}
\qquad
g = \begin{bmatrix} \text{consonant} \\ \text{nonsonorant} \\ \text{interrupted} \\ \text{lax} \\ \text{palatal} \end{bmatrix}
$$

6.5 Experimental Results

To evaluate the performance of the algorithm for generating hypotheses about plosive sounds, about 500 new samples extracted from continuous speech were analyzed.

Four of the speakers were male and one was female.

In 11% of the cases the right hypothesis did not reach the maximum degree of worthiness. To be able to use this component in a Speech Understanding System another type of evaluation was used.

Only one hypothesis was considered if there was only one phoneme hypothesis having a degree of worthiness at least 0.1 higher than the memberships of all the competing hypotheses, otherwise all the hypotheses whose memberships were within 0.1 of the membership of the most evident hypothesis were considered.

Under these assumptions, the following results were obtained:

absence of the right hypothesis: less than 1%;
generation of a single hypothesis: 80% of the cases;
generation of two hypotheses: 15% of the cases;
generation of three hypotheses: 5% of the cases.

An average number of 7 hypotheses was found to have a probability less than 1% of missing the right hypothesis. The right hypothesis was assigned the highest evidence in 85% of the cases. Some
improvement was achieved by refining the rules using the learning
algorithm described in Chapter 9.

Continuant and affricate sounds can be recognized by analyzing
the frication noise spectra under the control of context-independent
rules.

Centers of gravity and spectral peaks are the characteristic
acoustic cues of these sounds; a discussion of the algorithms for
their recognition has little relevance, and is omitted here for the
sake of brevity.

ROC curves for the nonsonorant sounds are similar to those obtained for the sonorant consonants and confirm the validity of the
rules.

For the plosive sounds appearing in consonant clusters and in
prevocalic or postvocalic contexts the rules are similar to those
derived for VCV cases but, of course, they contain only one vocalic
context. For the plosives which are between consonants, the burst
spectrum is the only valuable acoustic information from which acoustic cues may be extracted. Descriptions of the spectral shape of
the burst are the most suitable cues. Descriptions of this type
have been used by Blumstein and Stevens (1979) for manual recognition of plosives in CV syllables. Their results are fairly good
(about 70% recognition accuracy). Nevertheless context-dependent
cues remarkably improve the recognition rate, especially for lax
plosives in a prevocalic context. For these plosives the burst is
very often undetectable, at least in the Italian Language.

6.6 References

Recent interesting work on the analysis and recognition of nonsonorant sounds have been performed by Blumstein and Stevens (1978
and 1979), Searle et al.(1979), Siegel (1979), Rabiner and Sambur
(1977), Datta et al. (1980), Oka (1978), Fujisaki and Kunisaki (1978),
Fujisaki et al. (1980).

A paper by Cooper et al. (1955) describes the first extensive
synthesis experiments that have shown the importance of formant transitions in the perception of plosive sounds.

7. THE LEXICAL KNOWLEDGE SOURCE

7.1 Word Recognition
in Continuous Speech

The Lexical Knowledge Source and its controller, indicated in Fig. 1.6 of Chapter 1, will be described in this chapter. Their behavior can be compared with that of an operator which generates word hypotheses about a speech waveform, non-linearly coded into a data structure of phonetic features and phonemes.

These hypotheses are the major point of interaction between the concepts and the way they are coded into an acoustic message. The words of the spoken language are almost the same as those of the written language. Hence it may be argued that speech understanding has the same difficulty as understanding a written text, the only difference being that in the latter case words are represented by an orthographic alphabet, while, in the former, they are represented by a speech waveform carrying information which may be represented by a string of symbols belonging to a phonemic alphabet.

Unfortunately, the difference is not so small. In fact, while a word has always the same realization in a correctly written text, a correctly spoken word may be coded into a large variety of physical realisations (i.e., waveforms). Furthermore, the speech waveform of a sentence is an acoustic continuum where (Vaissière, 1980) the boundaries of successive words may not have clear acoustic correlates. Even when such cues exist, their detection may be inprecise or ambiguous.

Examination of sound spectrograms of fluent speech reveals that cues to word boundaries are often absent (Klatt and Stevens, 1973).

271

Furthermore, the absence of word boundary cues is responsible for many misperceptions of fluent speech (Bond and Garnes, 1980).

Mariani (1980) gives an example of such difficulties. The French sentence "j'ai mal aux pieds" (my feet hurt me), when unambiguously transcribed into a continuous string of correct phonemes may be segmented into more than 1500 possible sequences of correct French words. However, only very few of these alternative interpretations are syntactically and semantically correct.

The following are problem areas characterizing the difficulties of word recognition in continuous speech.

The first problem is the acoustic-phonetic non-invariance. As has been pointed out in the previous chapters, acoustic realizations of phonetic features are context-dependent. Even in the same context, the same phoneme, which is a discrete event, may have a large variety of acoustic realizations.

A second problem is that of speed. The same word may be pronounced at various speeds. Its discrete components may have different durations. Speed may affect the detection of some features, thus influencing the segmentation process and the hypotheses generated by the syllabic processor.

A third problem is related to the differences between speakers. Speakers differ in length and general shape of their vocal tract. Speakers may also use different phonetic features when pronouncing the same word. This may depend on the fact that the same language may be spoken in different ways in different geographic areas.

A fourth problem is the so-called phonological recoding (Klatt, 1979). It has been pointed out (Oshika et al., 1975) that the expected phonetic realization of a word depends on the sentence context in which it appears. For example, in the sentence "would you", the word-final phoneme /d/ and the word-initial /y/ are transformed into a phonetic segment which is represented by Oshika et al. as /j/. All these types of transformations, which may also occur inside words, are described by phonological rules. During speech production, such rules are assumed to operate on an underlying abstract phonemic representation of a sentence. This operation is called phonological recoding. Phonological rules must be taken into account when phonetic hypotheses are considered for producing lexical interpretations.

Cohen and Mercer (1975) have designed a phonological rule component for a speech understanding system. It contains a speaker performance model organized by taking a base-form selector, followed by a phonetic rule applier. A base-form for a word is a basic pronunciation of that word and is represented by a sequence of phonemes.

A phonetic rule applier is a process which transforms each basic form into a set containing every phonetic variant of the utterances in the basic form.

This is done by applying phonetic rules to the string of phonemes in the basic form.

A fifth problem area deals with recovering errors of the syllabic component.

These errors may be unambiguous insertions of a wrong segment or the loss of a segment, or the absence of the right hypothesis in a segment, and may be increased by the environmental noise.

The sixth problem concerns the use of suprasegmental cues such as fundamental frequency contour, pattern of segmental durations, and intensity contour. All these features are known as prosodic cues. Vaissière (1980) has shown that a word-final syllable length-ening and shapes of pitch contour may be cues for detecting word boundaries in sentences of the French Language.

A seventh problem is lexical representation.

A word is not only a complex relation between orthographic symbols and acoustic cues; it also has to contain information of syntactic and semantic nature. This makes a lexical item a complex data structure. Furthermore, lexical items have to be incorporated into a general lexical structure which should allow a fast access with high search speed. A reduction in storage can be obtained by avoiding duplication of information shared by many words.

The eighth problem concerns the strategy of lexical access.

This problem is related to representation. There may be two types of lexical access known as model-driven or top-down and data-driven or bottom-up.

Model-driven methods follow an analysis-by-synthesis model of speech perception (Halle and Stevens, 1962), consisting in using information drawn from phonetic features to search through the lexicon for possible sentence-initial words.

Hypothesized words are then rated by a word-verifying component and well-rated words are used to give syntactic and semantic con-straints for hypothesizing the following adjacent word. It is likely that this strategy is used in listening conditions where noise or distortions force the listener to rely more heavily on expectations and higher-level knowledge to hypothesize words.

Klatt (1979) believes that a bottom-up method of lexical access

is an essential part of the normal speech decoding process. He
reports that listeners can transcribe nonsense names embedded in
sentences and obeying the phonological constraints of English with
better that 90% phonemic accurancy. Cole et al. (1980) report that
an expert spectrogram reader, without knowing anything about the
words that are present in a sentence, can produce a broad phonetic
transcription that agrees with a panel of phoneticians from 80% to
90% of the time, depending on the scoring method used.

Erman and Smith (1981) believe that top-down methods are too
slow if many words have to be matched in each place in the utterance;
time is wasted matching words which may be likely syntactically but
have little acoustic support.

Examples of top-down systems are the HEARSAY-I (Erman, 1974a),
the HARPY (Lowerre, 1976), the IBM system (Bahl et al., 1978). Also
the HWIM system (Woods et al., 1976) is oriented towards a top-down
approach while the HEARSAY-II word hypothesizer (Smith, 1976) and
the LAFS system (Klatt, 1979) mostly rely on a data-driven lexical
component. Efficiency is improved in the LAFS system by precompiling phonological knowledge into a network of expected phonetic
sequences. Each phonetic event is represented by prototype spectra
to be compared with the spectra of the unknown sentence. In the
NOAH system (Erman and Smith, 1981), a recent version of the
HEARSAY-II word hypothesizer, knowledge representation is based on
knowledge acquired in a learning phase with the efficiency advantages as in LAFS but with more flexibility.

(NOAH is the name of N. Webster, the author of a famous dictionary.)

Klatt (1979) has pointed out the advantages of combining
suitable recognition strategies into a perceptual model. Modeling
efforts can serve to unify seemingly disparate facts into a cohesive
theory, to detect gaps in the knowledge available to support any
model, and to define testable alternatives to mechanisms described
in models.

In this chapter the extension of dynamic programming algorithms
to the recognition of connected words will first be described.
Then, recent approaches for generating and scoring word hypotheses
when the speech is asynchronously segmented and coded into states
will be reviewed. Finally, an original view will be introduced
regarding the generation of lexical hypotheses.

In this new apprach, the generation as well as the verification of lexical hypothesis is seen as a complex problem to be
solved by the development of a complex plan. Lexical representation
is obtained by a problem-reduction representation, where subproblems

involve the evaluation of syllabic hypotheses and the detection of
acoustic cues.   Each subproblem, in turn, is represented
by a graph of subproblems. A lexical problem, is solved when some
of its subproblems are solved. A degree of solution for each
elementary problem is defined and rules are provided for combining
degrees of solution in order to obtain the evidence of the lexical
hypothesis which is generated when the corresponding lexical problem
has been solved.

The following are the main perceptual findings which have been
found useful for the design of the model described in this chapter.
They come from a paper by Cole et al. (1980).

First, in many cases, acoustic cues to word boundaries are
absent and the listener must rely on the use of higher order knowledge to segment speech into words.

Second, recognition is very fast when word choice is highly
constrained by the context of the preceding words.

Third, speech perception normally proceeds word by word.

Fourth, word candidates are normally accessed from sounds
which begin a word.

Fifth, acoustic features are more prominent in stressed syllables (Umeda, 1977) and target phonemes are detected faster in
stressed syllables (Cutler and Foss, 1977).

7.2 Dynamic Programming for Matching Word
Patterns of Quasi-Continuous Feature Vectors

In this section, the concept of dynamic programming pattern
matching, introduced in Chapter 2, will be considered in more
detail in view of its application to connected words.

According to Sakoe (1979), the first attempt to extend the
dynamic programming method to patterns of connected words will be
made by considering "quasi-continuous" patterns. These patterns
are sequences of vectors of features extracted from the speech
signal with a relatively short (of the order of 10ms) and constant
sampling period.

An example of a feature vector is the set of samples at the
output of a filter bank corresponding to a 10ms interval.

Recognition of limited sequences of connected words belonging
to a limited size vocabulary (up to a few hundreds of words), can

be done by an algorithm that will be referred to as "two-level-dp-matching". This algorithm does not require a preliminary segmentation of the unknown message into words, but performs the matching between an unknown pattern and a sequence of prototypes in two steps, corresponding to the word and the phrase levels. The algorithm tries to match an unknown sequence of feature vectors with a reference pattern obtained by concatenating word reference patterns.

Let:

$$P = p(1), p(2), \ldots, p(i), \ldots, p(I) \qquad\qquad (7.1)$$

be the pattern of the unknown phrase and

$$A = a(1), a(2), \ldots, a(j), \ldots, a(J) \qquad\qquad (7.2)$$

be a prototype pattern.

Let:

$$W(n) = w(n, 1)\ w(n, 2)\ldots w(n, k)\ldots w(n, K(n)) \qquad\qquad (7.3)$$

be a word of a lexicon; $w(n, k)$ is the k-th feature vector of $W(n)$.

A prototype pattern A for a phrase W can be generated starting from a sequence A' of words as follows:

$$A' = W(1)W(2)\ldots W(n)\ldots W(N) \qquad\qquad (7.4)$$

Each word, in turn, is represented as a sequence of feature vectors. The whole sequence A' of feature vectors may be transformed into the prototype pattern A by applying some word-junction phonological rules.

Let PTH be the set of admissible phrases.

Pattern matching is performed by computing, for every possible prototype A in PTH, a distance D (A, P), and selecting the phrase A* such that:

$$D(A*, P) = \min_{A \in PTH} (D(A,P)) \qquad\qquad (7.5)$$

Calculation of all the distances necessary for obtaining A* requires a large amount of computation. For example if the set PTH of phrases is made of every possible concatenation of three digits, and each digit has a single prototype, the cardinality of PTH is 1000.

   Furthermore, for each prototype A, a calculation of distances, one for every allowed warping function, should be performed, as for the case of isolated words discussed in Chapter 2. In order to reduce the computational complexity, the matching process is divided into two levels, namely: the word level and the phrase level.

   Let $W(n)$ be a word of a lexicon L, $n \in (1, N(L))$, $N(L)$ is the size of the lexicon. A prototype A' is of length k if it consists of k words. This is indicated as $|A'| = k$.

   Let $L(k) \in PTH$ be the subset of phrases of length k. A phrase $F(n)$ in $L(k)$ is identified by a sequence of word indices

$$IN(k) = [n(1), \ldots, n(x), \ldots n(k)]$$

The best matching phrase A* is the one for which:

$$D(A*, P) = \min_{IN(k) \in L(k)} [D(P, W(n(1)) \, W(n(2)) \ldots W(n(x)) \ldots W(n(k)))] \quad (7.6)$$

where $IN(k)$ is every possible vector of k indices.

Now assume that P consists of z segments, that is:

$$P = P(h(0), h(1)) \, P(h(1), h(2)) \ldots P(h(z-1), h(z)) \quad (7.7)$$

with

$$h(0) = 0 \text{ and } h(z) = I, \text{ the duration of P.}$$

   Furthermore, assume that the total distance between two patterns A and P is defined as follows:

$$D(A, P) = \min_{m \in M} \sum_{i=1}^{I(m)} [D(i(m), j(i, m))] \quad (7.8)$$

where $D(i, j)$ is the vector distance between $p(i)$ and $a(j)$, and the minimum is taken on the summations computed along every path m corresponding to a warping function joining the points in a plane having coordinates representing pairs of vectors $(p(i), a(j))$ and satisfying certain constraints. These constraints may be represented by a window defined by the relation:

$$|i-j| < r$$

where r is a constant value (see Sakoe and Chiba, 1978, for more details).

If the pattern A consists of two segments, namely A(1) and A(2), the following relation holds:

$$D(P, A) = \min_{h} [D(p(0, h), A(1)) + D(P(h, l), A(2))] \quad 1 \le h \le l \quad (7.9)$$

Generalizing the above relation one gets:

$$D(A, P) = \min_{NX(k)} \sum_{x=1}^{k} \min (D(P(h(x-1), h(x)), A(n(x)))) \quad (7.10)$$

Each distance of the summation is computed according to (7.8).

Let us now devote a few words to explain (7.11). Equation (7.11) assumes that P and A have been broken into k segments. For each value of k, many sequences of break-points for P may be chosen. Let:

$$NX(k) = (h(0), h(1), \ldots, h(x), \ldots, h(k)) \quad (7.11)$$

be a generic expression for such sequences

Given a sequence NX(k) of break-points of P, to each segment P(h(x-1), h(x)), a segment A(n(x)) is associated for matching.

Given A, P and NX(k), the best matching is the one for which A is split into k segments A(n(x)), (x=1,2,...,k), and a generic segment A(n(x)) is such that its distance to P(h(x-1), h(x)) is minimum for every x. Let D(A, P, NX(k)) be the minimum distance in such a case.

This type of distance calculation has to be repeated for all k and IN(k) in order to identify the prototype A* that best matches P.

In order to avoid considering all possible word sequences with continuously varying break-points h(x), matching can proceed sequentially.

A first attempt can be made to match the beginning of the input pattern with the words in the lexicon. Only those prototype words with low distance from the input pattern are retained. Attempting to match the second word, the calculation of the distance between a vector frame of the input pattern and a vector frame of a prototype can be constrained to be within a fixed range around the best path so far, i.e., the local minimum (Rabiner et al., 1978). Then, the next word hypotheses can be assumed to begin in a region around

the end of the first word (Rabiner and Schmidt, 1980). The distances
of the second word matches can be accumulated with the first ones
and partial strings of candidates can be obtained. Again, only the
partial two-word strings with low accumulated distances are retained
and matching of the third word will be attempted in a region around
the ends of the partial strings. A partial string will be considered
to be an interpretation of the input when the end of its last word
matches a vector frame near the end of the input pattern.

     This type of search is a sort of "beam search" through the
space of all possible word sequences. Search methods will be
discussed in Chapter 8.

     These techniques allow high accuracies (>90%) to be achieved
for strings of digits with k<5 with an acceptable computational
complexity.

     Notice that prosodic cues may help in predicting the intervals
where words may be located. Moreover they may reinforce word
boundary hypotheses generated by the DP matching algorithm. Vaissière (1980) discusses these points in view of an application to
automatic speech recognition.

     The dynamic programming algorithm, although very good from the
theoretical point of view, is, in practice, very time-expensive.
In order to make it applicable, constraints are imposed by heuristic
considerations on the pairs of the vectors $(p(i), a(j))$ between
which the distance is computed. Furthermore, additional heuristic
constraints have to be imposed in order to make the algorithm
applicable to phrases.

     In practice, using "quasi-continuous" feature vectors, the
two-level DP-matching algorithm is applicable to vocabularies of
the size of a hundred words and phrases obtained by the concatenation of less than ten words. Furthermore, prototype patterns are
speaker-dependent and a large collection of them has to be provided
in order to obtain acceptable performances with many users.

     Attempts to use speaker-independent prototypes for the approach
described in this section have resulted so far in drastic reduction
of the vocabulary size to about 20 words.

     A special processing should also be performed to obtain phrase
prototypes from word prototypes in order to account for word
boundary effects.

7.3 Matching Speech States

7.3.1 Minimum-distance models

A great number of local distances have to be computed in order to obtain a degree of similarity between an unknown pattern and a prototype when they are represented by uniformly sampled feature vectors. Such a great number can be reduced if a preprocessing is performed which transforms sequences of feature vectors into speech states. Speech states may have different durations and may be represented by symbols or by a set of features extracted after segmentation. Segmentation of quasi-continuous sequences of feature vectors into speech states plays an important role in the recognition process.

In these approaches, segmentation is data-driven. It may be based on labels assigned to every centisecond feature vector. These labels may be further processed by a consolidator which transforms a sequence of them into a single symbol with a duration proportional to the number of labels transformed (Silverman and Dixon, 1980). Another method consists in matching feature vectors of fixed duration with speaker-dependent spectral prototypes. Only when a matching is successful, i.e., a low distance to a prototype is found, is a label of a phonetic feature generated.

Different labels can be provided to represent acoustic realizations of the same phonetic feature.

This method, proposed by Klatt (1979), allows one to conceive a network of spectral labels for each word of the lexicon. The distance between each word and a pattern of feature vectors is the sum of the distances between each feature vector (which is a spectral prototype) and a sequence of selected spectra in the input pattern.

Both the systems require a careful learning of spectral templates for each speaker. Other approaches propose to detect transients or stationary zones in the input pattern and to use these features for segmentation. Kawaguchi (1971), De Mori (1973), Liénard et al. (1974), Goldberg (1975), Kohonen et al. (1979) have proposed different approaches to this type of segmentation.

Let:

$$L=[l(j)], \; j \in (1, \; J),$$

be the set of labels with which the speech states may be coded; J is the cardinality of L.

Let:

$$X=[(x(1), t(1)), \ldots, (x(i), t(i)), \ldots (x(I), t(I))]$$

be the input pattern, represented by a sequence of pairs $(x(i), t(i))$, where $x(i)$ is a symbol of L and $t(i)$ is the time at the end of the interval in which $x(i)$ has been recognized.

The reference patterns can be represented by sequences of pairs:

$$(j(ks), d(ks))$$

where $j(ks) \in (1,J)$ is the index of the s-th label $l(j(ks))$ of the k-th template and $d(ks)$ is its maximum allowable duration. Thus the k-th prototype pattern $A(k)$ will be represented by the following sequence:

$$A(k) = ((j(k1), d(k1)), \ldots, (j(hs), d(hs)), \ldots, (j(ks(k), d(ks(k))))$$

where $s(k)$ is the number of symbols of $A(k)$.

The distance $D(X, A(k))$ is computed as follows:

$$D(X,A(k)) = \min_{S(k)} [\sum_{t=1}^{T} D[x(t), (l(k,t)]] \qquad (7.12)$$

where:

$$x(t) = x(1) \text{ for } t < t1,$$

$$x(t) = x(2) \text{ for } t1 \leq t < t2,$$

etc.

Assuming a synchronous sampling with sampling period of n ms, $0<t<T$, nT is the duration of X. Furthermore: $l(k,t)$ is the t-th sample of the realization of $A(k)$ made by starting with the following sequence of indices:

$$S(k) = k1, k2, \ldots, ks(k)$$

from which samples of $A(k)$ are obtained by the following rule:

$$l(k, t) = l(j(j1)) \text{ for } t < T1 < d(k1)$$

$$l(k, t) = l(j(k2)) \text{ for } T1 \leq t \leq T2: \quad T1 < T2 \leq T1+d(k2)$$

etc.

T1 and T2 are time instants satisfying the just introduced constraints.

$D[x(t), 1(k,t)]$ is the distance between the t-th sample of the input pattern and the t-th sample of the prototype.

The interpretation of X is the pattern class corresponding to $A*(k)$, computed as follows:

$$D(X, A*(k)) = \min_{k} D(X, A(k))$$

Vintsjuk (1976), Silverman and Dixon (1980) have proposed solutions to this problem.

A word or a sequence of words may be represented by a large set of strings of labels for speech states. In this case, a word item W can be represented by a grammar $G(W)$. Let this language be $L(G(W))$. Evaluation of word hypotheses is now the calculation of a distance between a phrase X, describing an input pattern, and a language $L(G(W))$, that is:

$$D(X, W) = D(X, L(G(W))) \qquad (7.13)$$

This distance is the minimum distance between X and the words generated by $G(W)$.

Let:

$$X=[x(1), x(2), \ldots, x(i), \ldots, x(I)]$$

be the sequence of speech state labels in X. The calculation of the distance

$$D(X, L(G(W))) = \min_{Z \in L(G(W))} d(X,Z) \qquad (7.14)$$

is a classical item in the theory of error correcting parsers.

It would be too long to recall this theory here. The reader may find a clear presentation of it in Fu and Lu (1977).

One can account also for the duration constraints by considering $G(W)$ to be an attributed grammar, i.e., a grammar capable of generating constraints on numerical attributes together with the labels of a string. This generation is performed by semantic rules associated with the syntactic rules of a grammar. K.C. You and K.S. Fu (1979) have described how to use attributed grammars in pattern recognition.

When phrases of connected words have to be recognized, constraining the successive calculations of distances is mandatory, otherwise the amount of computation would be tremendous.

An attempt to reduce the amount of computation is proposed by Nakatsu (1980) and consists in performing a forward DP-matching, followed by a backward DP-matching. Only those candidates for which the two matches agree at the middle of the phrase are retained and combined in the calculation of the total distance between each prototype and the input pattern.

A complex algorithm which attempts to partition a sequence of symbols into words was proposed by Kawaguchi (1971).

The algorithm is based on the definition of a familiarity measure of a word sequence and a degree of fitness with the data.

7.3.2. Stochastic models

A stochastic model for words is proposed by Jelinek (1976).

This model is used by a system with an acoustic front-end which generates a string of phone labels. A stochastic finite-state automaton describes the characteristics of an acoustic processor as a generator of symbols (acoustic model). For each input phone, it gives the probability that an output symbol $x(i)$ is a deformation of a pronounced phoneme, or that it has been inserted by an error of the acoustic processor, or the probability that a pronounced phoneme has been erroneously deleted by the acoustic processor. For each word W of the lexicon, another stochastic automaton describes its possible forms corresponding to alterations of the form derived from the orthographic prototype. Alterations may be due to pronunciation speed, speaker characteristics, etc. The latter stochastic automaton describes the possible phonetic realizations of a word and is called phonological model. These phonetic realizations, called surface forms, have associated a probability. Furthermore, for each surface form, a sequence of stochastic automata, one for each phoneme, describes the variety of strings with which the form can be coded by the acoustic processor.

A higher level decoder has to find the best interpretation of A, given a sequence X of symbols produced by the acoustic processor. The best interpretation is the one for which the following a posteriori probability is maximum:

$$Pr(A/X) = \frac{Pr(X/A)\,Pr(A)}{Pr(X)} \qquad (7.15)$$

If $Pr(X)$ is the same for every candidate A, the comparison between candidates may be based on the measurement of the following quantity:

$$m(A, X) = Pr(X/A) Pr(A) \qquad (7.16)$$

The quantity $Pr(A)$ can be computed from a language model which gives the probabilities of word sequences in a language. $Pr(X/A)$ can be obtained from the acoustic and the phonological model.

Nakagawa (1976) suggests accounting for the vagueness in phoneme recognition as follows.

Let A be a word and X be a sequence of acoustic or phonetic cues, then:

$$Pr(A/X) = Pr(A/PH) Pr(PH/X) \qquad (7.17)$$

PH is a sequence of phone or phoneme symbols:

$$PH = f(1), f(2), \ldots, f(k), \ldots, f(K).$$

In the approach proposed in the previous chapters, X may be seen as a sequence of vectors. Each vector is assigned a time value corresponding to the end of the signal segment it describes. Each vector contains the degree of evidence and the label of each phoneme hypothesized by the syllabic knowledge.

If $X = x(1), (x2), \ldots, x(k), \ldots, x(K)$

and

$$PH = f(1), f(2), \ldots, f(k), \ldots, f(K)$$

then:

$$Pr(PH/X) = \prod_{k=1}^{K} \frac{Pr(x(k)/f(k)) \, Pr(f(k))}{Pr(x(k))} \qquad (7.18)$$

because phoneme evidences are context-independent.

Furthermore:

$$Pr(A/PH) = \frac{Pr(PH/A) Pr(A)}{Pr(PH)} \qquad (7.19)$$

The absolute probabilities of the phonemes $Pr(f(k))$ and the words $Pr(A))$ can be computed and represented in the language model. The probability $Pr(PH/A)$ can be obtained from a phonological model which accounts for mispellings and errors of the acoustic decoder. $Pr(x(k)/f(k))$ can be described in an acoustic model representing the performances of the syllabic knowledge source in generating phoneme hypotheses. $Pr(x(k))$ can be obtained in the following way:

$$Pr(x(k)) = Pr(x(k)/f(1))Pr(f(1)) + Pr(x(k)/f2))Pr(f(2)) + \quad (7.20)$$
$$+ ... + Pr(x(k)/f(i(k)))Pr(f(i(k))) + ...$$
$$+ Pr(x(k)/f(I(k))Pr(f(I(k)))$$

where $f(i(k))$ is the $i(k)$-th phoneme which may generate $x(k)$ and $i(k) \in (1, I(k))$.

If the syllabic knowledge source is very selective (i.e., it generates, on the average, few hypotheses), the calculation of $Pr(x(k))$ is not complex.

7.4 Word Detection by the Hypothesize-and-Test Paradigm

When the lexicon contains thousands of items, special care has to be taken in order to avoid unnecessary computation and to save memory space. For this purpose, solutions for the generation of lexical hypotheses based on the hypothesize-and-test paradigm have been proposed. They consist in generating word candidates by a fast algorithm and then evaluating the evidence of the most promising of them by a verification process.

Most of the systems using this paradigm (Woods et al., 1976, Erman and Smith, 1981, Klatt, 1979) have a lexicon represented by a tree of phonemes.

Fig. 7.1. shows a simple example of such a tree. The root of the tree is R ; R is followed by nodes labelled by phonemes $f(11)$, $f(12)$, etc., with which surface forms of words may begin. Phoneme $f(22)$, for example, may be followed by phonemes $f(32)$ and $f(33)$, and so on. At the end of a concatenation of phonemes corresponding to a word, there is a dashed line representing a pointer to a lexical item. For example, the sequence of phonemes $f(11)$ $f(21)$ $f(31)$ corresponds to the surface form of the word W1.

There may be many paths pointing to the same lexical item. All the nodes labelled with phonemes which may be at the end of a word are linked with the root R because it is assumed that every word may be followed by every other word (which is a worst-case situation). These links allow one to apply phonological rules to every possible sequence of phonemes which may appear in continuously spoken phrases, including sequences across the word boundaries.

In the HWIM system (Woods et al., 1976), for example, the

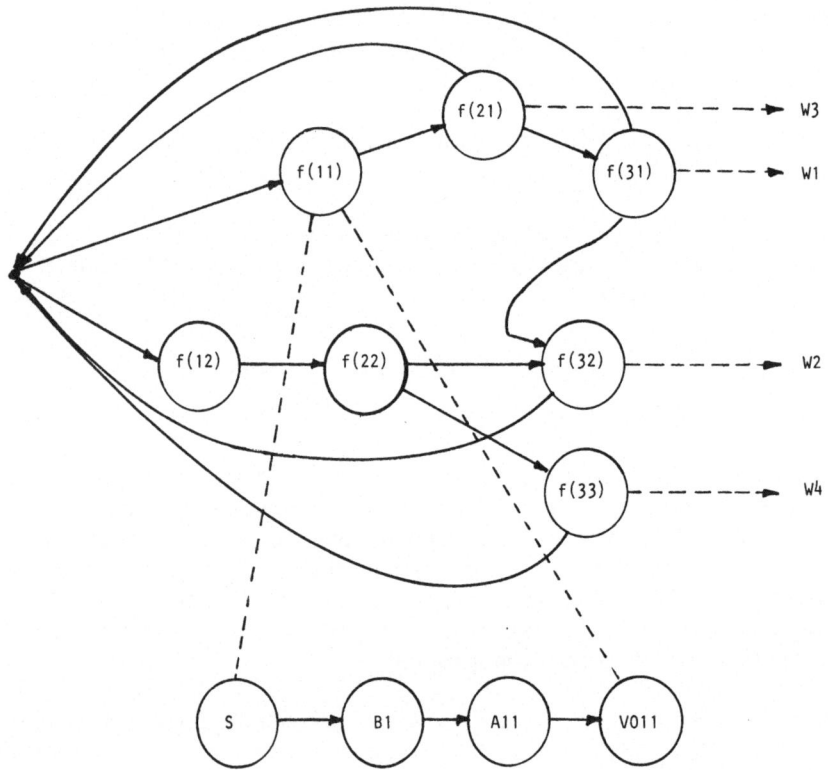

Fig. 7.1 Tree representation of the lexicon

lexical tree is accessed through a lattice of phoneme hypotheses which are generated by a bottom-up process. An algorithm for accessing such a lexicon is described in the following.

The algorithms for hypothesizing word candidates use two pointers, P1 which moves on the lexical tree and P2 which moves on the phoneme lattice (Fig. 7.2.).

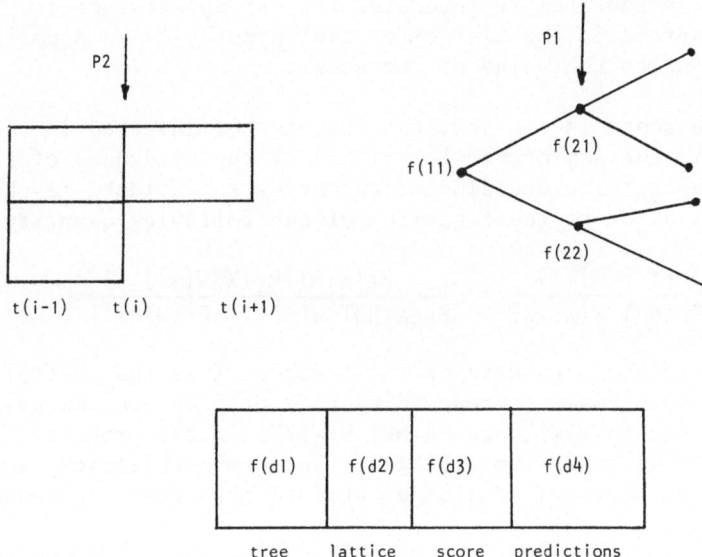

Fig. 7.2. Pointers used by an algorithm for lexical access

At the beginning, P1 points to the root R of the lexical tree
and P2 points to the initial time t0 of the input sentence. All
the phonemic hypotheses beginning at t0 and having a corresponding
label in the states of the lexical tree immediately following the
root cause the generation of an item in a queue of partial lexical
hypotheses. Each item of the queue contains a field f(d1) with a
pointer to the lexical tree, a field f(d2) with a pointer to the
phoneme lattice, a field f(d3) with a score of the partial hypothesis and a field f(d4) with the predictions of the phonemes which
may follow that partial hypothesis.

Items in the queue of partial hypotheses are ordered according
to their scores. The node at the top of the queue is extracted
and, if some of its predictions match some elements of the phoneme
lattice in a time interval following the time indicated by the

pointer in f(d2), the node of the lexical tree indicated by the pointer in f(d1) is accessed and becomes an active node. For every possible move from an active node whose predictions match with the input data, a new item is generated with an updated score. This item is inserted in the list of partial hypotheses in a position corresponding to the value of the score.

If the score is too low, the item is not inserted in the queue. The score is the sum of the logarithms of the evidences of the phoneme labels. The decision of accepting a candidate may be based on the evaluation of the logarithm of the following quantity:

$$RW = \frac{Pr(W(w)/X)}{Pr(NOT\ W(w)/X)} = \frac{Pr(W(w))Pr(X/W(w))}{Pr(X/NOT\ W(w))(1-Pr(W(w)))} \qquad (7.21)$$

where W(w) is the w-th word of the lexicon, X is the partial string of phoneme hypotheses corresponding to a path on the lexical tree which may lead to W(w), and Pr(NOT W(w)/X) is the probability of having NOT W(w) given the string X. These probabilities can be computed from a priori knowledge obtained in a learning phase.

Furthermore, generations of phoneme hypotheses may be thought of as independent events; thus the a priori probability of a string X given W(w) may be obtained as the product of the a priori probabilities of the phoneme labels in X.

Notice that in this model it is very hard to account for word boundary effects. For this reason, in the HWIM system (Woods et al., 1976), a preliminary matching is performed without considering word-boundary effects (this is referred to as pre-wbe strategy).

Klatt (1979) has introduced an interesting improvement proposing a lexical structure where each phone-label is substituted by a network of acoustic states. Coarticulation instances are accounted for in the network. To each state in the network is associated an acoustic cue represented by one or more spectral templates. A state following a chain of states representing acoustic cues which have been previously recognized is reached if one of its spectral templates matches a spectrum of the input sentence. This spectrum must be located in a given interval following a pointer which scans the spectra of the input signal. If the match is successful, the pointer advances just after the matched spectrum.

Fig. 7.3. shows a sample of the phone network corresponding to an unvoiced plosive preceding a vowel.

A silence is represented by a state S.

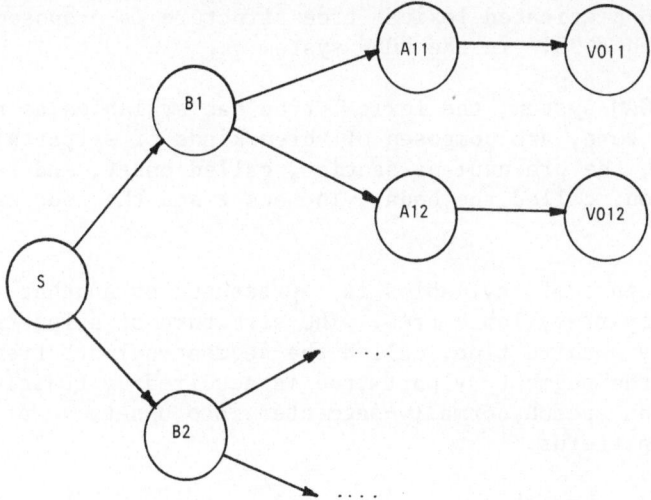

Fig. 7.3   Example of phone network

The following states B1, B2,... represent possible burst
spectra in different contexts.   States A11, A12,... represent
aspiration spectra following a burst of the type B1 and in different
prevocalic contexts.   Finally, states VO11, VO12,..., represent
voice onset spectra before a vowel.   Fig. 7.1. shows how this network can substitute the phone state f(11).

Crucial problems with these models arise because template
spectra are speaker-dependent and there are no criteria for deciding which templates are really archetypes of acoustic cues.
Furthermore, it is hard to incorporate in these models prosodic
cues and to account for deletion, insertion, and substitution
errors.

Once candidate hypotheses have been generated using a lexical
tree, one possible method for their verification is that of generating word spectra using a synthesis-by-rule program and computing
the distance between the segment of the input corresponding to the
hypothesized word and the data of the synthesized word.   In the
HWIM system (Woods et al., 1976) the distance is evaluated using a
metric introduced by Itakura (1975) and presented in Chapter 2,
Section 6.

A more sophisticated lexical tree structure is proposed by Erman and Smith (1981) in the NOAH system.

In the NOAH system, the lexical tree has syllables as nodes. Syllables, in turn, are composed of three kinds of sylparts: the vowel nucleus, the pre-nucleus section, called onset, and the post-nucleus section, called the coda. The onset and the coda may be optional.

The structure of syllables is represented by another tree called the sylpart-syllable tree. The structure of sylparts is represented by a third tree, called the segment-sylpart tree. The knowledge in the segment-sylpart tree is acquired by training the hypothesizer on speech normally-segmented into onset, vowel, and coda segment patterns.

A candidate hypothesis is generated by matching the lattice of phone labels attached to segments of the input sentence, using the method described by Goldberg (1975), with the segment-sylpart tree.

Sylpart candidates are generated with a time reference associated and a rating which estimates the likelihood that the sylpart candidate occurs at a given place in the utterance. Candidate syllables are then generated from candidate sylparts and candidate words are generated from candidate syllables.

Evaluation of hypotheses is performed with an algorithm proposed by McKeown (1977). Scores are the logarithms of the probabilities of the phone labels given the acoustic evidences.

The scores of the segments used for making a word hypothesis are added in order to obtain the whole score of a word hypothesis.

A weakness of this system is that all the coarticulation instances as well as the possible errors of the acoustic front-end have to be learned in a blind and rigid way, which does not account for the different a priori possibilities of the candidates.

An interesting tree representation of the lexicon is contained in the IBM system (Bahl et al., 1980). A word item is represented by a graph of the type shown in Fig. 7.4. This graph is obtained after phonological rules are applied to the base forms of a word. The arc labels $l(1)$, $l(2)$,... correspond to phonemes. The graph can be subdivided into portions called CLINKS (Confluent node LINKS) containing all arcs between the successive points which appear in every possible path in the graph. Two such points are A and B in the graph of Fig. 7.4. These points delimit a "CLINK" made of the arcs labelled by $l(1)$, $l(2)$, $l(3)$, $l(4)$. A word $W(i)$ is then represented by a single sequence of CLINKS: $C1(i)$, $C2(i)$,...,$Cn(i)$.

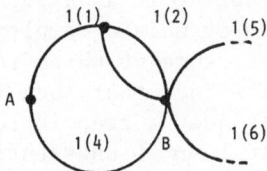

Fig. 7.4 Graph of a piece of a word obtained after application
          of phonological knowledge

The lexicon is a tree in which each node corresponds to a
"CLINK". Each "CLINK" is represented by a Markov source which
generates strings of centisecond labels and gives the probabilities
of insertions, deletions, and substitutions. Given a path W in
the tree and a string of input symbols s, the a priori probability
P(s/W) can be obtained and used in the acoustic model of the decoding algorithm.

Alignment problems are solved by using the Viterbi algorithm
which turns out to require a large amount of computation steps.

For this reason several heuristics have been introduced to
reduce the search space in the lexical tree.

A description of the Viterbi algorithm will be given in Section
3 of Chapter 8.

Alternative solutions to the lexical retrieval problem have
been proposed by Kobayashi and Niimi (1979) and Lennig and Mermelstein (1980) who proposed a model in which lexical structures and
verification actions are incorporated into a Language model which
is an Augmented Transition Network.

A good review of recent work on lexical access has been
provided by Haton (1980).

Other work on lexical representation for the French Language
has been published by Perrenou and Tep (1979), Meloni (1979) and
Vives (1976).

7.5   The Lexical Component
as a Problem Solver

A general view of the lexical controller is followed in this
book and consists in considering it as a problem solver. Grasping
the meaning of a spoken sentence is a complex problem which is
decomposed into subproblems, some of which require matching acoustic
information with words. The fact that such information is coded
into phonetic hypotheses is just a step in reducing the problem
complexity by an attempt to discard unessential information and
make the essential one emerge. The lexical problem solver may be
seen as a filter constraining the word hypotheses which may be
generated at a given point of the unknown sentence. The characteristics of this filter depend on model-driven constraints
(syntax, semantics, pragmatics) and data-driven constraints (the
already hypothesized syllables and the new ones which may be
hypothesized in accordance with the acoustic data).

The structural knowledge of the lexicon has the complexity of
a regular grammar and can be expressed by a Problem Reduction
Representation. The main difficulty at the lexical level is that
of achieving a fast and effective knowledge access. This is done
by decomposing a lexical access problem into plans which have to
be executed in order to achieve a committed task.

Plans are encoded into a network whose nodes may contain both
procedural and structural information.

Interesting general problem solving systems have been developed
in the last fifteen years. The General Problem Solver (GPS) of
Ernst and Newell (1969), the STRIPS of Fikes et al. (1977), and
the Nets of Action Hierarchies (NOAH) of Sacerdoti (1977) are among
the most important of such systems.

(Notice that NOAH here and below has a different meaning than
in the system of Erman and Smith (1981) mentioned in the previous
section.)

Special programming languages such as PLANNER (Hewitt, 1972).
CONNIVER (McDermott and Sussman, 1972), SOUP (see Sacerdoti, 1977
for references), QLISP (Wilber, 1976) have been proposed for problem
solving.

At the lexical level, the problem of controlling the generation
of lexical hypotheses is far more complex than that of establishing
structural rules between words, syllables, and suprasegmental
acoustic cues.

One problem in designing the lexical component of a SUS is
the automatic generation of all information associated with a lexical

item. A natural way of doing this is to consider the generation
of each item of word knowledge as a complex plan which is developed
under the control of a set of generation rules.

A second problem is that of generating word hypotheses belonging to a set predicted by the higher level KS, from the input data.

Again, the generation of hypotheses in a predicted subset

In order to construct a plan efficiently, the combinatorial
explosion of possible sequences of actions has to be contained.
The approach followed here is on the line of Sacerdoti's NOAH and
is based on hierarchically organized planning.

Most of this planning concerns the generation of the lexicon
itself and the procedures for its use.

Notice that the definition of this task is quite fuzzy because
the lexical problem solver has to generate the lexical hypotheses,
attempting to miss the right one as rarely as possible and with
the minimum amount of false alarms.

According to Sacerdoti (1977), the network is a graph structure whose nodes represent actions at various levels of detail,
organized into a hierarchy of partially ordered time sequences.
An action results in the generation of new states in the State
Space Representation of the problem to be solved. NOAH system is
conceived as a general purpose problem solver, and the lexical
component is described here as an application for which specific
knowledge simplifies the design. In the lexical component each
node in the network may refer to a set of child nodes that represent more detailed subactions. Associated with each node is a
triplet containing an action, the code of a program for expanding
the node, and a list in which are stored the changes to the interpretation caused by the action that the node represents. For the
sake of simplicity, only the action descriptions will be reported
in the following.

The main advantage of using plans for the lexical knowledge
is that a plan is non-linearly expanded into more detailed plans
only if some conditions are met. This is useful for successively
constraining the test and development of word hypotheses. Furthermore, even if the generation of a set of word hypotheses is not yet
complete, some CRITICS can be applied to the plans. CRITICS are
procedures that are executed to stop the execution of those plans
which are not expected to be completed successfully or are in
contradiction with other, more promising plans.

The lexical problem solver as described in this chapter
contains a Problem Reduction Representation generated by a
hierarchy of plans and a strategy for creating a State Space
Representation for hypothesizing words in a given time interval
of the spoken sentence. This strategy, embodied in the lexical
control KS, is also organized as a hierarchy of plans.

The main course of the research described in this chapter is
to construct a model of the lexical access in which a very large
set of word candidates, predicted by a language model as possible
interpretations at a given time of the input signal, can be progressively reduced to a few hypotheses by the use of acoustical
evidences and rules derived from perception models.

The algorithms that will be presented have been conceived for
a large vocabulary with only the attempt to minimize the number of
times the right hypothesis is missed and to keep as small as possible the number of words that are wrongly hypothesized.

Computer storage requirements and program complexity are also
important parameters for evaluating the efficiency of an algorithm.
Nevertheless, they won't be considered here because they mostly
depend on the system architecture. New architectures like a LISP
machine developed at MIT (Bawden et al., 1977; Steele and Sussman,
1980) can allow an efficient implementation of apparently complex
algorithms such as the ones described here. In other words, the
focus here is on the logical decomposition of a plan and on its
relevance for modelling a speech understanding function rather than
the attempt to get an efficient implementation on a classical sequential machine.

Fig. 7.5. shows the first level of the plan for generating the
lexical knowledge.

The plan description starts with a PLANHEAD node.

This plan is expanded into a split node (OR S) followed by
detailed plans regarding the generation or the updating of the
structural knowledge of the lexicon; both the subplans are followed
by a join (OR J) node.

The presence of the OR nodes is an example of the possibility
the system has of decomposing a problem in a non-linear way, which
is a peculiar characteristic of Sacerdoti's approach.

The generation of the lexical knowledge corresponds to the
setting of a first level of states in the State Space Representation.
First, a state is generated for each word, then, for each word, a
subtree is generated, corresponding to its possible alternative

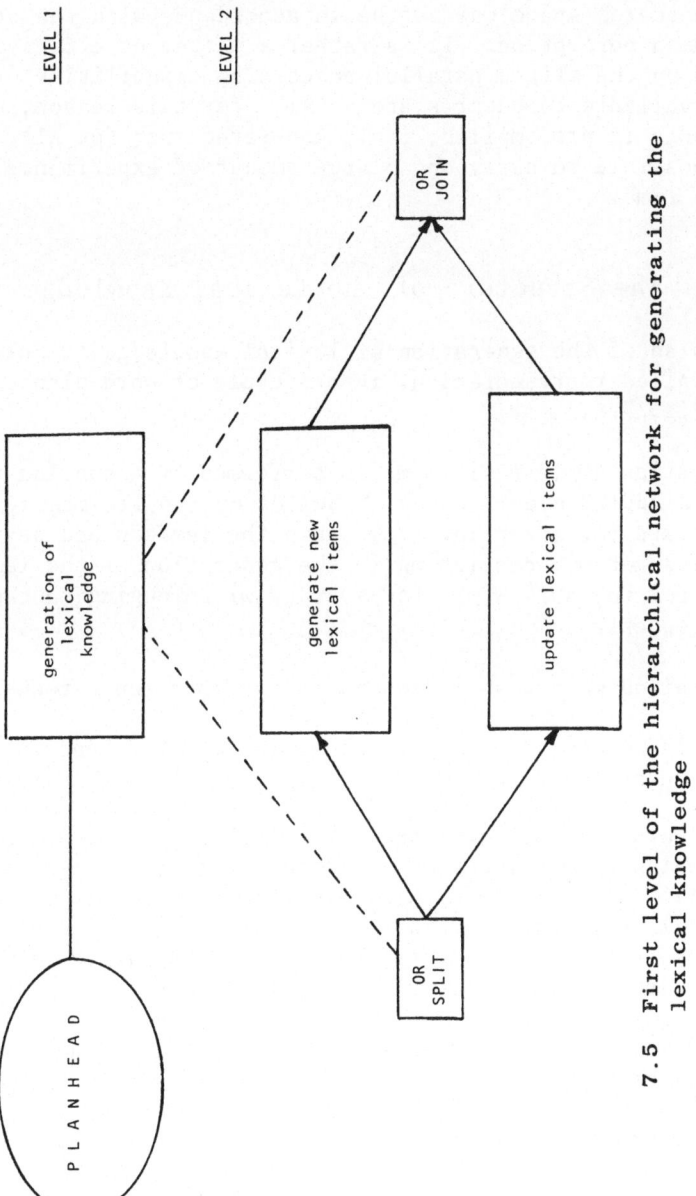

7.5 First level of the hierarchical network for generating the lexical knowledge

pronunciations. Then the search space in the state space is reduced
by focusing the attention on model-driven hypotheses and filtering
them progressively by using data-driven constraints.

Expanding the whole lexicon before applying constraints to
reduce the search space may not be in accordance with the strategy
used in human perception. It is rather a matter of efficiency which
is imposed by the slight parallel processing capabilities of the
actually available computer systems, Also for this reason, most of
the knowledge is precompiled, i.e., generated once for all. This
makes it possible to carry on a large amount of experiments in a
reasonable time.

7.6. The Structure of the Lexical Knowledge

The plan of the generation of lexical knowledge is split, in
a more detailed representation, into a cycle of word plans, as
shown in Fig. 7.6.

Generation of lexical items is performed by a subplan represented by a PBUILD node-type which builds up the lexicon. It takes
input requests for inserting items into the lexicon and performs
an iterative action, consisting in the generation of the lexical
knowledge for the word W(w), for every W(w) belonging to the set
of input requests.

A detailed word plan is generated for each input request.

The plan decompositions are shown in Fig. 7.6. and in the
following ones.

Each word plan enriches the structural knowledge of the
lexicon, which is a collection of representations of lexical items.
Let L =[W(1), W(2) ,...,W(w),..., W(n)] be such a collection. Each
element W(w) is a complex data structure representing the root of
a word. Updating the structural knowledge is simply an addition
of information to some items of the collection L.

The corresponding subplan will not be described here, for
the sake of brevity.

The generation of the syll-type tree prepares a data-structure
which will be enriched by the setting of the preconditions of each
word.

The structure of the syll-type tree will be described later
in this section.

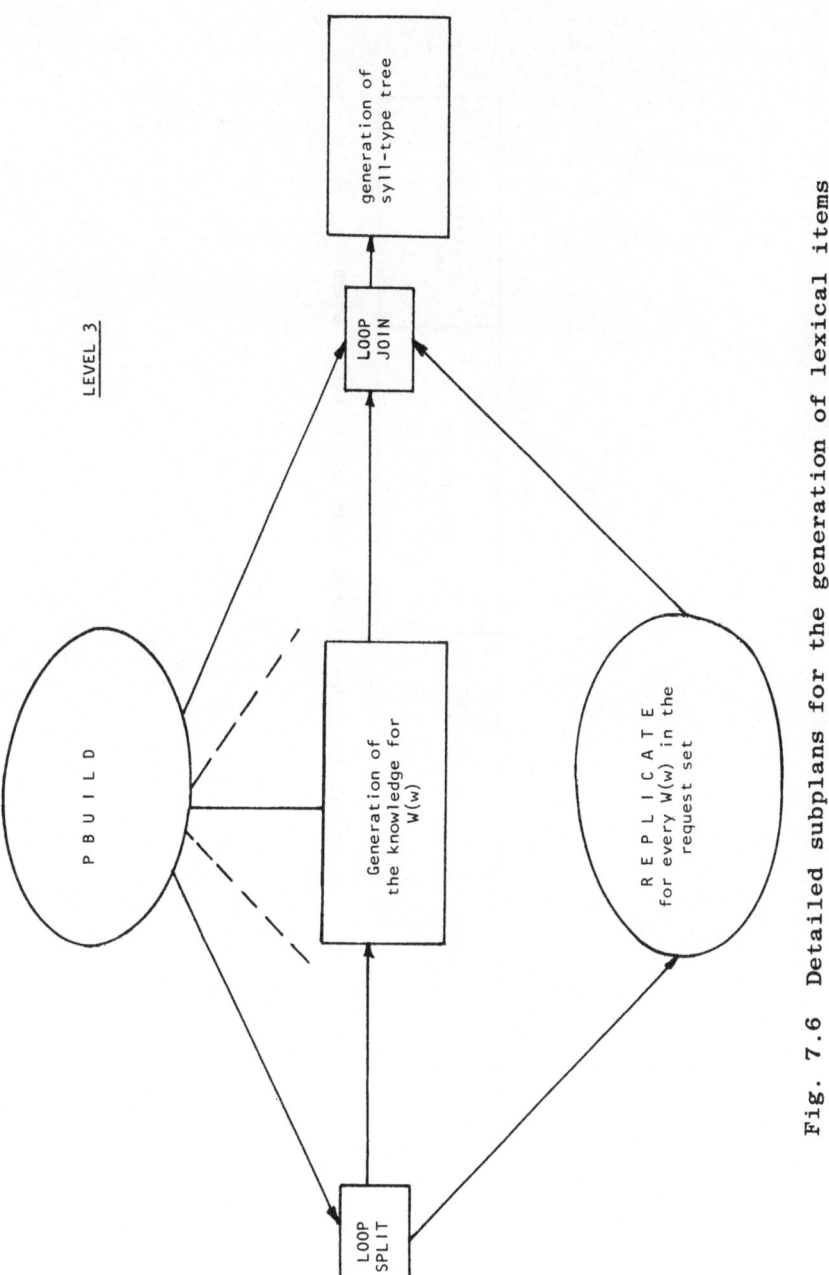

Fig. 7.6 Detailed subplans for the generation of lexical items

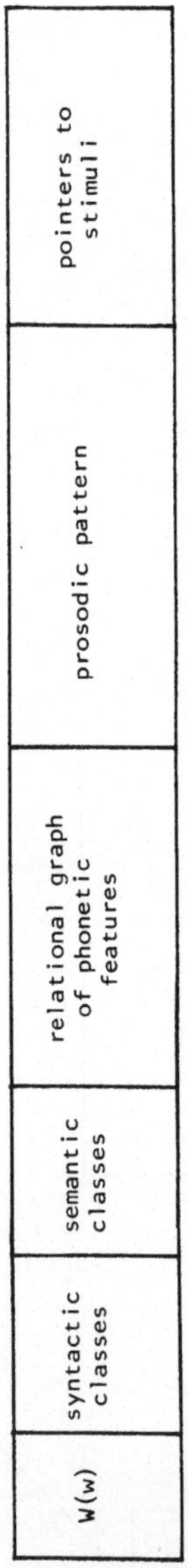

| W(w) | syntactic classes | semantic classes | relational graph of phonetic features | prosodic pattern | pointers to stimuli |

Fig. 7.7 Subplan for the generation of a word item

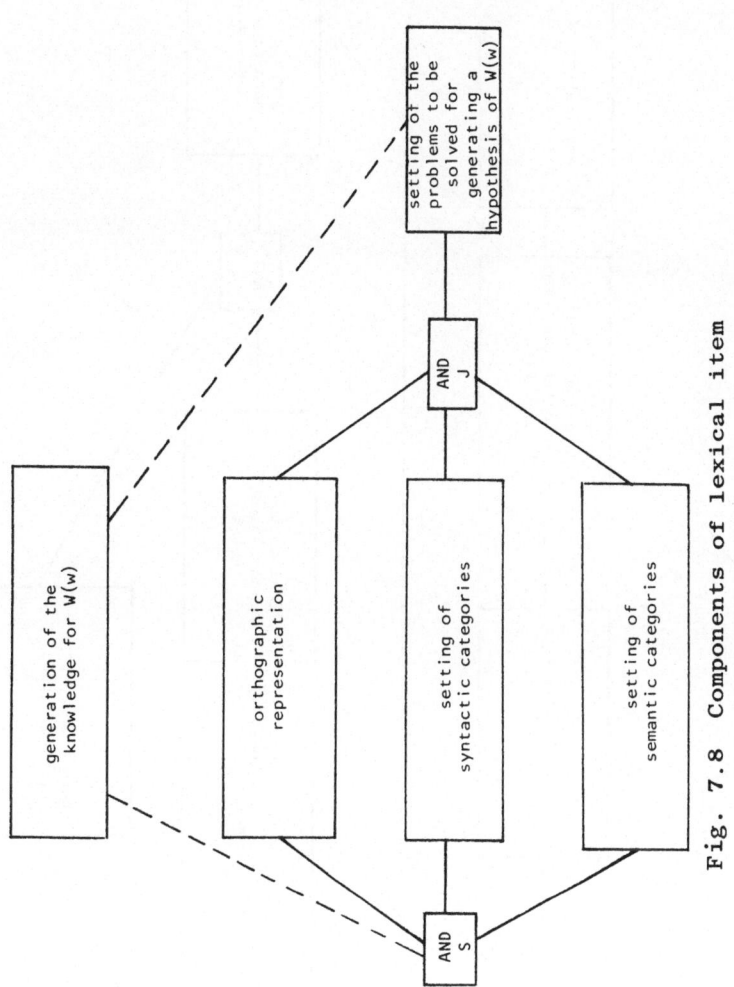

Fig. 7.8 Components of lexical item

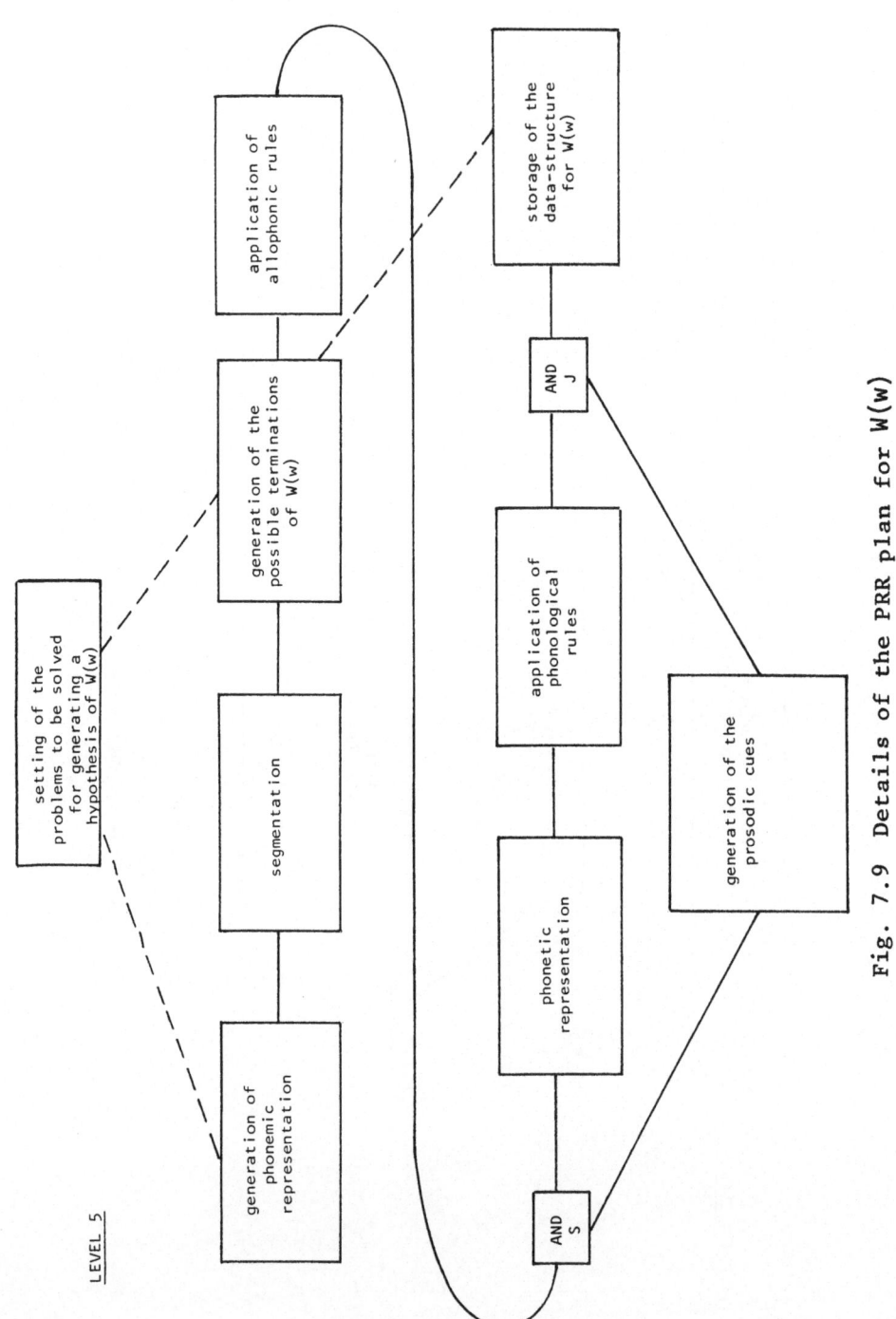

Fig. 7.9 Details of the PRR plan for W(w)

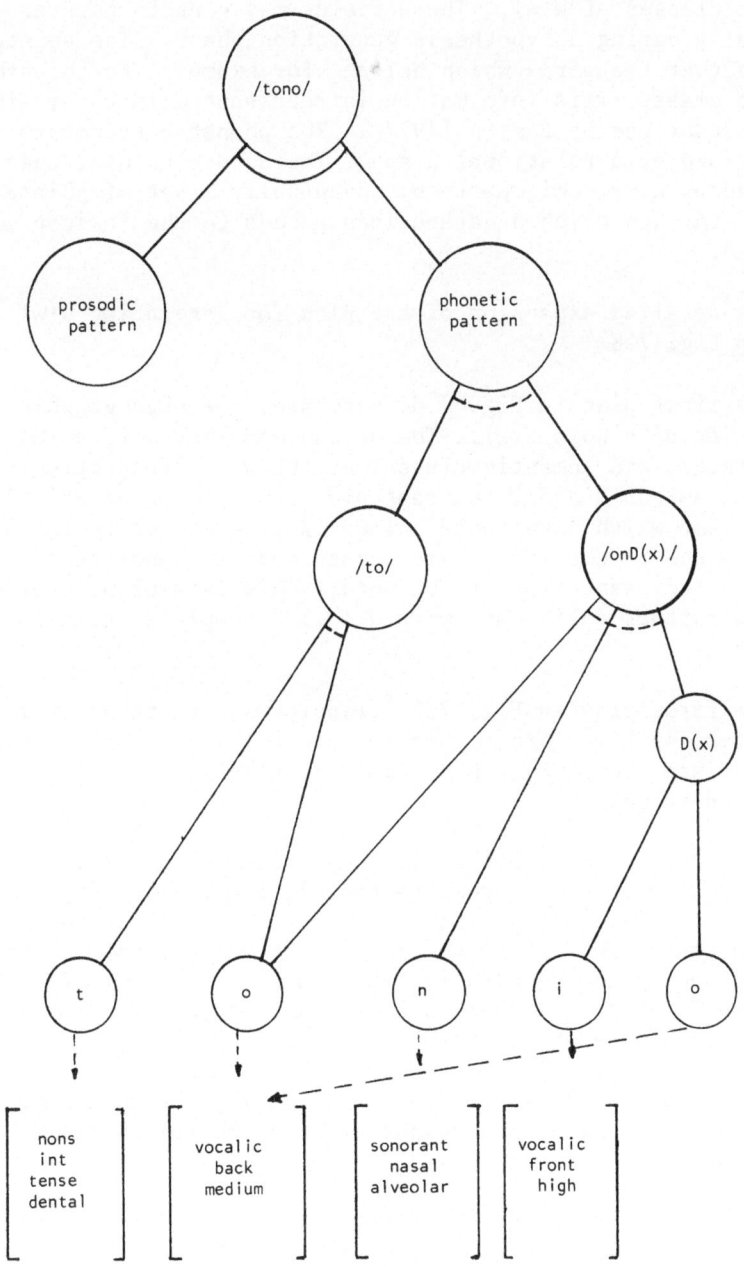

Fig. 7.10 PRR for the Italian word /tono/ ("tone" in English)

An element in L has the components shown in Fig. 7.7. The
first component is the orthographic transcription of the word W(w),
the second and the third describe respectively the syntactic and
semantic classes of W(w). These fields may contain pointers set
dynamically during a hypothesis generation phase. The pointers
link together the words which belong, for example, to the same
semantic class. This information is represented in a way similar
to that described by Paxton (1977). The phonetic structure of W(w)
is described by a relational graph. The suprasegmental cues are
described by a prosodic pattern. Eventually a set of pointers is
provided through which a data-driven access to the lexicon is
performed.

The detailed expansion of the plan for generating W(w) is
shown in Fig. 7.8.

The first plan in Fig. 7.8. generates the orthographic representation of a word W(w). The second and the third plans set
the syntactic and semantic classes of the words respectively.
After the AND JOIN node, the last plan generates a description of
the problems which have to be solved in order to write the hypotheses
W(w) into the blackboard. This corresponds to generating a Problem
Reduction Representation of the word. This last plan, called PRR
plan, is further split into more detailed subplans as shown in
Fig. 7.9.

The first plan in Fig. 7.9. transforms the orthographic representation of W(w) into a phonemic one. The next plan segments
the word into Pseudo-Syllabic-Segments by applying the rules
introduced in Chapter 4.

The third plan in Fig. 7.9. computes all the possible terminations of the word and represents them in a phonemic form.

The generation of all the possible terminations of a word
(singular and plural, masculine and feminine) is controlled by an
Augmented Transition Network Grammar whose detailed are omitted
here, for the sake of brevity.

For those nouns, adjectives or verbs which have standard
terminations, links to sub-graphs of standard terminations are
established.

For example, all the regular verbs of the first conjugation
are represented only by the graph of the possible coarticulations
of their roots with the terminations. The remaining syllables of
the terminations are represented in a structure which is common to
all the regular verbs of that conjugation.

The fourth plan adds to the phonemic representation of W(w) all its possible allophonic variations due to the influence of the speaker dialect and personality. All these variations are basic pronunciations of a word, called basic forms.

After the AND SPLIT node, a plan transforms each phoneme of the representation into a set of phonetic features and applies phonological rules. Phonological rules introduce the description of word boundary effects and reduce the phonetic representation to a sequence of minimal sets of features sufficient for characterizing the word. The representation obtained in this way is a collection of surface forms of a word.

Another plan sets a description of the prosodic cues for W(w) and, after an AND JOIN node, a final plan stores the data structure which has been generated into the storage of the lexical knowledge.

Fig. 7.10 shows as an example the graph which represents the phonetic composition of the word /tono/ ("tone" in English) together with its plural /toni/. The dashed arc represents a concatenation in time with the vowel /o/ shared by two successive PSSs. Phonological rules and their complexity are language dependent; while there are many such rules in languages like English and French, they are less complex, introducing few alternatives in the surface form of a word in Italian.

The prosodic cues specify the level of stress for each vowel. For example, /tono/ has the first stress on the first vowel, and the first vowel has to be rather longer than the second vowel. A detailed discussion on the prosodic cues which may be useful for word recognition may be found in a book chapter by Vaissière (1980).

The last plan of Fig. 7.9, which adds the lexical data to the lexical KS, also creates pointers and structures for accessing the word W(w) when proper stimuli are present in the syllabic blackboard. A detailed composition of this last plan is shown in Fig. 7.11.

The main problem in accessing a large lexicon is that of avoiding trying to match all the words with the entire blackboard of syllabic hypotheses.

This extensive trial can be avoided by attempting to detect only words of a set corresponding to concepts or syntactic classes which are under the focus of attention of the syntactic and semantic knowledge. Even in this case, it would be too complex to attempt to match all the words of the focused set with the syllabic blackboard. For this reason, a set of primary stimuli is selected for each word. Word stimuli contain sufficient conditions for calling the attention of the lexical controller to a word, provided that

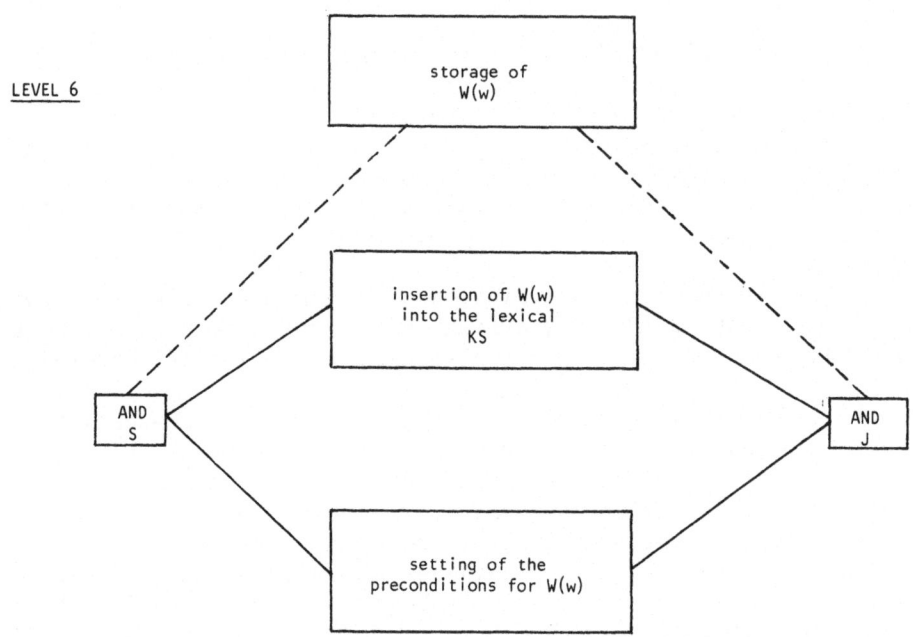

Fig. 7.11 Plan for storing W(w)

the focus of the higher level KS is in a lexical subset containing that word.

There are three types of primary stimuli, namely syll-type sequences, first complete syllables, and prosodic preconditions.

All the words having the same first complete syllable (the first syllable not affected by coarticulation with the previous word) are linked to a "partition of preconditions" by a plan which is shown in Fig. 7.11.

The syllable-type tree (syll-type tree) has syllable-types such as ((VB)(SON)(VB)) as nodes and pointers on the leaves to the partitions of preconditions.

The activation of a word candidate belonging to a lexical subset focused by the semantic knowledge depends on three types of preconditions, namely:

1)   The detection of an acceptable sequence of phonetic
features in the syllabic blackboard; this is done by partially
matching a fact-tree, represented by some hypotheses in the syllabic
blackboard, and the goal-tree, which is a part of the syllable-type
(syll-type) tree.

2)   An acceptable matching of more detailed features (phonemes)
than those of the syll-type tree between a syllable at the beginning
of a word candidate accepted by the syll-type tree and the data in
the syllabic blackboard.

3)   Congruence of the ratios of successive vowel durations
between the prosodic cues of the word candidate and the syllabic
hypotheses on whose basis that word has become a candidate.

Further details on the plan for generating word preconditions
will be discussed in the next section.

Some remarks are in order at the end of this section.

Once a word belonging to a data-driven set of hypotheses is
stimulated by an even partially hypothesized syllable, the remaining
process for establishing its congruence with the acoustic data is
performed from left to right.

The pointers between nodes of the data structures can be
avoided if a language allowing pattern matching (like LISP) is
used. This is particularly effective if implemented with a suitable
system architecture.

The application of phonological rules is performed as in the
LAFS system described by Klatt (1979).

Tree organization of the lexicon like the ones described by
Woods et al. (1976), Erman and Smith (1981), and Klatt (1979),
although particularly effective if a sequential computer is used,
do not allow incorporation of complex prosodic cues or relations
between word components. Furthermore they do not allow focussing
attention on a word using a "softly matched", partially corrupted
syllable as stimulus.

For this reason, three types of preconditions have been introduced, for each one of which only a sort of "soft-match" is required
in order to keep active a word candidate.

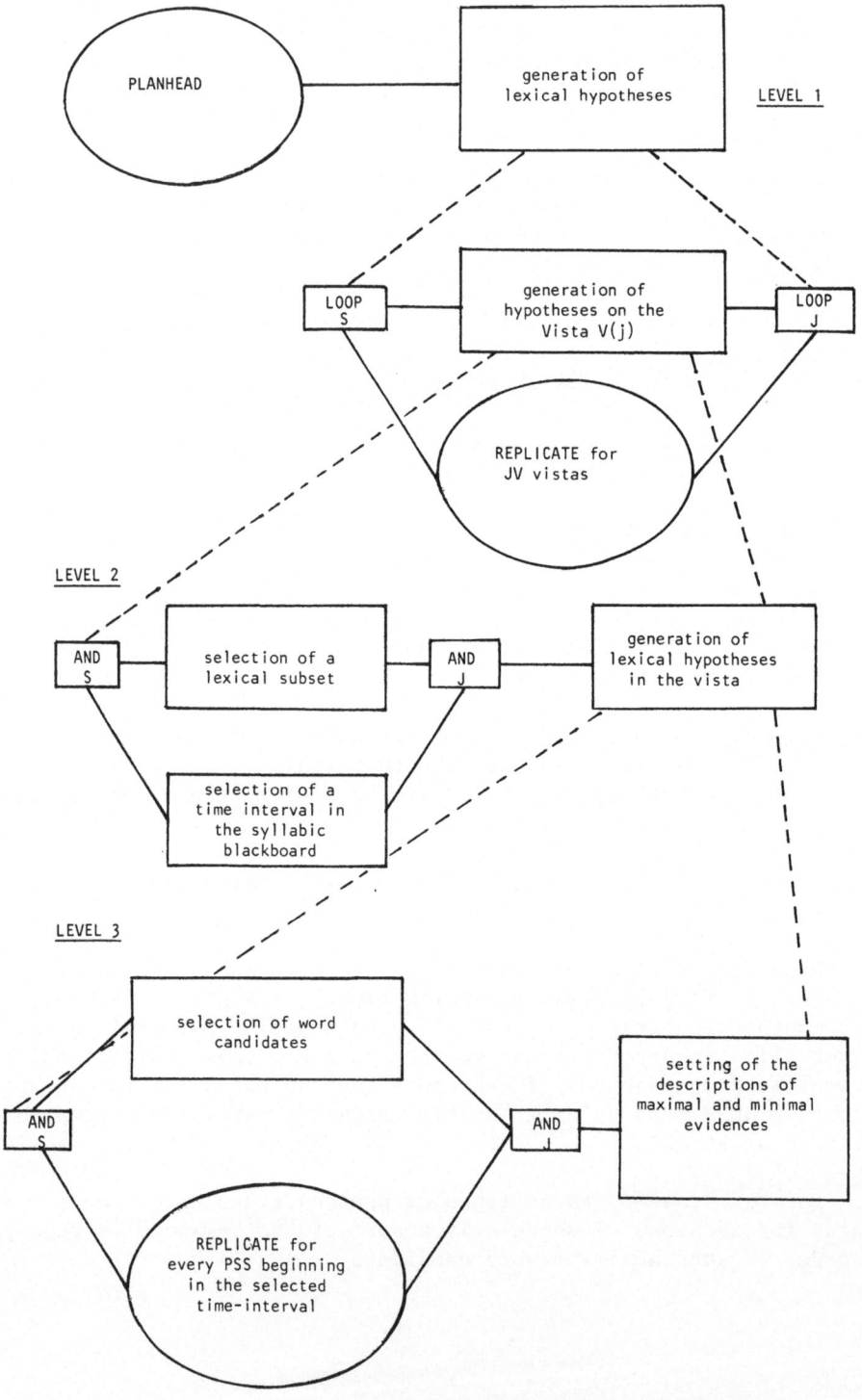

Fig. 7.12 Plan for generating lexical hypotheses

7.7   Strategies for Lexical Access

7.7.1   Top-down constraints

Fig. 7.12 shows a decomposition of the plan which generates lexical hypotheses. The REPLICATE node together with the LOOP S and LOOP J nodes represents the repetitions of the plan for generating hypotheses on a lexical subset.

The model-driven constraints reduce the set of lexical hypotheses that can be generated to a subset of the lexical items. This subset is called "Vista". When there are no constraints, the vista covers the entire lexicon. Furthermore, hypotheses belonging to a vista are looked for in a time interval which covers at most the duration of the sentence to be interpreted. Let $V(j)$ be the j-th vista requested by the higher level knowledge and let $(t(bj), t(ej))$ be the times of the beginning and the end of the interval in which the words of $V(j)$ are looked for. A vista may contain several sets of words belonging to different syntactic and semantic classes. Words may be arranged in sets   each one of which is assigned a degree of expectation $E(jx)$, $j \epsilon (1,JV)$, $x \epsilon (1,X(j))$, where JV is the number of vistas activated for the interpretation of a sentence and $X(j)$ is the number of sets of expectation in the vista $V(j)$.

Setting a vista $V(j)$ corresponds to establishing a link between the word $W(jw)$ of the vista $V(j)$, $w \epsilon (1,N(Xj))$, $N(Xj)$ being the number of lexical items in $V(j)$. Whether the links are pointers or labels to be matched depends on the language used, and the efficacy of the implementation depends on the system architecture. The vista interval $(t(bj), t(ej))$ is compared with the entries of the directory of syllabic hypotheses and the times $t(ji)$ which are inside the interval $(t(bj), t(ej))$ are considered as possible starting points for generating candidates of hypotheses in $V(j)$.

Generation of hypotheses in $V(j)$ anchored at $t(ji)$ can be represented as a tree rooted at $V(j)$ in a State Space Representation; each node of the tree corresponds to a time instant $t(ji)$ and is labelled as $V(ji)$ (Fig. 7.13). Based on the model-driven constraints, each word $W(jw) \epsilon V(j)$ can be hypothesized with an anchor at $t(ji)$. Thus, the node $V(ji)$ can be expanded by an operator which generates the words of $V(ji)$ and associates to each word an expectation value.

This tree is expanded by the attempt to match each word against the acoustic data in order to select the most evident hypotheses. Expansion of $V(ji)$ is constrained by imposing a hierarchy of preconditions to be checked before a verification process is executed on a word candidate. Notice that this verification process is computationally expensive because it involves the composition of evidences.

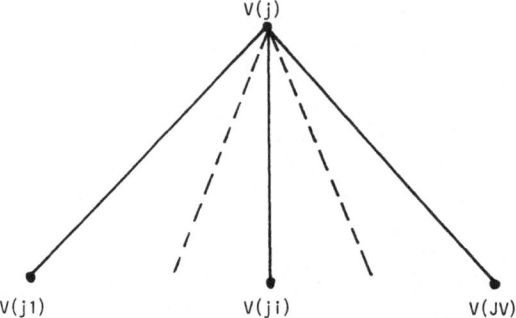

Fig. 7.13   SSR for a set of vistas

7.7.2   Preconditions based on the first syllable

Preconditions of a lexical item are of various types. It has been shown in the previous section that one of them concerns the word initial syllable and could be extended to the stressed syllable or some degradation of them. The selection of the precondition or data-driven access constraints to the lexicon is language dependent. Perceptual experiments (see for example Cole et al., 1980) support a model of human perception in which lexicon access is primarily based on the first syllable.

For the Italian Language and for the system described so far, preconditions for a word are all the possible initial "complete" syllables. A complete syllable of a word is a PSS which does not begin with a phoneme which may be affected by coarticulation by another phoneme at the end of the previous word. For example, the word /tono/ starts with a complete syllable /to/ and does not have other allophonic variations at the beginning; thus /to/ is the first complete syllable. Some monosyllabic words or bisyllabic words with the stress on the first syllable have preconditions on the first syllable even if there may be some coarticulation effects with adjacent words.

Addressing a lexical item based on a syllable implies that some components of this syllable have been detected even if their detection is affected by a degree of vagueness. Because of the segmentation algorithm, consistent errors in segmentation mostly depend on a fault in hypothesizing the presence of a vowel. System performances show that these errors have a frequency of occurrence less than 1%. Nevertheless, this possibility is taken into account

by introducing secondary preconditions, derived from the primary
preconditions by assuming that one vowel may be missed in the
entire word, and the syllable is extended according to the rules of
the segmentation grammar.

For example, for the word /tono/, the secondary preconditions
are derived by assuming that the first /o/ is missed, allowing the
syllable to be extended as /tno/ or /tno (postvocalic-sonorant)/
and for the plural to /tni/ or /tni (postvocalic-sonorant)/.

The expressions in parentheses represent phonetic classes that
the first phoneme of the following word may belong to. These
phonemes may be detected in a single pseudo-syllable segment
together with the rest of the word in which the vowel /o/ has been
missed. This possibility is very rare because the /o/ is stressed
and the system usually does not fail to detect stressed vowels.

The secondary preconditions are not allowed to be stimuli for
lexical access unless no hypotheses of a vista can be generated with
high evidence or there are high-evidence hypotheses; the higher
level controller considers that all of them can be composed of
terminations and beginnings of expected word sequences, and the
maximum evidence that can be obtained by the hypotheses stimulated
by the secondary preconditions may lead to a hypothesis more evident
than those of other vistas.

This item will be clarified in the next section, where details
on hypothesis evaluation will be given. For the moment, let us
consider how a word can be accessed by standard preconditions.

7.7.3 Precondition degradations

The plan which generates the lexicon determines, for each word
W(w), the first complete syllable FSw and enters W(w) in a partition
of preconditions corresponding to FSw.

In order to account for the possible degradations with which
the syllable may be hypothesized, the precondition field of W(w)
is linked with several syllable classes belonging to different
levels of detail.

Notice that the syllabic KS is designed in such a way that
the probability of failure to hypothesize a syllable at its first
level of detail is practically negligible and the probability of
failure to generate higher levels of detail increases, but is kept
small, except in the case of consonant clusters. In these cases,
the conditions associated with arcs of the syllabic ATNG do not
allow the system to proceed along the network up to the generation
of phonemic hypotheses.

Shikano and Kohda (1978) report that extensive protocols may be designed for the Japanese language using words made only of VCV syllables. In these cases the system proposed here would be able to generate all the phonemic hypotheses in a purely bottom-up way with a small probability of failure to generate the right phonemic hypothesis.

If it is true that in human perception the lexicon is accessed by the first syllable of a word, this syllable has to be rather stable and the case of allophonic variations or word boundary effects should be rare.

This is the case of the Italian Language, for which most of the words have only one possible first complete syllable.

7.7.4 The lexicon as a content-addressable memory

The preconditions are considered as disjoint primary problems for the general problem solving activity, which is word hypothesization. If none of them are satisfactorily solved for a word, the problem of hypothesizing that word has to be considered unsolvable for the time interval of the vista.

It has been mentioned that a syllable which is a precondition may have been only partially hypothesized. For example, for /to/ only the following phonetic subproblems may have been solved: [<nonsonorant-interrupted-tense consonant> <vowel back>]. These admissible degradations and the evidence of the preconditions depend on the level and the possibility of solution of their subproblems but not on the words the preconditions belongs to.

For example, the admissible degradations of /to/ are the admissible degradations of a <pseudo-syllabic-segment> consisting of a <consonant-interrupted-tense-consonant> followed by a <vowel back>. They do not depend on the fact that /to/ is in /tono/. The reason for this is that a syllabic segment has a context-independent structure. Furthermore it has been found for the Italian Language that the statistic of the degradations of a syllabic segment depends only on a subset of the phonetic features characterizing the segment. For example /to/ has similar degradations as /po/, the degradations being dependent on the fact that the first consonant is an unvoiced plosive, regardless of its particular identity.

The data structure of the partitions of preconditions for the first complete syllable contains a partition for every syllable class $(SC(j))$ $(j=1,2,...,SM)$, SM is the number of syllable classes. Notice that in this subsection j refers to a partition and not to a vista.

Syllable classes are obtained by applying the segmentation grammar to generate all the admissible Pseudo-Syllabic-Segments with the following set of symbols: [UN, SON, V, PREVSON, POSTVSON]. The partition of the j-th syllable class SC(j) has the following structure:

$$(SC(j) \ (PHC(j1)/L1(j) \ (FS(j11)/K1(j1) \ (W(j111)/(I1(j11)...)))$$

$$(FS(j12)/K2(j1).........................)$$

$$.$$
$$.$$
$$.$$

$$(.....................:.....................))$$

$$(PHC(j\ell)/L\ell(j) \ (.......................................)$$

$$.$$
$$.$$

$$(FS(j\ell k)/Kk(j\ell) \ (...W(j\ell ki)/Ii(j\ell k)...))$$

$$.$$
$$.$$
$$.$$

$$(............................))$$

$$.$$
$$.$$
$$.$$

$$(..))$$

To each syllable class SC(j) is associated a set of Phonetic Compositions (PHC(jℓ)). For example, if SC(j) is ((UN)(V)), a phonetic composition PHC(jℓ) may be ((NIT)(VB)); another may be ((NCL)(VF)). To each phonetic composition PHC(jℓ) is associated a condition Lℓ(j). This condition is met when at least one word beginning with a syllable of type SC(j) has a syllabic structure for which a match has been obtained between the data in the syllabic blackboard and the syll-type tree. The use of this tree will be described later.

To each phonetic composition PHC(jℓ) is associated a set of phonemic syllables FS(jℓk).

A phonemic syllable FS(jℓk) contains a possible phonemic transcription of PHC(jℓ). If, for example, PHC(jℓ) is ((NIT)(VB)), then one possible PHC(jℓk) is /to/. To each phonemic syllable FS(jℓk) is associated a condition Kk(jℓ) which acts in a similar way as Lℓ(j).

To each phonemic syllable is associated a set of words beginning with that syllable. To each word W(jℓki) is again associated a condition Ii(jℓk) which is set by an algorithm using the syll-type tree, the syntactic and the semantic predictions.

Example 7.7.1

For example, ((UN)(V)) can have the following composition:

(((UN)V)((NIT)(VB)/L1(1) ((to)/K1(1 1) (tono)/(1 1 1)

                                    ((po)/................)

                        .

                   (......................))

          ((NCL)(VB))/L2(1) ((vo)/..........................)

                        (..............................))

(..))

7.7.5. The syll-type tree

The tree of syllable types has a root corresponding to every possible syll-type. The descendants of the root are all the syllable types SC(j), with a specification of the places of articulation for the vowels in SC(j). For example ((UN)(V)) is represented by the following nodes: ((UN(VB)), ((UN)(VC)), ((UN)(VF)). Each such node has a set of descendants containing all the possible syllable types made from the symbols UN, SON, and the places of articulation of the vowels. A leaf of the syllable tree identifies a set of words whose syllabic structure is compatible with the sequence generated by moving from the root to the leaf. For example, the root /ton/ of the word /tono/ will be in the set associated with the leaf corresponding to the path (((UN)(VB))((VB)(SON)V)).

A path is characterized by a sequence of nodes $N(p) = (h(g1), h(g2),...,h(gb))$. Each path ends with a node which is associated to a data structure DATSTR(N(p)) containing pointers to the conditions in the data structure of the first syllable. The structure of DATSTR(N(p)) for the case of a two node path: $h(g1) = j$ and $h(g2) = n$ is the following:

DATA-STR(j,n) = (j(11(j,n)(k1(j,n11)(ij(j,n,11,k1)........)

                             (.......................))
              (1z(j,n)......)(i1(j,n,1z,k1)...........)
                            (ky(j;n,1z)(i1(j,n,1z,ky).....)
                              :   (...............))
              (.................................))

ℓz sets a condition $L\ell(j)$ in SC(j), ky sets a condition $Kk(j\ell)$ and ix sets a condition $Ii(j\ell k)$ when the leaf of N(p) is reached.

Let $(SC(j)(t1, t2))$ be a syllable type hypothesized in the syllabic blackboard. It identifies a node in the syll-type tree. By examination of the syllables beginning at t2 in the blackboard, some descendants of $(SC(j)(t1,t2))$ are identified in the syll-type tree. Proceeding in this way, certain leaves of the syll-type tree are reached and certain conditions of the first syllable data structure in the partition corresponding to $SC(j)$ are set.

Now, more detailed hypotheses about $(SC(j)(t1,t2))$ are looked for in the syllabic blackboard. These detailed hypotheses identify some phonetic and phonemic compositions of syllables which allow certain word hypotheses in $SC(j)$ to be reached.

These words are those associated with phonemic syllables whose associated conditions are set by the algorithm of the syll-type tree.

Example 7.7.2

The word /tono/ is identified by a leaf of the syll-type tree containing the nodes $(((UN)(VB))((VB)(SON)V))$. The word /tono/ is identified by a set of conditions for the first syllable data structure of the Example 7.7.1. If $((UN)(VB))$ is the j-th syllable type and the word /tono/ is the first word addressed by $SC(j)$, then the following set of conditions appear at the leaf of /tono/ in the syllable-type tree:

$j=1$

$\exists z/\ell z(1,n) = L\ell(1)$

$\exists y/ky(1,n,\ell z) = k\ell(1,1)$

$\exists x/ix(1,n,\ell z,ky) = Ii(1,1,1)$

The syll-types in the tree contain details on the place of articulation of the vowels even if these details are not present in the syllable types of the first syllable partitions. The reason is that the place of articulation of the vowels is detected practically without error (even if with some false-alarms) and more details on the features detected in a bottom-up way allow a better preselection of word hypotheses.

Among the words in the partition $SC(j)$ whose conditions are met, only those which are consistent with prosodic, semantic, and syntactic predictions are considered for generating verification processes.

Notice that the tree structures used in this approach have

broad phonetic features as node labels. The probability that these
features are not hypothesized by a data-driven stimulation is
negligible.

Based on this fact, only a few erroneous configurations are
inserted into the trees in order to account for the very unlikely
errors of phonetic feature hypothesization. Thus, the algorithm
that uses the tree does not have to attempt substitutions, insertions, or deletions, with a great advantage in speed and simplicity.

Learning is also simplified because it is merely reduced to
the few possible errors that depend only on the syllabic type and
not on the phonemes or the words.

The syll-type tree can be seen as a constraint which avoids
associating a word hypothesis with a sequence of input syllables
absolutely incompatible with that word.

7.7.6 Precondition evidences

Evidences of preconditions based on the first complete syllable
are evaluated as follows.

Let SILB be a PSS. Analyzing several realizations of SILB,
different levels of problem solution can be observed. The level of
problem solution is represented by a pair composed of the cardinality
of a set of terminal subproblems, sufficient for the solution of the
general problem, and the number of such subproblems which have been
solved by the bottom-up analysis of a particular utterance of SYLB.
For example, /to/ has only one set of sufficient subproblems
(shown in Fig. 7.10); this set contains eight subproblems; if only
seven of them have been solved, the level of solution is the pair
(8,7). Levels of solutions are represented by discrete points
along the horizontal axis in Fig. 7.14.

Good statistics can be obtained if it is accepted that evidence
assignments can be based only on a subset of the sufficient sets of
subproblems. For example, the same rule for assigning evidences
can be applied for all the syllables of the type ((NIT)(VB)).

Evidences of preconditions are defined over a partially
ordered set which is obtained by first ordering the sets of subproblems according to the values of the second number in the levels
of solution, and by performing a partial ordering inside each level
based on the possibilities of the solved problems. The evidence
distributions are settled subjectively after analysis of the recognition performances of syllable types extracted from continuous
speech. This assignment is made in such a way that evidence judgements better or equal to a value qualitatively described as "evident"
affect levels of solution covering at least 80% of the observed cases.

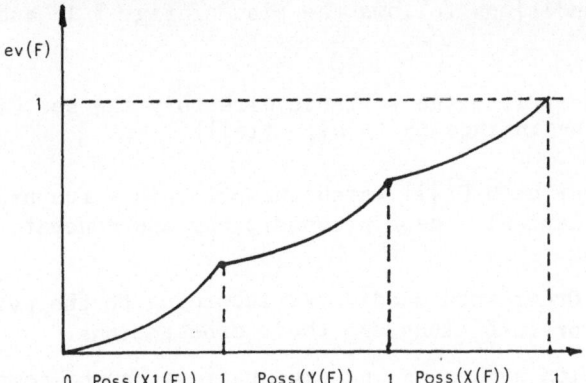

Fig. 7.14 Relation between evidence, level and possibility of
          solution

Notice that the evidence values can belong to an ordered
set of linguistic variables. This implies that evidence evaluation
is not accurate. The same probably happens in human perception.

Because this evidence assignment is very crude, we do not
need to know precise statistics on syllable degradations in order
to decide how to do it. Good statistics on a hundred syllable
types can be enough to decide how to assign evidences to the preconditions of all the words of the Italian Language.

The underlying concept of the above statement is that the
evidence of a precondition depends on the experience of frequencies
of occurrence of events belonging to the same broad class of phonetic
features and on the level of problem solution they correspond to.
It is not difficult to verify that an information-theoretical
approach to this problem, based on the precise computation of
probability values, would be impractical.

A further selection, after discarding word candidates of low
evidence, is performed by controlling that the ratio of the durations of the first two hypothesized vowels of a word is inside an
interval which is one of the prosodic features of each word.

7.7.7 The algorithm for lexical access

Given a vista $V(j)$ $(t(bj), t(ej))$ and a data structure of
phonetic hypotheses, the algorithm which generates word candidates

based on preconditions follows the plan of Fig. 7.15 and is summarized below:

> STEP 1: Sort from the syllabic directory the set $\Gamma(ji)$ of
> syllable beginnings in $(t(bj), t(ej))$.

> STEP 2: For each $\Gamma(ji)$, match the syllabic structure
> with the lexical access preconditions and generate
> candidates.

> STEP 3: Order word candidates according to the evidences
> of their preconditions and their expectations.

Precondition evidences and expectations can be composed using
a fuzzy algebra. The main motivation for this choice is that fuzzy
algebra is very flexible and evidences do not have to be expressed
with high accuracy.

Using the first complete syllable as a precondition for lexical
access is in accordance with the results of recent psychoacoustic
experiments (Cole and Jakimik, 1980). Using the syll-type tree in
the same way as Erman and Smith (1981) allows elimination of those
candidates for which some basic phonetic feature does not appear
in the data.

The use of prosodic cues is in accordance with the suggestion
by Vaissière (1980).

Accessing the lexicon with three types of preconditions has
potential advantages over other solutions (HWIM, NOAH, LAFS) for
which the access is strictly left-to-right and limited to only one
type of precondition.

7.8 Selection of Candidates
and Hypothesis Evaluation

7.8.1 Evaluation of precondition evidences

Let $H(jik)$, be the k-th candidate hypothesis at time $t(ji)$,
$k \in (1, Kji)$, Kji is the number of candidate hypotheses at time $t(ji)$.

A candidate hypothesis is generated when all the preconditions
are met with enough evidence. A candidate hypothesis represents
a collection of subproblems whose solution has to be attempted in
order to find the best hypotheses, if any, of $V(j)$ to be written
onto the blackboard.

The solution for subproblems may require a large amount of
processing. For this reason a further reduction of candidates or,

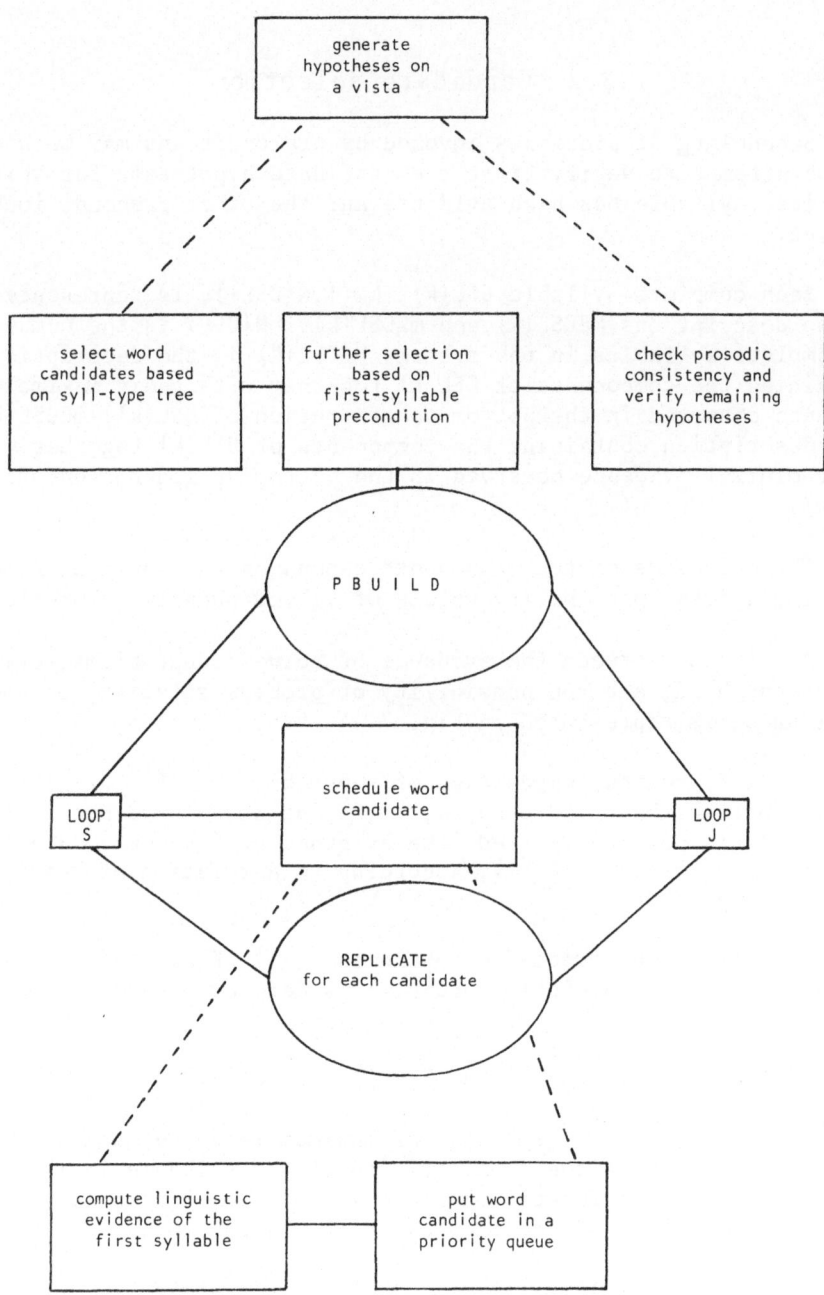

Fig. 7.15 Subplan for generating hypotheses on a vista

at least, a scheduling of the process invoked by them may be
required before evaluation of word hypotheses.

7.8.2 Candidate selection

Scheduling of processes invoked by preconditions may be useful
in the attempt to verify first the candidate hypotheses for which
the first syllable has high evidence and the other preconditions
are met.

Each complete syllable $CSL(k)$, $k \in (1, N(CSL))$, is represented
by two descriptions $MECSL(k)$ and $mECSL(k)$. $N(CSL)$ is the number
of complete syllables in the system; $MECSL(k)$ is the description
containing the components of $CSL(k)$ together with their maximal
evidence observed in the bottom-up generation of $CSL(k)$; $mECSL(k)$
is a description containing the components of $CSL(k)$ together with
their minimal evidence observed in the bottom-up generation of
$CSL(k)$.

The evidences of the components depend on the level of problem
solution and the possibility values of solved phoneme subproblems.

A relation between the evidence of solved subproblems, $ev(F)$,
on the one hand, and the possibility of problem solution, on the
other hand, is represented in Fig. 7.14.

Let $Y(F)$ be the composition of subproblems $X1(F)$, $X2(F)$ and
$X3(F)$. Assume that, for example, $X1(F)$ and $X2(F)$ have been solved
and $X3(F)$ has not been solved, the evidence of F can be evaluated
on the basis of $Poss((X2(F))$, according to a relation of the type
shown in Fig. 7.14.

Let $CSL(k)$ be composed of phonemes $F(k1)$, $F(k2)$, $F(k3)$. The
maximal evidence description $MECSL(k)$ is defined by the following
set:

$$MCSL(k) = (Mev(F(k1)), Mev(F(k2)), Mev(F(k3))), \qquad (7.22)$$

where $Mev(F(ki))$ $(i = 1,2,3)$ is the maximum value reached by the
evidence of $F(ki)$ in the i-th position of the syllable $CSL(k)$
after a set of experiments. Analogously, the minimal evidence
description $mECSL(k)$ is defined by the following set:

$$mECSL(k) = (mev(F(k1)), mev(F(k2)), mev(F(k3))) \qquad (7.23)$$

where mev $(F(ki))$ is the minimum value reached by the evidence of $F(ki)$ in the i-th position of the syllable $CSL(K)$. In case none of the subproblems of a phoneme are solved, the corresponding value of $mev(F(ki))$ is zero.

Maximal and minimal evidence descriptions can be learned automatically by considering together all the syllables of the same type and assigning the same description to them.

Example 7.8.1

Let us consider, as an example, the syllable /to/. It is composed of two phonemes /t/ = F1 and /o/ = F2. The sub-problems of the two phonemes are:

X1(F1) = nonsonorant	X1(F2) = vocalic
X2(F1) = interrupted	X2(F2) = back
X3(F1) = tense	X3(F2) = medium
X4(F1) = dental	

As the syllables of the type (NIT)(Vocalic) can be hypothesized with full evidence by a bottom-up process, the maximal evidence description of the syllable /to/ is:

$$MECSL(/to/) = (1,1)$$

because

$$Mev(F1) = Max\ Poss\ (F1) = 1$$
$$Mev(F2) = Max\ Poss\ (F2) = 1.$$

The minimal evidence description is:

$$mECSL(/to/) = (0.3,0.7)$$

For every phoneme $F(ki)$ an evidence ratio $ER(ki)$ is defined as follows:

$$ER(ki) = \frac{ae(F(ki))-mev(F(ki))}{Mev(F(ki))-mev(F(ki))} \tag{7.24}$$

where $ae(F(ki))$ is the evidence of the phoneme $F(ki)$ in the syllable under analysis. Of course, only positive values of $ER(ki)$ are accepted.

Let us further define the following quantities:

$$N(ki) = ae(F(ki))-mev(F(ki))$$
$$D(ki) = Mev(F(ki))-mev(F(ki))$$

(7.25)

which will be used later.

If a PSS contains a single vowel and a single consonant, the evidence ratio is computed for both. Let ERV be the evidence ratio of the vowel and ERSC be the evidence ratio for the consonant. A global evidence judgement represented by a linguistic variable is obtained according to the rules shown in Fig. 7.16 by putting VWE=ERV and SCE=ERSC.

Example 7.8.2

Let /to/ be the first complete syllable of a word for which the preconditions are met. Let /t/ be recognized as <nonsonorant-interrupted-tense> with a global possibility of 1. Using a diagram like the one in Fig. 7.14 an actual evidence ae = 0.75 is obtained. For the vowel /o/, an actual evidence ae = 1 is obtained by a similar procedure.

Using the maximal evidences defined in the previous example, the following evidence ratios are obtained:

$$ERV = \frac{1-0.7}{1-0.7} = 1$$

$$ERSC = \frac{0.75-0.3}{1-0.3} = \frac{0.45}{0.7} = 0.64$$

Thus the linguistic evidence of the syllable is "evident" according to Fig. 7.16.

If, in a syllable, there are two vowels, the global vocalic evidence VWE is obtained as a ratio between the sum of the numerators (ki) of the two vowels and the sum of the denumerators D(ki) of the two vowels.

The same operation is performed for the consonants and the global consonantal evidence SCE is obtained. Then, the evidence of the whole syllable is obtained from Fig. 7.16.

Fig. 7.17 shows the details of the subplan for setting the descriptions of maximal and minimal evidences.

This plan is the main component of the plan for updating the lexical knowledge.

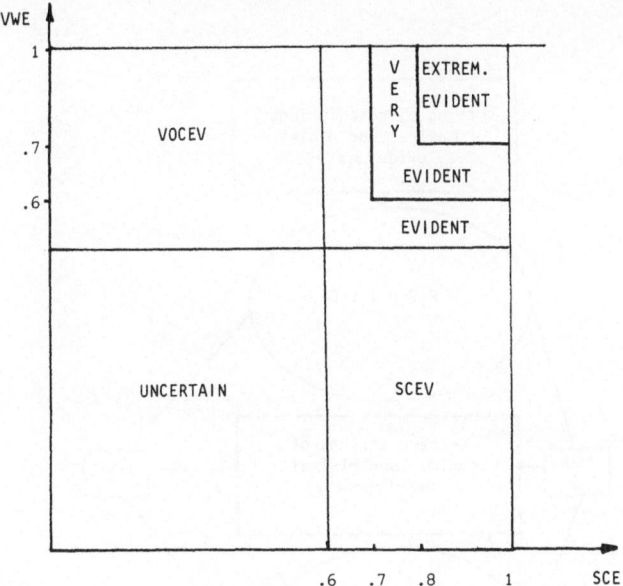

Fig. 7.16 Definition of linguistic evidences

7.8.3   Other possible methods
for hypothesis evaluation

The syllabic KS codes the speech waveform into an asynchronous sequence of sets of symbols.

Let $t(i)$, $i \in (1, I)$, be the i-th instant of time written in the syllabic directory, at which one or more syllabic hypotheses start.

A vector $L(i)$ of symbols can be associated with each $t(i)$. Each symbol in $L(i)$ represents a syllabic hypothesis and its evidence. Now, if a word hypothesis W is generated in the time interval (t_i, t_{i+k}), a sequence of vectors $SL(ik)$ can be associated with the hypothesis W. $SL(ik)$ contains all the vectors $L(j)$ corresponding to tj such that:

$$t_i \le t_j \le t_{i+k}$$

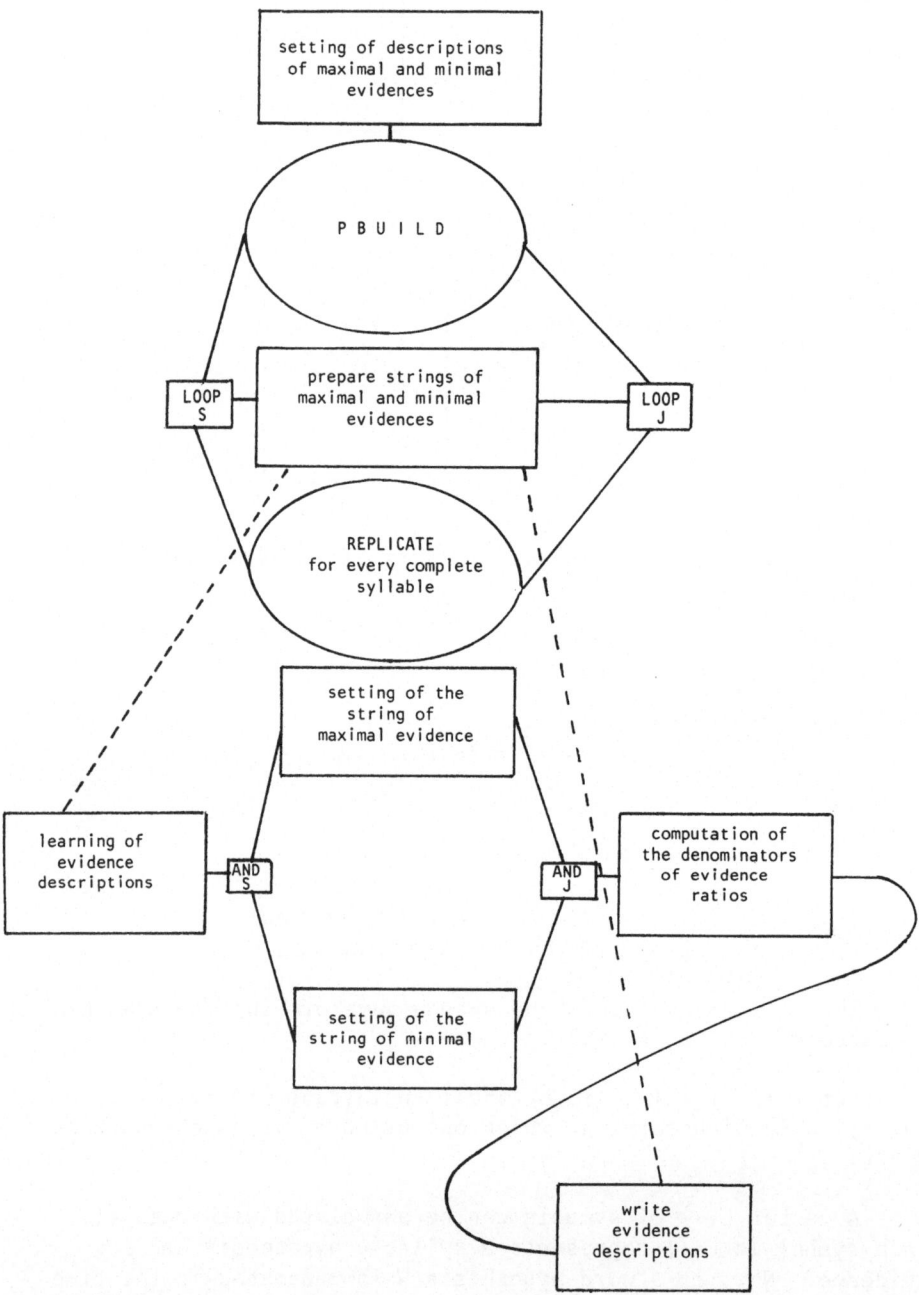

Fig. 7.17 Subplan for setting descriptions of maximal and minimal
          evidences

Now the a posteriori probability:

$$Pr(W/SL(ik)) = \frac{Pr(SL(ik)/W) \; Pr(W)}{Pr(SL(ik))} \qquad (7.26)$$

can be used as a measure of evidence. The probability (7.26) can be obtained by multiplying the a posteriori probabilities of the vectors in SL(ik) and an admissible search strategy, based on short-fall densities of logarithms of probabilities, can be applied here as in the HWIM system (Woods, 1977).

Another possibility is to use a sequential decoding method, following Jelinek (1976). To apply the sequential decoding algorithm, only the product $Pr(SL(ik)/W)Pr(W)$ has to be computed. This product can be computed by a Language Model in which the probability $Pr(W)$ may account for the expectation of the word W in a particular vista and the probability $Pr(SL(ik)/W)$ may also account for the probability of a base form in W. The calculation of the latter probability may be obtained by inferring a finite-state machine for each word. If $W = SYL1 \; SYL2 \; SYL3$, where $SYLi(i=1,2,3)$ are syllables and SL(ik) is composed of three vectors containing the three syllables in the right sequence, $Pr(SL(ik)/W)$ can be easily computed as follows:

$$Pr(SL(ik)/W) = Pr(L(1)/SYL1)Pr(L(2)/SYL2)Pr(L(3)/SYL3) \qquad (7.27)$$

where L(i) is the i-th vector of evidence symbols. Otherwise, probabilities of syllable insertion and deletion have to be considered.

Furthermore, let us assume that the evidence of the hypothesis SLYi is EV(i). The vector L(i) may be represented as follows:

$$L(i) = ((SYLi)(EV(i))) \; (ALPHA(i)) \qquad (7.28)$$

ALPHA(i) is the vector of the syllables competing with SYLi and their evidences.

With this in mind one gets:

$$Pr(L(i)/SYL(i)) = Pr(EV(i)/SYLi)Pr(ALPHA(i)/(EV(i) \\ SYLi)) \qquad (7.29)$$

The term $Pr(EV(i)/SYLi)$ can be obtained by a statistic of the evidences with which SYLi is recognized. The calculation of $Pr(ALPHA(i)/(EV(i)SYLi))$ is quite complex and can be obtained with some approximation. This approximation is required because the cardinality of the set of vectors L(i) which can be generated by the syllabic component is very high.

This approximation is the price paid by the attempt to have
an efficient acoustic processor. By contrast, an acoustic processor
can be used which is forced to generate a single phonemic or phonetic
symbol at a time, perhaps after performing a low level phone con-
solidation process.

Such a processor cannot take into account coarticulation effects
and acoustic ambiguities even if error probability distributions
can be computed at its output. The knowledge of error probabilities
of a system cannot lead to understanding a sentence if the errors
cannot be recovered from the redundancy of the message. The same
comment applies to those systems which are capable of generating
several phoneme hypotheses but with low average rates for the right
hypotheses.

The hypothesis evaluation method used in the project proposed
in this book is based on the opinion that statistical calculations
are always approximate and unnecessarily complex. Moreover, evidence
evaluations are more qualitative than quantitative, it is unlikely
that the human system is capable of carrying several hypotheses in
parallel (at most it can hold very few of them), expectations and
evidences are different dimensions of the problem represented as an
AND/OR graph of subproblems (time sequences can be considered as
conjunctions), and each terminal problem can be partially solved
with a degree of solution.

For these reasons, qualitative evidences will be used, based
on the evaluation of possibilities of the solution of phoneme subproblems and statistical methods will be used to decide which
hypotheses have to be kept or discarded.

7.9   Strategies for the Generation
of Lexical Hypotheses

After having met the preconditions and having selected the word
candidates, a set of word verification processes, one for each word
candidate, is activated. A verification process of a word candidate
verifies if the problem corresponding to such a word can be solved
with enough evidence by the syllabic hypotheses already written into
the blackboard and those which can be written by some syllable
verification processes whose instantiation is invoked by the lexical
KS.

Word verification consists in trying to match a goal-tree represented by the graph of the word and a fact-tree represented by
the structure of the hypotheses in the syllabic blackboard.

The rules for deciding the generation of word hypotheses are
given in the following.

Let MWj be the word hypothesis of V(j) which has achieved the maximum evidence; let EV(MWj) be such evidence. A word hypothesis W(w) is accepted if its evidence EV(W(w)) satisfies this relation:

$$EV(MWj)-DE<EV(W(w))<EV(MWj) \qquad (7.30)$$

where DE is a logical threshold defined on the basis of errors and false alarms statistics.

Fig. 7.18 shows an SSR for the vista V(j). The state tree represents the evolution of the problem solving activity from the configuration shown in Fig. 7.13.

The generation of lexical hypotheses starts from the active states generated by the matching of the preconditions. As there is a difference between the partial word recognition represented by

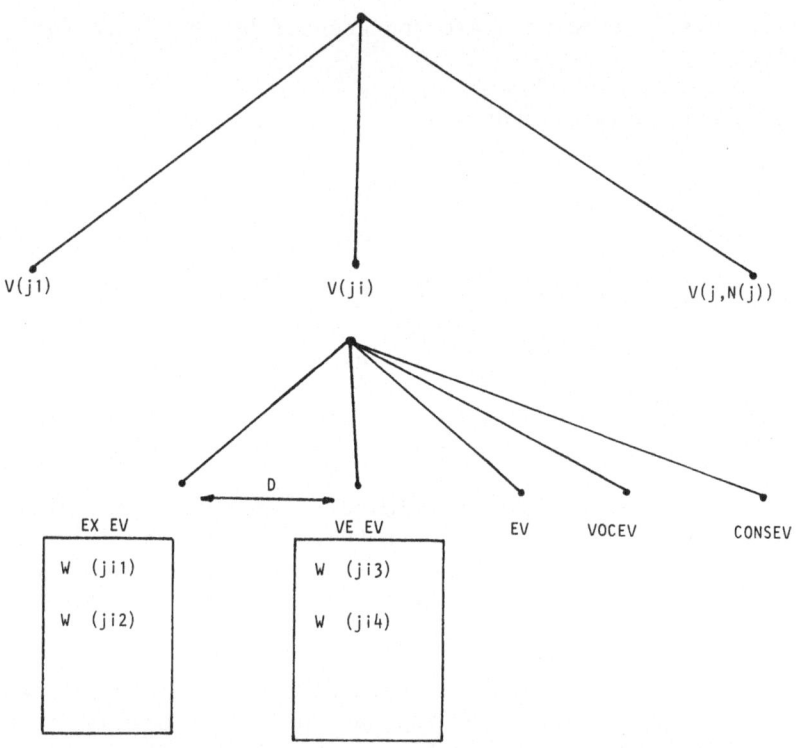

Fig. 7.18   Expansion of the state V(ji) in the SSR

the open nodes of the tree and the goal which is a sort of matching
between the words of $V(j)$ and the data structure of syllabic hypotheses, operators are applied to the partial recognitions represented
at the nodes in the SSR to attempt to reduce the difference between
the situation represented by the nodes and the goal. Each word may
be considered as a goal tree which is progressively reduced by a
"reducer" when its subproblems are solved. The syllabic data-
structure is a fact tree which may be expanded by an operator which
performs lexical-dependent generation of complex syllabic hypotheses.

The control strategy tries to apply operators and reducers
until the goal is reached. This approach, proposed by Nilsson (1977)
for theorem proving, is extended here to a multiple-task problem.
In fact, after each reduction of a goal tree, the evidence associated
with it is computed and a new node is generated in the state space.
At a given moment, only a subset of states is active, namely those
whose evidence is between the maximum evidence and this maximum
minus DE. To each node in the state space is also associated the
position of the goal tree which still has to be reduced.

Fig. 7.18 shows the expansion of the state labelled $V(j)$ in
the lexical SSR, which is a data structure associated with the lexical
control KS.

The linguistic evidences are represented by the following
abbreviations:

EX EV : extremely evident

VE EV : very evident

EV   : evident

VOCEV : only vowels are evident

CONSEV: only consonants are evident

The ordering of evidences corresponds to the order of possibility
that a word hypothesis is true, given an evidence judgement.

To each node labelled by an evidence judgement is associated a
table of word candidates. Only those tables for which (7.30) holds
are reported in Fig. 7.18. Further application of operators will
reduce the word candidates, leading to the word hypotheses shown at
the leaves of the SSR.

As it will be too complex to store goal trees at every step
of the generation of hypotheses, the sequence of solutions of the
subproblems of a lexical item is fixed and all the subproblems are

numbered. Thus a reduced goal tree is represented by a number corresponding to the first subproblem which is still unsolved. Furthermore, lexical-dependent operators are applied only when required by a subproblem which has to be solved.

Deletion or insertion errors are treated as follows. It is practically impossible that a syllable is completely deleted or inserted. Rather, it is possible that a vowel has been missed and a deformed syllable is produced by the analyzer. This is taken into account by considering deletions of vowels as allophonic variations of a word. No more than one vowel deletion per word is allowed, since the probability of more than one deletion in a word is negligible due to the good performance of the syllabic component. Other distortions may appear inside syllables and are represented as syllable degradations.

Because of the good performance of the syllabic component, only few candidate hypotheses usually remain active after application of the two types of preconditions described in the previous section. The main goal of the lexical problem solver is to write hypotheses into the blackboard only if they have reached a sufficiently high global evidence.

In order to obtain the global evidence of the word hypothesis, the sum of all the numerators of evidence ratios of the vowels is divided by the sum of the denumerators of the evidence ratios of the vowels, yielding a global vocalic ratio ERV. By doing the same with the consonants, a global consonant evidence ratio ERSC is computed. Then, using the diagram of Fig. 7.16, the evidence of the whole word can be obtained.

It is also possible that, given two words W1 and W2 such that W1=BETA W2, where BETA is a string of phonemes, the evidences of W1 and W2 turn out to be both very high. This is a peculiarity of the method, in accordance with what one may expect in such situations.

The system also has learning capabilities because the subplan for updating the lexical knowledge can improve the knowledge of each word and add new configurations in the syll-type tree. Furthermore, it can modify the duration interval which acts as a prosodic precondition for a word and the configuration of maximal and minimal evidences.

7.10 References

Basic papers for reviewing the problem of the lexical representation, lexical access, and relations with speech perception are those by Klatt (1979), Haton (1980), Erman and Smith (1981).

The structure of plans in a network and the comparison of such an approach with previous ones can be found in the book by Sacerdoti (1977).

The possible use of prosody for the word detection is discussed in a book chapter by Vaissière (1980).

Perceptual considerations can be found in the book edited by Cole (1980).

Content addressing is treated in a recent book by Kohonen (1978).

8. ON THE STRUCTURE AND USE OF
TASK-DEPENDENT KNOWLEDGE

8.1 Introduction

Task-dependent knowledge contains syntax, semantics, and
pragmatics. Its purpose is the solution of ambiguities and recovery
from errors which still remain at the lexical level. Furthermore,
task-dependent knowledge may be used to transform the surface
structure of a recognized sentence into its conceptual representation.

Notice that even in a sequence of perfectly recognized words
a word may have several meanings.

This set of meanings can be filtered by using the task-dependent
knowledge.

The use of task-dependent knowledge to interpret a sentence
shares with important items of Artificial Intelligence (AI) the
need to create programs capable of performing tasks normally considered to require intelligence (Charniak, 1976).

In the case of speech understanding, the "intelligent program"
has to use a knowledge for performing inferences in order to ask
the question, "What is the meaning of the sentence carried by the
speech signal?" The system uses knowledge which is represented at
various levels.

At each level the data to be used should be represented in a
proper way in order to trigger the proper inference rules.

Problems of knowledge representation, inference, triggering,
organization of the rules to be applied, inference mechanisms

specifying the type of interference rules, and world representation are common to question-answering and speech understanding systems.

A further level of complexity characterizes speech understanding systems. A large variety of speech patterns may lead to the same interpretation and several speech patterns may not have enough redundancy to compensate for imprecision, giving rise to more than one interpretation.

Thus, inference triggering, i.e., the problem on when and why inferences have to be made, has a natural solution in the framework of concurrent and cooperating processes. Furthermore, inferences cannot be based merely on binary logics and the inference mechanism has also to provide rules for combining evidences of hypotheses.

From the point of view of theorem proving (Nilsson, 1977) the lexical hypotheses are facts, and some of them can be seen as the elements of a fact-tree which has to be matched with a goal-tree representing an interpretation.

The complexity of the speech understanding task is high because for a single spoken sentence, there are many possible "ill-defined" fact-trees which should be matched with a large variety of goal-trees, some of which can also be ill-defined.

It is too difficult, at the moment, to propose a general solu-tion to the speech understanding problem. Rather, valuable solutions have been proposed for restricted tasks.

Task-dependent knowledge may be represented in several ways, depending on many factors, the most important of which is its complexity. The simplest way is to represent it by a finite-state automaton whose transitions are labelled by a word symbol.

If the task is simple, the network of states and transitions is small.

In this case, each word can be represented by a network whose transitions are labelled by phone symbols. This network is a concise representation of the Problem Reduction Representation (PRR) of the understanding task.

The inference rules are organized in such a way that inferences proceed along the network when a direct matching is performed between an arc label and a hypothesis on the blackboard. Because of impre-cision and ambiguities, several chains of inference may be active at a given moment.

A State Space Representation (SSR) of the problem is a tree whose nodes correspond to partial paths of the network. To each state is associated a score which may be a probability, a degree of possibility, or a purely heuristic measurement indicating the degree of matching between the input data and the interpretation represented by the path of the network corresponding to the node. The goal state in the SSR is the one corresponding to the best scored path in the network going from the initial to the final state of the automaton.

When evidences are measured by probabilities, the automaton can be a stochastic one, allowing a priori probabilities to be computed for each sequence of symbols corresponding to a path. This automaton is often referred to as language model (Jelinek, 1976).

Search in the state space can be performed by several strategies which will be briefly reviewed in this chapter.

Another section of this chapter will be devoted to the evaluation of language complexity.

For very complex protocols, task-dependent knowledge has to be kept separate from the phonetic and phonemic knowledge and, perhaps, syntax, semantics, and pragmatics have also to be kept separate. In this case we are faced with problem solving in domains which must use large amount of diverse, error-prone, and incomplete knowledge in order to search in a large space. For these problems, "opportunistic" strategies must be used because there may not be a computationally feasible definition of the solution space. The task of understanding the data is accomplished at various levels of analysis. At each level, the units of analysis are the hypothesis elements.

Integrated knowledge models, like those used in the HARPY system (Lowerre, 1976) or the IBM system (Jelinek, 1976) have given better performances than the systems with distributed knowledge, like the HEARSAY-II (Reddy et al., 1977). This comparison has been based on a very constrained protocol having a lexicon of about 1000 words.

Even if no sufficient and significant experiments have been made on these systems, one may argue that integrated systems may perform better on very constrained protocols. Generally they require a complex tuning whenever the protocols or the speaker change. Distributed knowledge systems, although more complex, are potentially more powerful and suitable to be easily adapted to different speakers and protocols. Moreover, they represent the only available model for very complex tasks and their investigation is valuable research, even for models with simple protocols.

The most interesting aspects of the solution proposed in this chapter are the organization of the use of semantic knowledge and evaluation of priorities based on Approximate Reasoning.

Semantic knowledge is organized in a Semantic ATNG whose arc transitions are conditioned by a direct match between key words or phrases and lexical hypotheses.

A direct match may cause a "pattern directed invocation" of a semantic subnetwork. This invocation can be repeated recursively until an action is performed which invokes the syntactic knowledge (which is also represented by an ATNG) for filling up, mainly with function words (such as prepositions), the gaps in the interpretation of the spoken sentence which have been left after the application of the semantic rules.

Learning can also be performed by updating the semantic and syntactic ATNGs.

8.2   Finite-State Language Models

A simple finite-state language model is a finite-state machine whose states correspond to sets of partial interpretations and whose arc labels are words.

Fig. 8.1 shows an example of a finite-state language model. The initial state I corresponds to the beginning of the set of acceptable sentences. A sentence may begin with word W_i, $i \in (1,6)$. The words W_1, W_2, W_3 define a set of words $W'(01)$.

The automaton can be seen as a model for generating the language L which can be recognized using the automaton as language model. Let us consider a sentence $SEN = SW_1\ SW_2...SW_k$.   If SEN belongs to L, SW_1 must belong to $W'(01)$ or $W'(02)$.   If $SW_1 \in W'(01)$, the automaton, after generating SW_1, is in the state S_1 and the next word it may generate has to belong to either $W'(11)$ or $W'(12)$.   Thus SW_2 must belong either to $W'(11)$ or $W'(12)$. Assume that to each word of each set is associated a proper arc. In other words, if $W'(ij)$ contains $N(ij)$ words and is the label for the arc joining state S_i with state S_k, $N(ij)$ arcs can be established between S_i and S_k with a one-to-one correspondence between these arcs and the words in $W'(ij)$. Under this assumption, there will be a one-to-one correspondence between sentences in L and paths joining the initial and final states of the automaton. Ignoring for a moment the coarticulation problem, or reducing its complexity by a simplifying assumption, a word can be represented by a finite state automaton whose arc labels are phone symbols belonging to a small alphabet.

Fig. 8.2 shows, as an example, a very simple language model (Fig. 8.2a).

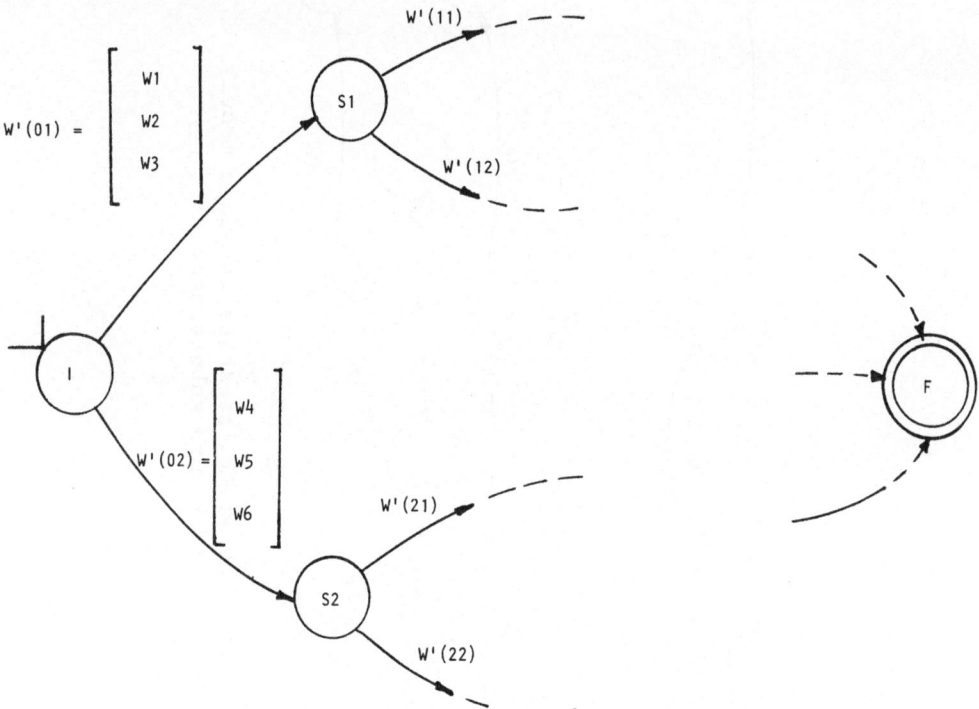

Fig. 8.1 Finite-state language model

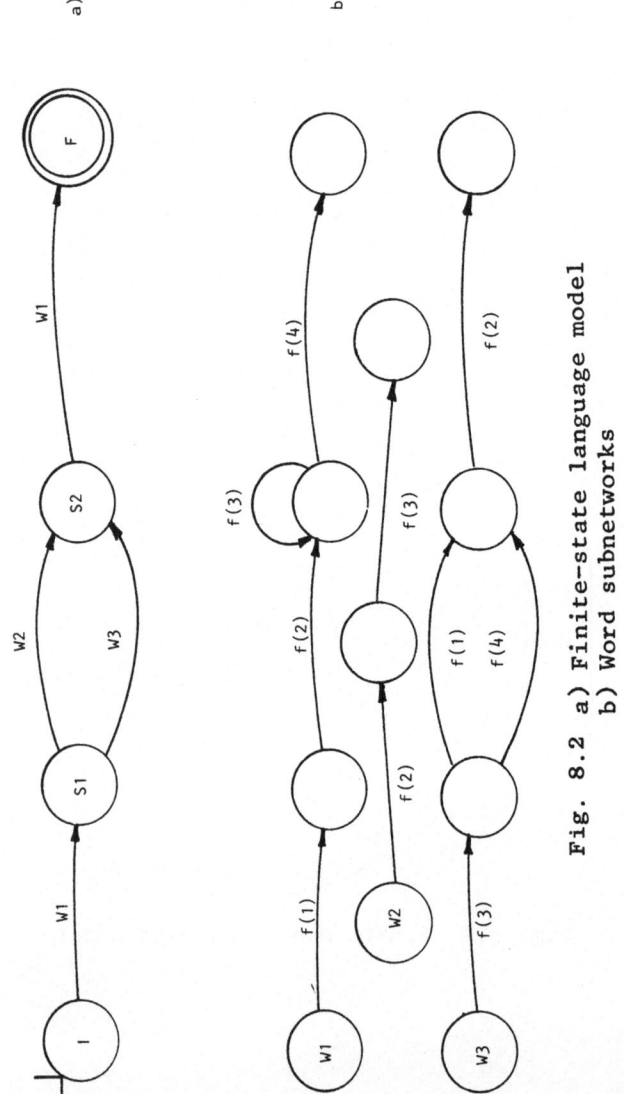

Fig. 8.2 a) Finite-state language model
         b) Word subnetworks

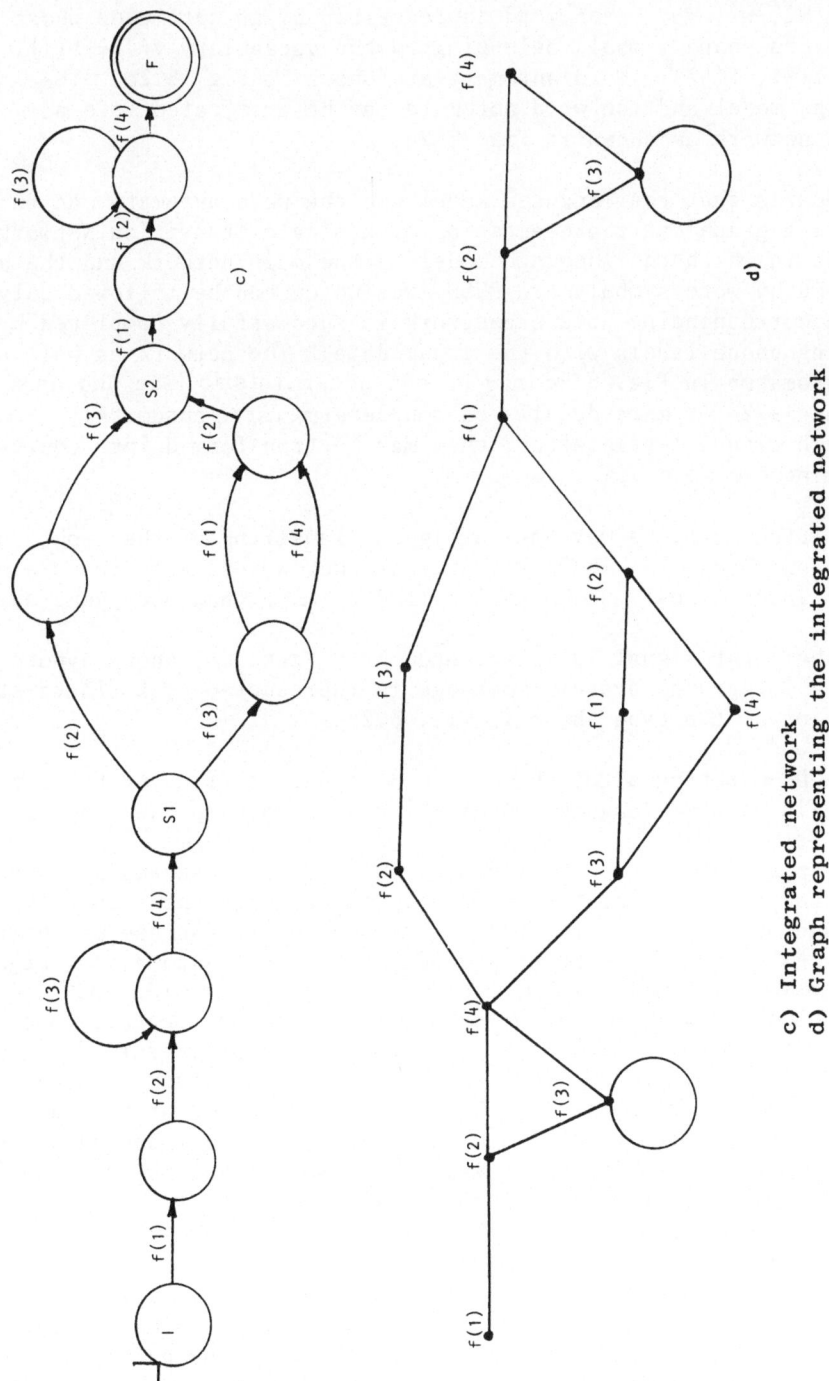

c) Integrated network
d) Graph representing the integrated network

   The language model is capable of generating two sentences:
SEN1 = W1 W2 W1 and SEN2 = W1 W3 W1. The lexicon VW contains three
words: W1, W2, W3. Each word is described by an automaton whose
labels are phone symbols belonging to the vocabulary Vf = {f(1),
f(2), f(3), f(4)}. Word automata are shown in Fig. 8.2b. The
language model and the word automata can be integrated into a
single network as shown in Fig. 8.2c.

   Notice that the language model and the word automata can be
seen as a graphical representation of a single Transition Network
Grammar in which the language model is the main network and the arcs
labelled by word symbols are PUSH arcs which can be followed only
if the corresponding word subnetwork is successfully completed by
matching phone labels with the input data. The network as well as
the automaton in Fig. 8.2c may be non-deterministic (see Aho and
Ullman, 1972 for more details on non-deterministic automata),
although with a little effort they may be transformed into equivalent
deterministic automata.

   Furthermore, we may have ambiguous labelling of the input data
with phone symbols. There may be, in other words, time intervals
of the input signal which are labelled by more than one phone symbol.

   The input signal is represented by a lattice of phone hypotheses
and the integrated system knowledge is represented by a finite-state
automaton of the type shown in Fig. 8.2c.

   A DP-matching algorithm can be designed for finding the path
in the integrated language model which best matches the input lattice.

   For this purpose, the language model is represented by a graph
whose nodes have a one-to-one correspondence with the arcs of the
language automaton and there is a one-to-one correspondence between
sequences of nodes in the graph and paths in the automaton. Figure
8.2d shows the graph equivalent to the automaton of Fig. 8.2c.
Figure 8.3a shows a simple graph representing the language model;
Fig. 8.3b represents a lattice of phone hypotheses describing an
unknown input signal and its representation by a graph. The nodes
of the language graph $f(i)$, $i \in (1,F)$, and the nodes of the lattice
of phone hypotheses graph $l(j)$, $j \in (1,J)$ can be represented along
two orthogonal axes in such a way that a set of points SM is defined
containing the points of coordinates (i,j) $\forall i \in (1,F)$, $j \in (1,J)$.
Let $h(i,j) \in$ SM be a point in SM. Fig. 8.4 shows an example of
such points. Every point $h(i,j)$ for which $j = 1$ or $j = 2$ or $j = 3$
and $i = 1$ can be a starting point of a path in the (i,j)-plane.
Every point $h(i,j)$ with $j = 6$ or $j = 7$ and $i = 6$ can be an end
point for a path. Let S be the set of the starting points and E be
the set of end points.

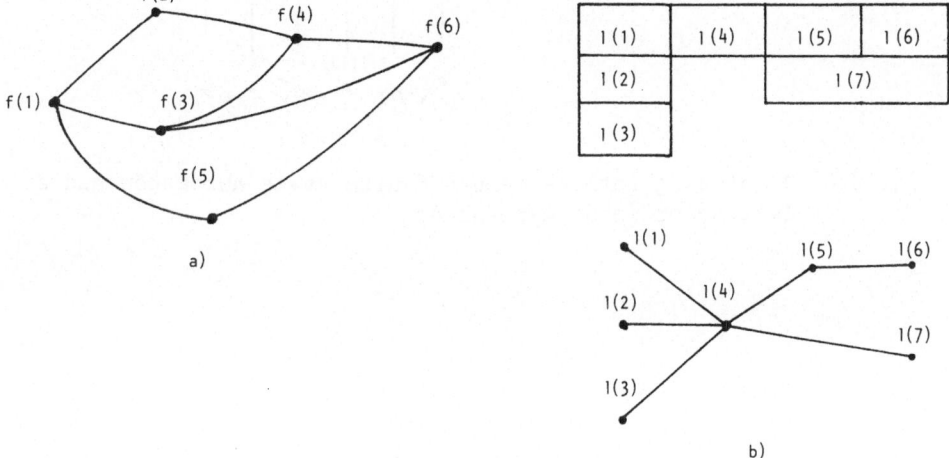

Fig. 8.3   a) Simple graph representing a language model
           b) Lattice of phone hypotheses and its corresponding
              graph

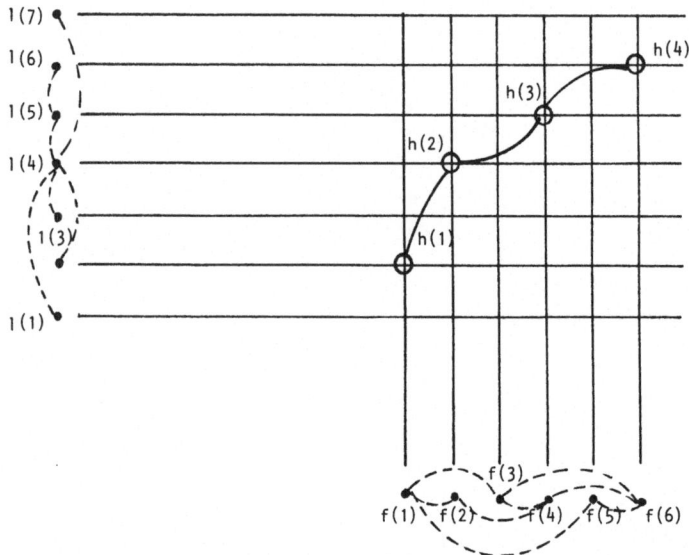

Fig. 8.4   A matching path between a finite-state automaton and a
           lattice of input hypotheses

An interpretation H of the input signal is represented by a
path in the (i,j)-plane such that:

$$H = ((i(1),j(1))(i(2),j(2))...(i(k),j(k))...(i(K),j(K))) \qquad (8.1)$$

where:

8.2.1   the sequence $(i(1), i(2)...i(k)...i(K))$ corresponds to a
        path from the initial node to the final node in the language
        graph.

8.2.2   the sequence $(j(1), j(2)...,j(k)...,j(K))$ corresponds to a
        path from an initial node to a final node in the graph of
        the input lattice.

Notice that the above two conditions imply that:

- $h\ (i(k),\ j(k)) \in SM \forall\ k,$

- $h\ (i(1),\ j(1)) \in S,$

- $h\ (i(K),\ j(K)) \in E.$

For the sake of simplicity let:

$h\ (i(k),\ j(k)) = h(k).$

A path H can be represented as:

$$H = h(1)\ h(2)...h(k)...h(K). \tag{8.2}$$

Each admissible path H can be assigned a degree of evidence which may be assumed to be the sum of the scores assigned to each point of the path. These scores can be, for example, degrees of similarity or logarithms of a priori probabilities.

For example, the following path is an admissible one in the situation shown in Fig. 8.4:

$$H = h(1)\ h(2)\ h(3)\ h(4)$$

where:

$h(1) = h(1,2)$

$h(2) = h(2,4)$

$h(3) = h(4,5)$

$h(4) = h(6,6).$

The path corresponds to matching the input sequence:

$$I = (1(2),\ 1(4),\ 1(5),\ 1(6))$$

with the sentence of the language:

$$SEN = (f(1),\ f(2),\ f(4),\ f(6))$$

The score $EV(k)$ associated with the node $h(i(k),\ j(k))$ is the degree of similarity between $f(i(k))$ and $1(j(k))$.

The constraint that H has points whose coordinates correspond to a path in the language graph and to another path in the lattice graph is motivated by the fact that the input lattice should contain the correct sequence of phone bounds even if it may also contain wrong bounds.

If the performances of the acoustic processor are not good enough for the correct sequence of phone bounds to be present in the lattice, constraints 8.2.1 and 8.2.2 have to be removed.

In this case, sequences of points in the (i,j)-plane may have other types of constraints.

A constraint may be that the total number of insertions and deletions of symbols in the input lattice has to be lower than a fixed percent of the maximum length of the sequences in the input lattice.

If insertions and deletions are admitted, the evidence $EV(k)$ of the k-th point in H is obtained by the composition of the similarity

$$D(k) = D(f(i(k)), 1(j(k)))$$

and the loss $L(k)$ due to an insertion or a deletion. If $D(k)$ and $L(k)$ are the logarithms of probabilities, composition corresponds to an algebraic summation.

The case of deletion is the one in which passing from $h(k-1)$ to $h(k)$ implies that one label $f(i) \in FD(k)$ is skipped in a path of the language graph. $FD(k)$ is the set of symbols which can be deleted at the k-th point in H.

Furthermore, assume that the probability of two consecutive deletions is negligible.

Let $L(i,k)$ be the loss due to the deletion of $f(i) \in FD(k)$ between $f(i(k-1))$ matched with $1(j(k-1))$ and $f(i(k))$ matched with $1(j(k))$. Let LK be the set of the indices of $f(i) \in FD(k)$.

Assuming the values of the loss function to be negative numbers, one gets:

$$L(k) = \max_{i \in LK} L(i,k) \qquad\qquad (8.3)$$

Analogously, we may have an insertion if a symbol $1(j)$ is skipped when passing from $h(k-1)$ to $h(k)$. In this case, $L(k)$ will correspond to the maximum score due to the insertion of a symbol $1(j)$ chosen among those which are between $1(j(k-1))$ and $1(j(k))$ in the graph of the input lattice.

8.3 Measuring Evidences

An evidence measure which has been widely used in SUS is derived from the evaluation of the a posteriori probability (Jelinek 1976). We will show here how this approach can be applied in the framework proposed in this book. When words have been hypothesized and written into the blackboard, we may assume that the input signal has been coded into a sequence $D(S)$ of discrete events:

$$D(S) = a(t(1)), a(t(2)),\ldots,a(t(j)),\ldots,a(t(J(S))),$$

where each $t(j)$ is the beginning of a time interval in which one or more phonemic hypotheses have been generated.

Assume that at time $t(j)$ the phonemic hypothesis:

$$a(t(j)) = f(j,1), f(j,2),\ldots,f(j,k),\ldots,f(j,K(j)) \quad j \ (1,(J(S)),$$

has been generated. Each phoneme hypothesis $f(j,k)$, $k \in (1,K(j))$ is assigned a degree of membership $\mu(j,k)$. Let $M(j)$ be the vector of the $\mu(j,k)$, $k \in (1,K(j))$.

Now, let the speech understanding system have an integrated knowledge represented by an automaton whose arcs are labelled by phoneme or phone symbols. If we are interested in dealing with possible deletions, insertions, and substitutions, we may transform the automaton into a stochastic one, in which all the real situations which may cause the transitions from one state to another are represented, together with their a priori probabilities.

What is obtained in this way is a Markov source, which is basically a collection of states and transitions with a probability associated with each transition between two states. Furthermore, the source can be seen as a model for generating all the possible sequences $D(S)$ of discrete events describing the speech waveform together with a degree of evidence $\mu(j,k)$ for every phoneme $f(j,k)$ of every discrete event $a(t(j))$. Thus, to each transition of the Markov source between state $S(i)$ and state $S(n)$ may be associated a probability:

$$Pr((\mu(1), \mu(2),\ldots, \mu(\ell),\ldots, \mu(L))/(S(i) \to S(n))) \qquad (8.4)$$

where $\mu(1)$ is the degree of evidence of the phoneme $f(1)$, $1 \in (1,L)$ and L is the number of different phonemes in the language.

Now, for each event $a(t(j))$, $j \in (1,J(S)$, the syllabic knowledge source generates a vector:

$$M(j) = \{\mu(j,1),\ldots,\mu(j,k),\ldots,\mu(j,K(j))\}$$

For each vector M(j) and for each transition S(i) → S(n), a probability value

$$Pinj = Pr(M(j) /(S(i) → S(n)))$$ (8.5)

can be obtained.

The sequence D(S) can be represented by the sequence of vectors M(j), j∈(1,J(S) which can be seen as a message generated by a Markov source like the one shown in Fig. 8.5.

The Gin are the a priori probabilities of moving from state S(i) to state S(n) with a sequence of output vectors M = M(1)..., M(j),...,MJ(S).

The probability of having M(j) at the output of the Markov source during the transition S(i) → S(n) is Pinj. Dashed lines represent transitions with no output; their associated a priori probabilities are Din.

In a language model having a single integrated knowledge source, states may represent a partial interpretation of a sentence, while transitions correspond to phoneme labels. Fig. 8.6 shows an example taken from Bahl et al. (1980). The automaton shows the phonological model of the word "band". The model can be enriched by adding loops corresponding to insertions and dashed lines indicating transitions with no outputs (deletions).

Let us now try to show how the algorithms for phoneme hypothesization, introduced in Chapters 4-6, could be applied in a Markov model.

For each transition labelled by a phone symbol, the probability:

$$Pr((μ(1), μ(2),..., μ(ℓ),..., μ(L))/(S(i)→S(n)))$$

refers to the evidences of all the phones or phonemes of the alphabet, given the pronunciations of the single sound corresponding to the label for the transition S(i) → S(n). In practice, μ(ℓ) can assume only a few discrete values and the a priori probabilities (8.4) can be inferred by experiments for all the combinations of these values. Given a vector M(j), the Pinj can be obtained from one of these combinations.

All the inference of the syllabic knowledge source has been done with the attempt of using context-dependent and speaker-independent rules for obtaining a reduced number of phoneme candidates with evidences that are independent of the context and of the speaker.

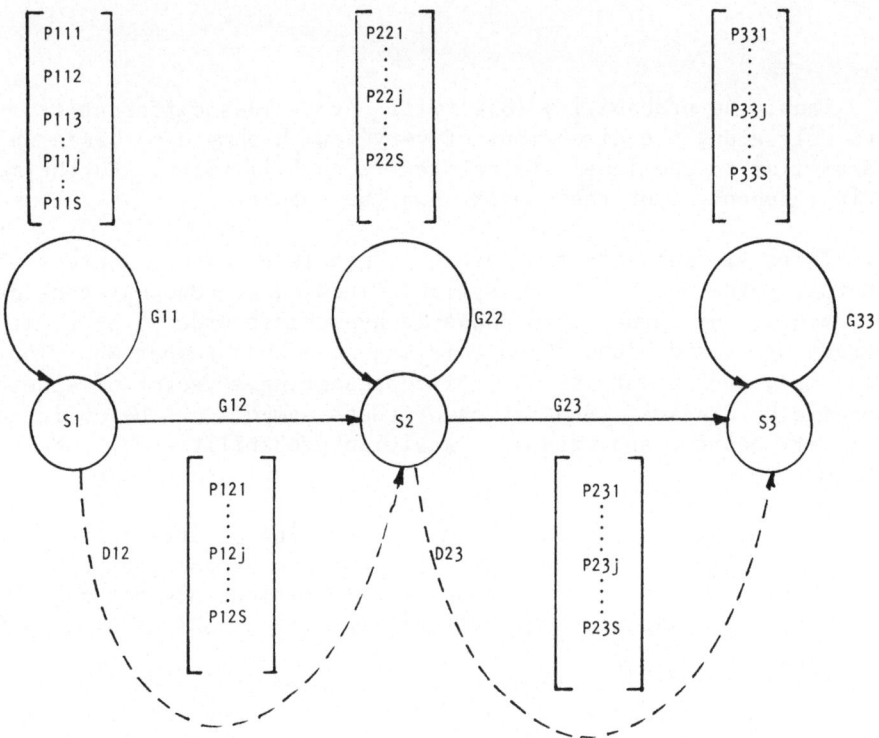

Fig. 8.5 Model of a Markov source

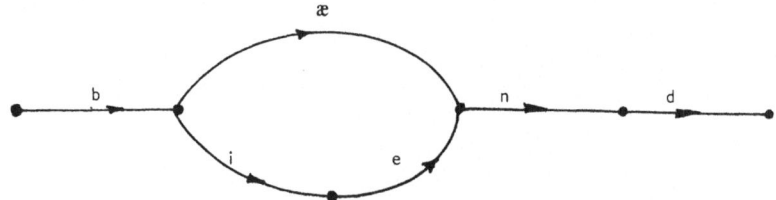

Fig. 8.6 Phonological model of the word "band"

Thus, the probability (8.4) will have values different from zero only along the dimensions of very few phoneme hypotheses which are similar to the label of the transition $S(i) \rightarrow S(n)$. Furthermore, it is independent of the context and the speaker.

There are many other alternative models of Markov sources for generating the output of the syllabic knowledge. One may consider, for each j, the sequence of phonemic hypotheses ordered by their degrees of evidence and bound this sequence by a number NB. In this case, each event is a symbol representing a vector of ordered phonemic hypotheses. Statistics of these symbols can be collected for every phoneme and used as transition probabilities of a Markov source.

Given a stochastic automaton representing an integrated model of the language and the generator of phonemic symbols, hypothesis evaluation may be performed using two statistical algorithms, one of which is derived from the Viterbi algorithm (Bahl et al., 1980).

The first statistical algorithm, denoted SAL1, consists in finding the probability of reaching a state through all the possible paths starting from initial states, given a sequence M of vectors. This sequence can be seen as the message generated by the source with symbols $M(j)$.

The algorithm SAL1 consists in constructing a lattice in which a node $ND(i,j)$ corresponds to a pair $(S(i),M(j))$, composed of a state of the source and a vector of phoneme evidences.

Each node $ND(i,j)$ can be assigned a probability $Pr(i,j)$. This probability is computed as follows:

$$Pr(i,j) = \sum_{b \in B(i,j)} Pr(ND(b)) \cdot Pr(ND(b) \rightarrow ND(ij)) \quad (8.6)$$

where b is a pair of indices and $B(i,j)$ is the set of the indices

of all the predecessors of $ND(i,j)$. The following boundary condition is assumed:

$$PR(S(0),\emptyset) = 1,$$

$\emptyset$ is the null symbol.

To each node $ND(i,j)$ in the lattice is associated a state $S(i)$ and
an input vector $M(j)$; we may assume that a state $S(b)$ is associated
with the node $ND(b)$. From these considerations one gets:

$$Pr(ND(b) \rightarrow ND(i,j)) = Pr(S(b) \rightarrow S(i)) \cdot Pr(M(k)/(S(b) \rightarrow S(i))) \quad (8.7)$$

where $M(k)$ is either the input vector $M(j)$ or the null vector $\emptyset$ if
$ND(b) = ND(1,j)$. $M(j)$ is indicated in the lattice diagram as the
symbol generated by the source in a transition from $S(b)$ to $S(i)$.

Example 8.3.1

Fig. 8.7a shows a Markov source generating strings with a
terminal alphabet:

$$V_T = \{x,y\}$$

States are labelled respectively $S(0)$, $S(1)$, $S(2)$.

This is equivalent to saying that there are only two types of
input vectors other than the null vector $\emptyset$, namely X, which reduces
to the symbol x, Y which reduces to the symbol y.

To each state transition is associated the probability:

$$Pr(S(u) \rightarrow S(v)); \quad u,v \in (1,2,3)$$

and a vector of probabilities:

$$Pr(x/(S(u) \rightarrow S(v)))$$

$$Pr(y/(S(u) \rightarrow S(v))).$$

Let $M = x\ x\ y\ y$ be an observed sequence of symbols. Fig. 8.7b
shows the lattice made by algorithm SAL1. To each transition in
the lattice is associated a probability computed according to (8.7).
The second statistical algorithm is denoted SAL2 and consists in
finding the best path PS from a state $S(0)$ to a state $S(n)$ when a
sequence M of vectors of evidences is observed. The best path is
the one for which the following a posteriori probability is maximized:

$$P(PS/M) = \frac{P(M/PS)\,P(PS)}{P(M)} \quad (8.8)$$

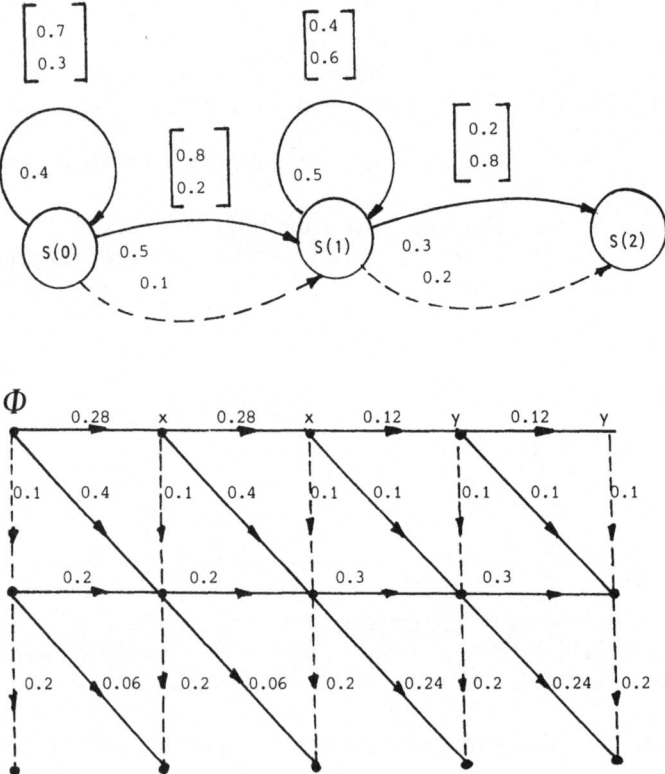

Fig. 8.7 Example of the application of the algorithm SAL1

As $P(M)$ is the same for all the paths PS examined, the best path is the one for which

$$P(M,PS) = P(M/PS)P(PS)$$

is maximized. Assuming, for a moment, that there haven't been insertions or deletions, there will be N-1 vectors M(j) and the likelihood of a path could be computed as follows:

$$P(M,PS) = P(S(0)) \prod_{j=0}^{N-1} P(M(S)/(S(j) \to S(j+1))) P(S(j) \to S(j+1)) \quad (8.9)$$

If insertions and deletions are also allowed, a lattice like the one in Fig. 8.7 has to be built and the best path in the lattice has to be looked for.

The algorithm for this search is the Viterbi algorithm, whose basic recursion is the calculation of the maximum likelihood at a node ND(i,j) of the lattice. Let LK(i,j) be such a likelihood; the basic recursion is:

$$LK(i,j) = \underset{b \in B(i,j)}{Max} LK(b) \ Pr(ND(b) \to ND(ij)) \quad (8.10)$$

LK(i,j) represents the evidence of the most likely sequence of transitions which generated the sequence M(1), M(2),...,M(j), taking the system to state S(i). This algorithm is particularly useful for aligning corrupted strings with prototype strings. B(i,j) is the same as for (8.6).

Example 8.3.2

Fig. 8.8 shows the values of LK(i,j) computed by (8.10) for the same input string and Markov source as shown in Fig. 8.7. From these data, one can conclude that the most likely generation of M = x x y y is (S(0) → S(0),x)(S(0)→S(1),x)(S(1)→S(1),y)(S(1)→S(1),y) which shows three insertions:

(S(0) → S(0),x) and twice (S(1) → S(1),y).

8.4   Search Strategies

8.4.1 Branch-and-bound algorithms

The search for the best path through the lattice of Fig. 8.8 is a dynamic programming problem.

The computational complexity involved in attempting to solve problems of this type becomes too large if the number of nodes is high. A reduction in complexity can be achieved by attempting to avoid visiting some of the nodes for which one can infer that they cannot belong to the best path. A type of algorithm for solving these problems is known as *branch-and bound*. A general presentation of these algorithms has been provided by Hall (1971).

φ	x	xx	xxy	xxyy
	S(0) x 0.28	S(0) x 0.08	S(0) y 0.009	S(0) y 0.001†
S(1) ∅0.1	S(0)∅ 0.028 S(0)x 0.4 ―――――― S(1)x 0.02	S(0)∅ 0.008 S(0)x 0.112 ―――――― S(1)x 0.08	S(0) ∅0.009 S(0) y 0.008 S(1) y 0.08 ――――――	S(0)∅ 0.0001 S(0) y 0.0009 ―――――― S(1) y 0.01
S(1) ∅ 0.02	S(1)∅0.08 ―――――― S(1)x0.006	S(1)∅0.08 S(1)x0.024	S(1)∅0.0067 S(1)y0.027	S(1)∅0.002 S(1)y0.008

Fig. 8.8 Example of the application of the algorithm SAL2

Consider the problem of minimizing a function h in some discrete set U, i.e.:

$$h: \quad U \to R \tag{8.11}$$

(R is the set of real numbers). We wish to find $u^* \in U$ such that:

$$h(u^*) \le h(u), \forall u \in U.$$

Suppose that we have available a function 1 which computes a lower bound for h in a subset $U' \subseteq U$. Thus:

$$1(U') \le h(u) \quad \forall u \in U' \text{ and } U' \subseteq U.$$

Furthermore, when U' is a single element a:

$$1(U') = 1(\{a\}) = h(a).$$

Notice that the following properties hold for bounding functions:

$$1(U') \ge 1(U'') \text{ if } U' \subseteq U''$$

$$1(U' \cap U'') \quad \max (1(U'), 1(U'')).$$

Let PAR be a *partition* of a set U.

PAR is a set of subsets of U which are mutually exclusive and exhaustive, i.e.:

$$PAR = \{U_1, U_2, \ldots, U_n / U_i \in U \text{ and } U_i \cap U_j = \emptyset \text{ if } i = j$$

$$\text{and } \bigcup_{i=1}^{n} U_i = U\} \tag{8.12}$$

$\emptyset$ is the null set.

The branch-and-bound algorithm sets a partition PAR on U, then it finds the subset U_k for which $l(U_k)$ is minimum. U_k is further partitioned into subsets. The subset for which the lower bounding function is minimum is then considered and partitioned. This cycle is repeated until one or more subsets are found containing a single element and for which the lower bounding function is minimum.

The algorithm is most efficient when l computes the greatest lower bound g*; in this case it makes no unnecessary partitions; furthermore, if l were constant for all sets of more than one element, the search would ultimately have to evaluate h at every point, making an exhaustive search.

One could even contemplate using any estimate of the greatest lower bound, recognizing that if the bound condition is violated a non-optimal solution may be returned, but the error could be no more than the error in the estimator.

The search can be forced to refine partitions until a possible solution point has been found and then to continue searching until the minimal point is found; one way to do this would be to subtract a depth factor from the bounds to bias the search towards depth.

If the search algorithm is conceived in such a way that an early termination does not prevent finding the optimum, the algorithm is said to be *admissible*.

Branch-and-bound algorithms can be applied to graph searching.

Knowledge representations used by a problem solver may be *weighted directed graphs* (WDG).

Definition 8.4.1

A *weighted directed graph* (WDG) is an ordered set (N,E,W), where N is the set of *nodes* of the graph, E is a set of ordered pairs of nodes called *edges* ; W is a *weight* function mapping E to the non-negative real numbers.

Nodes can be represented by points.

An edge (a,b) is a line going from point a to point b.

We may also say that $w(a,b)$ is the length of edge (a,b).

Definition 8

A *path* S between two nodes s and t is a sequence of nodes:

$$S(s,t) = (s = n(0),n(1),\ldots,n(k) = t)$$

and the length of S is defined as follows:

$$w(S(s,t)) = \sum_{j=1}^{k} w(n(j-1),n(j)) \qquad\qquad (8.13)$$

Using the absolute values of the logarithms of the probabilities as weights of the edges, the definition of the length of a path is increasing as the a priori probability associated with that path decreases.

The set U to be searched is the set of all possible loop-free paths from s to t and the subsets of paths which will be used in partitions of U will be those which start at s and have an initial part in common.

The function 1 will give the length of the path in common. Partitioning will be done by extending the initial path by one edge for all possible edges, eliminating those which introduce loops; each of these extended initial paths defines a subset of the refined partition. A search procedure of this type has been proposed by Dijkstra (1959).

A heuristic graph search can be performed with the branch-and-bound algorithm.

Suppose we have a heuristic function $h(a,b)$ which estimates the distance through the graph from a to b.

If:

$$h(a,b) \leq d(a,b)$$

where $d(a,b)$ is the true minimal distance through the graph from a to b, and using a bounding function:

$$1(s,...,a) = w(s,...,a) + h(a,t) \qquad\qquad (8.14)$$

we have a lower bound for the set of all paths initially along
(s,...,a) and ending at t.

When w is the magnitude of the log of an a priori probability,
Jelinek (1976) suggests using for h(a,t) the magnitude of the log
of the a priori probability, computed from the language model,
of reaching t through any possible path from a.

Graph search can be represented by a tree in which to each
node corresponds a set of paths going from s to t and having the
path (s,...,a) in common. To each node of the tree is associated
a score equal to the value of the bounding function 1 (s,...,a),
calculated on the common sub-path (s,...,a).

Details on search algorithms can be found in the book by
Nilsson (1971).

A *depth-first-search* is achieved by ranking the nodes according
to the values of the bounding function and expanding first the node
with minimum value of the bounding function.

Other search methods derived from branch-and-bound are the
breadth-first and the *depth-first*.

The *breadth-first* method consists in expanding first all the
level-1 nodes, i.e., those following the root, obtaining the level-
2 nodes, and so on.

The *depth-first* method consists in expanding first the first
unexpanded node of maximum level.

Fig. 8.9 shows examples of the best-first, breadth-first, and
depth-first methods.

Inside each node is shown the order of expansion, while close
to each node but outside there is the node score.

8.4.2   Non-admissible search algorithms

For those models for which the knowledge is integrated into a
single network, it is often very hard to compute the values of a
bounding function at each state because the number of states is very
high. For systems of this type, such as the HARPY system, non-
admissible search algorithms have been proposed. The non-admissible
search algorithms are those which do not guarantee that the optimal
solution of a problem is always found.

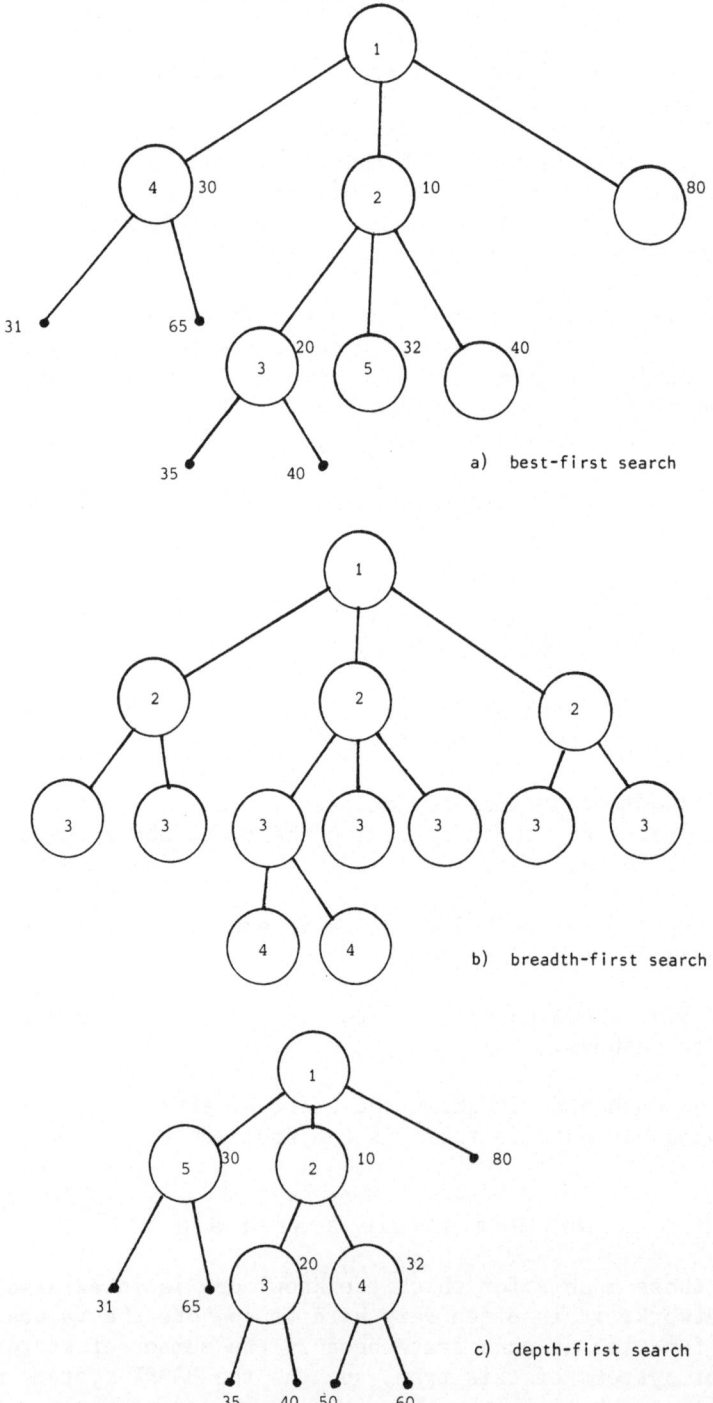

a) best-first search

b) breadth-first search

c) depth-first search

Fig. 8.9 Examples of best-first, breadth-first, and depth-first
          algorithms.

The HARPY system was designed by Lowerre (1976) and has recently been discussed in a framework of human cognition by Newell (1980).

Basically, the knowledge of HARPY is represented by a network in which to each state is associated a phone template for the label of the sound to be heard, if in that state, and a set of transitions to states that can immediately follow it in time. Taking the transitions to have associated probabilities and the template comparison process to yield a probability that the phone was heard at that point, the interpretation of the input utterance corresponds to the maximum likelihood path through the network. The following equation is applied iteratively:

$$P(S(i),(n+1)) = C(Ai,D(n+1))\underset{j \in \Sigma}{Max}\ P(S(j),n)P(S(j) \rightarrow S(i)) \qquad (8.15)$$

where: $P(S(i),(n+1)$ is the probability of being in state $S(i)$ at time $t = n+7$, i.e., after $n+1$ phone matches. $C(Ai, D(n+1))$ is the score of the comparison of the template Ai of state $S(i)$ and the acoustic data of the $(n+1)$-st input segment. Notice that input samples may not have the same duration. Σ is the set of states.

In the DRAGON system (J.K. Baker, 1975) the algorithm is a dynamic programming matching that sweeps across the entire rectangle of states and time segments like the Viterbi algorithm.

The computational complexity of the DRAGON system was high even with simple networks.

To reduce such complexity, a *beam-search* algorithm has been adopted in the HARPY system.

The algorithm consists in keeping active only a variable fraction of the states (the beam) at each step of the search.

Taking the magnitude of the logarithm of probabilities as the score and performing additions instead of multiplications, the states which are kept active are those whose score is below a threshold, computed by taking the minimum score of the nodes and adding a fixed value S.

Fig. 8.10   shows an example, in the State Space Representation of the beam search algorithm.

Quinton (1980) has investigated the performances of a modified version of the beam search algorithm. The modification consists in making variable the extension SB of the beam, thus accepting, at

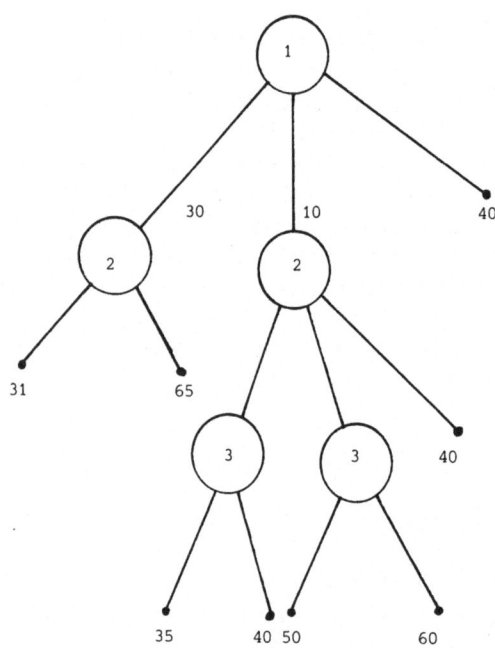

Fig. 8.10 Example of beam-search algorithm

each step, all the candidates whose scores SC satisfy the following relation (scores are cummulative distances from prototypes):

$$m \leq SC \leq m + SB$$

where m is the minimum score. The expression for SB is the following:

$$SB = \frac{So}{\sqrt{n}} \qquad\qquad (8.16)$$

where So is a constant and n is the number of words in the hypothesis. The main reason for this choice is that Quinton uses as score for a sentence, a weighted average of the scores of the words, the weights being the word lengths. Now, if σ is the standard deviation of the distribution of the word scores, the distribution of the scores for sentences of n words will have a standard deviation of the type:

$$\sigma = \frac{\sigma_o}{\sqrt{n}}$$

Thus, as the analysis advances, the score values become more dense around the average. A variable beam extension compensates for this increase in density.

Fig. 8.11 shows the percent of correct solutions, the percent of the examined space, and the standard deviation of the examined space for a beam search experiment.

All the curves in Fig. 8.11 are functions of So and refer to an experiment of 50 French sentences recognized by the KEAL speech understanding system (Mercier and Quinton, 1980).

Quinton compares the performances of the beam-search method with those of other non-admissible algorithms. The details of this comparison are omitted here for the sake of brevity. The main

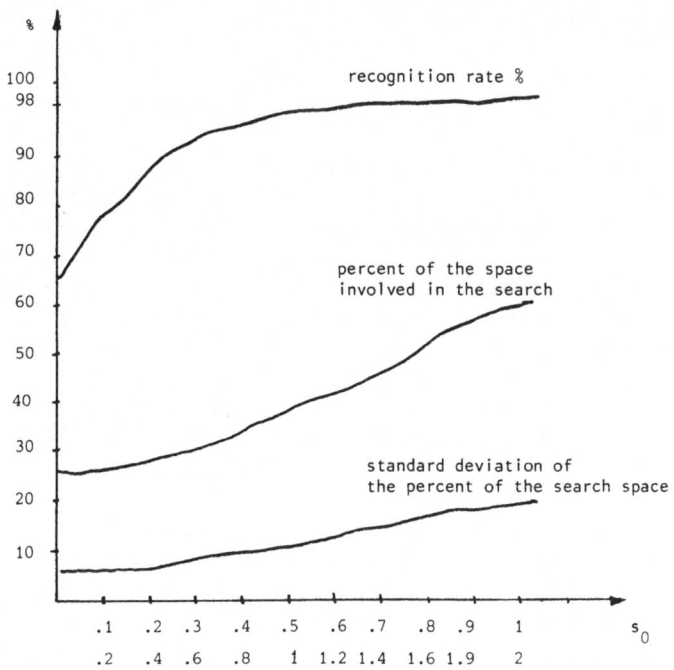

Fig. 8.11 Performances of the beam-search algorithm

result is that the beam-search algorithm gives the same performances
as other methods, but with the lowest percentage of space used for
the search and the lowest standard deviation of this percentage.

8.5   On the Use of Production Systems
for Problem Solving

When the task-dependent knowledge is very complex or is
incompletely specified, an integrated network cannot be used.

A sort of semantic knowledge, in addition to evidence or likelihood evaluations, greatly helps in focusing the search in a
problem space. The representation of semantics and a strategy for
its use in performing inferences is a fundamental problem. This
knowledge may be represented by a set of *production rules*. A
production rule is a statement of the form:

$$\text{IF } C \text{ THEN } A \tag{8.17}$$

where C is a conjunction of conditions, i.e.:

$$C = C(1) \wedge C(2) \wedge \ldots \wedge C(K)$$

and A is a set of actions, i.e.:

$$A = \{A(1), A(2), \ldots, A(n)\}.$$

Production rules may be seen as an integrated representation
of the structural knowledge and its corresponding controller.

Newell (1980) proposes a model of human cognition in which
production rules occupy a *long-term memory* and partial hypotheses
are written into a blackboard, which is a sort of *short-term memory*.

The short-term memory contains elements which are symbolic
structures. Each condition is represented by a template which has
to be matched with an element of the short-term memory.

A production is satisfied if all of its conditions find matching
elements in the short-term memory. Such a set of matching elements
constitues a possible *instantiation* of the production.

In systems like HARPY (Lowerre, 1976) or the IBM system
(Jelinek, 1976) the network representing the integrated knowledge
is a stochastic finite state automaton which can be described by a
set of stochastic rules of the type:

$$S(i) \xrightarrow{Pr(i,j)} aS(j)$$

$$\tag{8.18}$$

where Pr(ij) is a transition probability, S(i) and S(j) are state symbols which can be seen as nonterminal symbols of a linear grammar (Aho and Ullman, 1972).

The stochastic rule (8.18) can be seen as a piece of structural knowledge which can be transformed into a production rule of the type: IF (S(j) is an active state) *and* (a is in the blackboard) THEN compute the likelihood of reaching S(j) from S(i). Newell (1980) discusses the whole set of production rules which may describe the HARPY system.

The simple example we have followed so far shows how the rewriting rule of a grammar can be transformed into a production rule. With some effort, a collection of production rules can be derived from structural rules more complex than the ones used in a linear grammar and from the semantic rules associated with them, and useful for performing the composition of evidence measurements. Thus, in principle, production systems may describe Augmented Transition Network Grammars and Problem Reduction Representations and the rules of evidence composition embedded in these representations.

The total set of instantiations for all productions is called the *conflict set*. From it, an instantiation is selected and its action executed.

Conflict resolution is governed by several types of rules. The first type is *refraction* which inhibits a rule from being applied a second time on the same data. The second type of conflict resolution is *recency* which prefers instantiations that bind to elements that have more recently entered short-term memory. The third type of conflict resolution is *special case order* which prefers a production that is a special case of another.

When a production system is derived from a set of rules belonging to a grammar, the suitability of the application of a rule depends on the type of parsing which establishes how the structural rules should be used.

Recall that most of the knowledge of a speech understanding system may be described in terms of structural rules and, when these rules belong to a context-free grammar, they can be represented by an AND/OR graph. The terminal nodes of the AND/OR graph represent subproblems which consist in detecting some primitive features in the data and in measuring a degree of evidence from them.

Problem Reduction Representations may be established for each level of knowledge of the system and the terminal problems for a PRR of a level may be the hypotheses generated at lower levels.

Problem solving, given a PRR and the evidences of the terminal problems, can be performed by looking for the best scored path in the AND/OR graph.

Assuming now that scores are increasing with evidences, the evidence of a subproblem represented by an AND node is equal to the minimum of the evidences of the sons of that node. The evidence of a subproblem represented by an OR node is equal to the maximum of the evidences of the sons of the node.

Nilsson (1971) has described an algorithm, called the α-β algorithm for obtaining the best path through an AND/OR graph.

Stockman (1979) has proposed another algorithm, in which an ordered search is performed in a State Space Representation (SRR), corresponding to a mini-max procedure in the Problem Reduction Representation (PRR).

In Stockman's algorithm, a state in SRR encodes a path in the AND/OR graph.

Search in SRR is merit ordered and corresponds to breadth-first on OR alternatives and depth-first on AND alternatives.

In addition to his algorithm, which may perform better (i.e., visiting less states) than the α-β algorithm, Stockman proposes a method for restricting consideration to only those structures relevant to the sample data and not to those possible in the universe of samples. Restrictions are imposed by first solving in a bottom-up way some *primary problems* which are semantically key problems. These problems should either be easy to solve or have high value in focusing search. The set of primary problems is also the smallest set of problems which can be considered without losing completeness in the search for a solution to the initial problem.

Following these ideas, Stockman has developed a top-down, bottom-up, non-directional best-first parser, in which semantic considerations may determine a preferred sequence of parsing operations, trying to model the fact that the greatly flexible structural recognition capability of the human is largely due to his ability to evoke relevant models in the appropriate context of details.

Stockman's parser is described in his Ph.D. thesis (Stockman, 1977) and will not be described here, for the sake of brevity. An interesting application of such a parser was the automatic interpretation of carotid pulses. In this application, the knowledge of the system is represented as a PRR whose interface with the data is a *data scanner* which is called into action by the parser and attempts

to solve a problem (the detection of a primitive signal form) in a
given time-interval.

Many concepts from Stockman's parser can be applied in designing
the control KS at the various levels of the SUS. Notice that most
of them have to work both in data-driven (bottom-up) or in model-
driven (top-down) ways; context dependencies can be represented by
semantic rules.

Eventually, the structural knowledge and the rules of a non-
directional parser can be embodied into a production system.

Probably, in Speech Understanding, it is not enough to single
out primary problems of semantic evidence. Rather, there is a
hierarchy of importance of semantic problems. This hierarchy can
be embedded into a Semantic Augmented Transition Network Grammar,
as will be shown in Section 8.7.

McDermott (1978) has reported other methods than production
systems for representing and using knowledge in Artificial Intelligence.

So far, production systems have been the best suited for
handling together the use of knowledge and the rules for composing
hypothesis evidences.

Conceiving the system as a production system, a big problem
still remains open. The problem, which is discussed in a paper by
De Mori and Saitta (1979) is that in Speech Understanding Systems
conditions are verified with different degrees of evidences. These
degrees of evidences should affect the priorities with which actions
are performed.

Rules for relating items of evidence with items of priority
will be introduced in the next Section.

8.6. Scheduling of Interpretation Processes
Based on Approximate Reasoning

8.6.1 Background

The search for control strategies and methods for scheduling
interpretation processes is among the most difficult tasks involved
in designing Speech, Language, and Image Understanding Systems.

Most of the approaches use numerical probabilities or heuristic
scores to evaluate the hypotheses generated by the system and to

assign priorities to the processes that have to be executed to verify
new hypotheses or theories. Valuable as they are, all these approaches leave room for improvements.

A control system should be designed on the basis of inference
rules that are not probabilistic in nature. For example, the
assignment of a priority to an action process depends mostly on the
importance of the action in order to have a rapid convergence of
the interpretation process towards the goal of grasping the meaning
of the input message.

Nevertheless, every interpretation of the speech signal in
terms of phonetic, phonemic, or high-level features can be evaluated,
as suggested by Woods (1977), by a posteriori probabilities using
the Bayes theorem. But, even for these evaluations, a high degree
of imprecision is involved; this is mostly due to the approximations
made in extracting acoustic features, in evaluating a priori probabilities, and in the often questionable assumption of event indepen-
dence. Thus, the a posteriori probabilities obtained by the application of the Bayes theorem are affected by a degree of imprecision.

The imprecision with which the interpretation hypotheses are
evaluated and a need to have inference rules which are not exclusively
probabilistic have suggested the approach proposed in this Section.
Here, probabilistic qualification may be performed for hypothesis
evaluation using linguistic probability values. Other qualifications
such as evidence expectation or complexity may also be performed.
This leads to a proposition that is a premise for an approximate
reasoning. A suitable use of linguistic variables introduces an
approximation that does no harm, because the features extracted and
the rules for their extraction are vague and imprecise.

Let Ω be a segment of the speech signal or of its spectrogram;
using rules embedded in a knowledge source, an interpretation
$\mathfrak{J}_{1i}(\Omega)$ the i-th interpretation of Ω at level 1) is formed and its
evaluation is expressed by a proposition of the following type:

$$\mathfrak{J}_{1i}(\Omega) \text{ is } \lambda(1i) \tag{8.19}$$

where $\lambda(1i)$ may belong to a lexicon of probability values such as
likely, very likely, etc., or other types of evidence values.

If $\mathfrak{J}_{1i}(\Omega)$ is a precondition for an action $A(i)$, a process
"perform $(A(i))$" is created; in order to evaluate its priority, an
approximate reasoning is made, whose implication is an inference
rule of the form:

$$\text{IF } \mathfrak{J}_{1i}(\Omega) \text{ is } \lambda(i) \text{ THEN priority (perform}(A(i)) \text{ is high} \tag{8.20}$$

where $\lambda(i)$ is generally different from $\lambda(1i)$ and "high" is a fuzzy
linguistic variable belonging to a lexicon of priority values.
Inference rules like (8.20) are not probabilistic in nature because
the action $A(i)$ may be more, less, or just as suitable as other
actions, many actions may be highly suitable for exploring a few
interpretation possibilities in parallel. The inference rule (8.20)
and an evidence qualification may be combined by an approximate
reasoning to give the priority of $A(i)$, given its precondition.

In order to obtain an approximate conclusion from a premise
like (8.19) and an implication like (8.20), possibility distributions
are associated to the inference rules using the "bounded sum operation" on the membership of the linguistic evidence and priority
values.

Each linguistic priority value is assigned a queue where
candidate interpretation processes are inserted, following the
First-In-First-Out (FIFO) policy. When a processor is ready for
running an interpretation process, a supervisor selects the first
candidate of the queue with highest priority. A detailed discussion
on process scheduling can be found in a book by Brinch Hansen (1973).

8.6.2. On the use of truth functions and the fuzzy logic

Fuzzy logic is based upon a Lukasiewicz multi-valued logic
(see the book by Rescher, 1969, for details) and fuzzy set theory,
with the aim of generalizing logical reasoning based upon two-valued
propositional logic.

The most important contributions to fuzzy logic have been given
by Zadeh (1973, 1975, 1976), Bellman and Zadeh (1977), Baldwin (1979),
Gaines (1976), Lee (1972), Giles (1979), Baldwin and Pilsworth (1980),
and Dubois and Prade (1979, 1980).

The reader can find in these contributions an axiomatic presentation of fuzzy reasoning, motivations, and comparison with classical
logic. We shall be following and adapting to our purpose the approach
by Baldwin (1979) because it is the simplest, and developed enough to
cover the applications we are interested in.

The hypotheses which are generated by the lexical or the syllabic KSs have associated a degree of evidence defined in the
interval $[0,1]$.

Furthermore, we may have evidence propositions like: "a
hypothesis is very evident".

The truth value of a proposition is represented by a truth function v, defined as a mapping:

v: $[0,1] \rightarrow [0,1]$

In our case a proposition may be of the type:

p = W is λ-evident

which is represented by the following fuzzy restriction:

R(Evidence(W)) = λ (8.21)

We can further associate with the proposition p a fuzzy truth value τ and say:

(W is λ) is τ

We now give fuzzy labels for certain truth functions. In order to define the truth function "true" we have to respect the constraint that:

(W is λ) is τ

should be equivalent to:

W is λ *true*

In other word, the degree of evidence μ of W with respect to the restriction λ should be equal to the degree of truth γ_{true} for the application of the label λ to W, i.e.:

$\gamma_{true}(\mu) = \mu \quad \forall \ \mu \in [0,1]$ (8.22)

Similarly:

(W is λ) is *false*

should be equivalent to:

(W is $\neg \lambda$)

defining the truth function for *false* as:

$\gamma_{false}(\mu) = 1 - \mu$ (8.23)

Furthermore, we may define *fairly true* by the function:

$$\gamma_{\text{fairly true}} (\mu) = \gamma_{\text{true}}^{1/2} (\mu) \tag{8.24}$$

and *very true* by the function:

$$\gamma_{\text{very true}} (\mu) = \gamma_{\text{true}}^{2} (\mu) \tag{8.25}$$

Other truth functions are defined as follows:

absolutely true, $\gamma_{\text{a.t.}} (\mu) = \begin{cases} 1 & \text{if } \mu = 1 \\ 0 & \text{otherwise} \end{cases}$ \hfill (8.26)

absolutely false, $\gamma_{\text{a.f.}} (\mu) = \begin{cases} 1 & \text{if } \mu = 0 \\ 0 & \text{otherwise} \end{cases}$ \hfill (8.27)

undecided, $\gamma_{\text{undecided}} = 1 \; \forall \; \mu \in [0,1]$ \hfill (8.28)

The following relations hold between fuzzy truth values:

$(\text{very})^{x} \text{ true} \to \text{absolutely true as } x \to \infty$

$(\text{very})^{x} \text{ false} \to \text{absolutely false as } x \to \infty$

$(\text{fairly})^{x} \text{ true} \to \text{undecided as } x \to \infty$

$(\text{fairly})^{x} \text{ false} \to \text{undecided as } x \to \infty$

where $(k)^{x} h = k (k)^{x-1} h; \quad x = 2,3,\ldots$.

8.6.3 Priority assignment and approximate reasoning

A Speech Understanding System is organized in levels; let
$<l(1), \; l(2), \; \ldots, \; l(i), \; \ldots, \; l(LV)>$ be such levels. The procedural
knowledge of the system is represented by rules of the type:

$<\text{precondition} \to \text{action}>$.

The i-th rule contains a set of preconditions PR(i) and a set
of actions A(i). A precondition pr(ik) in PR(i) that allows the
action A(i) to be requested is expressed in terms of items $l(l1,i,k1)$
$l(l1,i,k2) \ldots l(l2,i,k1) \; l(l2,i,k2)\ldots \; l(li,i,k1) \; l(li,i,k2)$
$l(li,i,km)\ldots l(lLV,i,k1) l(lLV,i,k2) \; \ldots$ corresponding to hypotheses
(theories) already verified at the levels $l(1), \; l(2),\ldots,l(i),\ldots,$
$l(LV)$. Each item $l(li,i,km)$ is assigned a degree of evidence of
being a correct interpretation of the input data. This evidence
is represented by a linguistic value $\lambda(l(li,i,km))$. Combining the
evidences of the precondition items, it is possible to evaluate the

linguistic evidence of the whole precondition pr(ik); let $\lambda(ik)$ be
such a value. This assignment corresponds to the following proposition:

pr(i,k) is $\lambda(ik)$

Priority assignment is performed by the understanding controller
on the basis of two types of evaluation. The first refers to the
possibility of the success of an action, given the evidence of the
preconditions. The second refers to the suitability of the action
with respect to a knowledge like computational complexity and goal
satisfaction. The former evaluation is based on a proposition of
the following type:

IF pr(i,k) is evident THEN A(i) is suitable   (8.29)

where "suitable" is a fuzzy linguistic variable, defined over the
unit interval, belonging to a dictionary V of variables (suitable,
very suitable, more or less suitable, etc.).

Formula (8.29) is an example of the rules of inference of the
understanding controller. Such rules may be learned by experiment
or they may be initially settled by the designer.

During the understanding process, the controller will assign
a priority linguistic value to the process executing the action
A(i), according to the result of the following reasoning, which is
approximate if $\lambda(i,k)$ is different from "evident":

pr(i,k) is $\lambda(ik)$ (premise)

IF pr(i,k) is evident THEN A(i) is suitable (implication) (8.30)
_____
Execution of A(i) is σ (approximate conclusion)

In fact, $\lambda(ik)$ is a fuzzy restriction or a possibility distribution $\Pi(u)$ over the unit interval [0,1]; "IF pr)i,k) is evident THEN
A(i) is suitable" is a fuzzy relation R(u,v) or a two-dimensional
possibility distribution. The composition of $\Pi(u)$ and R(u,v) gives
a possibility distribution over the unit interval [$0 \leq v \leq 1$] on which
suitability linguistic values belonging to the alphabet V are
defined. The linguistic value σ V, which best approximates the
obtained distribution, is used to assign a priority to the execution
of A(i). Other, more sophisticated types of approximate reasoning
can be used if required by the complexity of the problem.

Once many action processes are competing for execution, the
degrees at which their execution is suitable can be compared, in
order to assign priorities to them. Linguistic variables are

assigned to the execution of the processes; these variables can be inserted directly into the implications, giving inference rules of this type (prec is for precondition):

IF prec is evident THEN priority (perform (A(i)) is high (8.31)

If prec has a linguistic value of evidence LEV(j) different from "evident", an approximate reasoning has to be performed, in order to obtain a linguistic value h(LEV(j)) for the height of the priority with which A(i) has to be performed.

A first step toward performing an approximate reasoning is the computation of the truth τ of the following conditional proposition:

(prec is *evident* / prec is λ j)

i.e.:

v(prec is *evident* / prec is λ j) = τ.

Following Baldwin (1979), the membership function of τ is obtained as follows:

$$\gamma_\tau(\mu) = \bigvee_{\mu \in [0,1]} \gamma_{evident}(\mu) ; \forall u \in [0,1]$$

$$\gamma_{\lambda_j}(\mu) = u \tag{8.32}$$

$\gamma_\tau(u)$ is a restriction representing the possibility distribution of the premise.

This restriction has to be composed with the fuzzy relation $R(u,v)$ representing the implication: IF *evident* THEN *high*.

Following the Lukasiewicz definition of implication, we have the following definition of membership function for the fuzzy relation $R(u,v)$:

$$\mu_R(u,v) = 1 \wedge (1 - u + v) \tag{8.33}$$

This "bounded sum" represents the fact that if *high* has a higher truth value than *evident*, the inference rule has membership equal to one, otherwise the "worthiness" of the relation decreases with the difference between the truth values of *evident* and *high*.

By composition of the fuzzy restriction of the premise and the fuzzy relation of the implication, we obtain a fuzzy restriction expressing the truth value τ (c) of the conclusion: priority is high.

This truth value is represented by the following function:

$$\gamma_{\tau(c)}(v) = \bigvee_{u\in[0,1]} \{ \gamma_{\tau}(u) \wedge [1 \wedge (1-u+v] \} \qquad (8.34)$$

Often, the premises may be a compound of statements. The compound may contain an evidence statement E of what has already been recognized, an expectation statement P related to the a priori probability of what has been missed during recognition, a duration statement D regarding the extent of an already generated hypothesis over the sentence to be interpreted, and a complexity statement C about the amount of computation which has to be performed in order to end up with a complete hypothesis. In other words, the premise may have the form:

If E *and* P *and* D *and* C, *then* H,

where H is a priority statement.

In a real situation we may find different premise statements like E' instead of E, P' instead of P, D' instead of D and C' instead of C.

In order to apply the algorithm for approximate reasoning, we have to evaluate the truth of the following conditional statement:

v {E *and* P *and* D *and* C/E' *and* P' *and* D' *and* C'}

$\qquad$ = v {(E/E') *and* (P/P') *and* (D/D') *and* (C/C')} $\qquad (8.35)$

whose truth is just the minimum of the truth of every single conditional statement for each value of u, $u\in[0,1]$.

Once a possibility distribution for the truth of the conclusion: "priority is high" has been obtained, the best approximation of this distribution with functions representing standard truth values can be found. Furthermore each priority queue can be associated with one of these truth values on the basis of a one-to-one correspondence.

8.7 Outline of a Semantically-Guided Use of Task-Dependent Knowledge

8.7.1 System organization

The system is organized in such a way that the interpretation of a sentence is seen as the accomplishment of a complex plan which is subdivided into subplans according to Sacerdoti's **NOAH** system (Sacerdoti, 1977).

Fig. 8.12 shows the decomposition of the plan "interpret sentence" into subplans. The first decomposition contains the conjunction of two subplans named "Bottom-up generation of syllabic hypotheses" (BUGSH) and "Set first vista" (SFV). When these subplans have been successfully achieved (unsuccess would mean the system did not detect the times of the beginning and end of a sentence) two other subplans are executed.

The first one consists in generating "seed-theories" which are seeds of interpretation hypotheses, based on words belonging to the first vista and recognized with sufficient evidence by the lexical KS. The first vista is a subset of the lexicon, consisting of key words which are expected in some time intervals of the speech signal. The composition of the first vista is mainly decided by the pragmatic knowledge. For example, in a protocol for an automated travel reservation and information system, the first vista consists of key words characterizing concepts such as "request of information", "reservation", "cancellation" etc. An automatic program can learn such words from written sentences.

As there may be more than one keyword characterizing a concept, attention is paid to selecting those which permit the shortest interval of expectation in a sentence.

After seed theories are generated, a plan for the dynamic growing of these theories is set.

The BUGSH plan is composed of the following subplans

- sampling and quantizing,
- pitch extraction and description,
- Fast-Fourier Transformation (FFT),
- estimation of Linear Prediction Coefficients (LPC),
- extraction and description of gross spectral features,
- segmentation,
- generation of hypotheses about the composition of syllables.

These subplans are not represented in Fig. 8.12 for the sake of simplicity.

Execution of the subplans mentioned so far corresponds to the creation of processes by the "general control strategy KS" shown in Fig. 1.6. A high degree of parallelism can be achieved in the execution of these subplans.

Nii and Feigenbaum (1978) have proposed an approach to problem solving activities based on a hierarchically organized control structure. The system design proposed in this book and shown in Fig. 1.6 is a hierarchically organized control structure which has

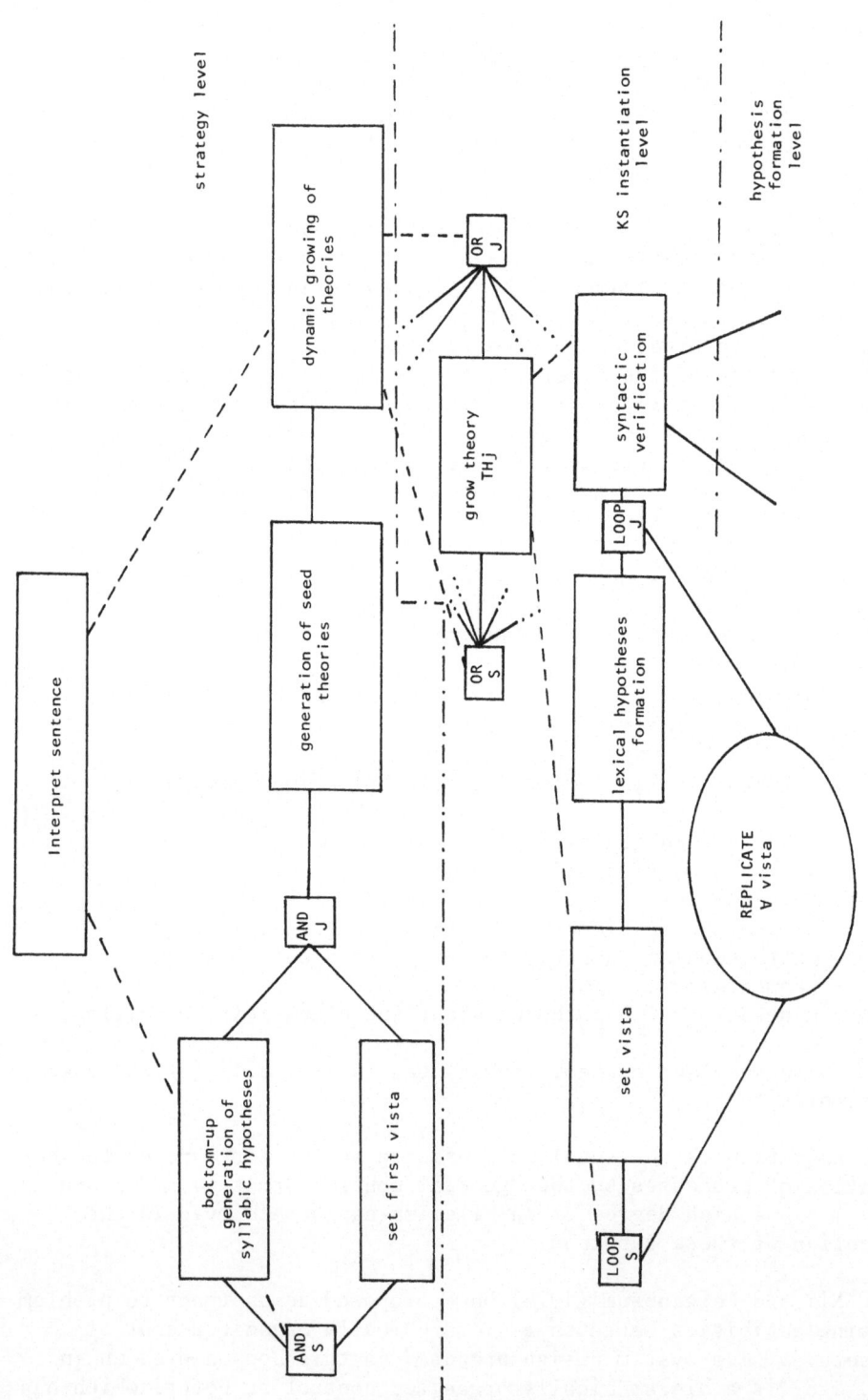

Fig. 8.12   Decomposition of the plan: "interpret sentence"

the general control strategy KS at the top of the hierarchy, corresponding to a "strategy level" (see also Fig. 8.12).

Below, in the hierarchy, there is a KS-Activation level which
corresponds to the KS control strategies associated with each layer
of the speech understanding system.

Each KS instantiation applies a piece of structural knowledge
which has built-in conditions (like the GETF X of the syllabic ATNG)
for controlling the generation of hypotheses. This "hypothesis
formation level" is the lowest one in the control structure.

The plan for the dynamic growing of theories is made of subplans,
one for each active hypothesis. The achievement of each one of these
subplans is under the responsibility of the control KS at the level
of syntax, semantics, and pragmatics and corresponds to a "KS
instantiation". A KS instantiation is a controller acting on a set
of data using a part of the knowledge. This results in the creation
of an interpretation process which executes an interpretation program
using a structural KS, having as input some lower level hypotheses.
The program code and the structural KS are used as "pure segments",
i.e., as information which cannot be modified. Several images of a
pure segment can be created for working, perhaps in concurrence, on
different sets of data and for generating different hypotheses.

Each one of such images is a process or a KS instantiation. The
book by Brinch Hansen (1973) offers a good background on the theory
of concurrent processes.

Each KS instantiation at the semantic level attempts to grow
a theory $TH(j)$, $j \epsilon (1, NTH(t))$, where $NTH(t)$ is the number of active
theories at time t.

An example of KS instantiation of the semantic knowledge, is
the creation of a vista.

Growing a theory corresponds to iteratively creating a "vista"
and attempting to generate lexical hypotheses about the words in
the "vista". This process uses a semantic network as structural
knowledge source and a non-directional parser as procedural knowledge. The semantic network and its parser will be described in
the next subsection.

Once a semantic hypothesis has been generated for the whole
sentence, a syntactic verification is performed. This verification
requires the evaluation of evidences of all the words in the sentence,
including the so-called "function words" like articles, prepositions,
etc.

Verification involves also a consistency analysis between the
syntactic structure of the hypothesized sentence and the pitch
contour.

If a verification is successful, a hypothesis about the sentence is generated.

Generation of lexical hypotheses begins with a request to the
lexical control KS, which attempts to perform a lexical plan according to the procedure described in Chapter 7. Again this is performed through KS instantiations and eventually the generation of
lexical hypotheses.

The control actions are only partially data-directed. Word
hypothesization is stimulated by the evidence of syllabic hypotheses
generated by the syllabic KS. Nevertheless, word hypothesization
is constrained by semantic predictions represented by "vistas"
selected by a non-directional parser, using a semantic network.
This prediction is not based on expected word adjacency but on conceptual dependency between words.

The general behaviour of the system respects a hierarchy of
control actions. The general control strategy is at the top of this
hierarchy. It may cause chains of KS instantiations. A KS instantiation, in turn, may result in hypothesis formation. This type of
hierarchy of control structures is in accordance with a proposal by
Nii and Feigenbaum for signal understanding systems.

From the system architecture point of view, the main controller
contains a sort of supervisor which decides to use the strategy KS
for interpreting a sentence. KS instantiations may be performed
sequentially or in parallel.

Following the scheme of Fig. 8.13, sentence interpretation is
controlled by a subsystem called "interpret sentence". This subsystem
controls the execution of the plans according to the scheme of Fig.
8.12. The subsystem is capable of sending and receiving messages
from other subsystems, each one of which corresponds to a control
KS.

Each KS controller reacts to a request by creating an interpretation process (a KS instantiation), which is submitted to a
scheduler for execution.

The scheduler evaluates the priority of each process, considering it as an action following a set of premises which have been
met with some degree of worthiness. The priorities are expressed
by linguistic values and are evaluated by Approximate Reasoning.
According to its priority, a process is queued in a list. There

may be lists for very high priority, high priority, fairly high
priority, average priority etc.

Processes are executed in a set of processors, which may be
real or virtual.

An Initiator/Terminator (IT) subsystem attempts to allocate
the processes in the priority queues to the available processors.

Once a process has been allocated to a processor, its execution
is initiated and the process becomes active. An active process may
advance a request to a KS controller, causing the generation of
other processes.

The process which advances a request may be kept in a "waiting"
state until the request has been honored.

When a process is terminated, after possibly writing some
hypotheses on the blackboard, a message is sent to the IT subsystem,
which provides for process deallocation and for the allocation of
another process extracted from one of the queues.

The queues are analyzed following a priority order. The IT
subsystem selects the first process from the first non-empty queue
for allocation.

Processes in each priority queue are ordered following a first-
in-first-served policy.

A highly parallel implementation based on Hewitt's "actor
system" has been proposed by Giordana et al. (1980). This system
has been simulated in the SIMULA-67 language on a DEC PDP10 computer.

From this preliminary experience it appears that is worth
having processors devoted to specific levels of the system. In this
case, a processor has access only to a specific structural knowledge
and to some specific levels of the blackboard.

The proposed structure is particularly efficient if the structural and procedural knowledge of the system are so rich that having
the right hypothesis on the highest priority queue is very likely
and the average number of competing hypotheses reduces rapidly as
the seed hypotheses grow.

8.7.2. The semantic knowledge

The semantic knowledge can be conceived in such a way that it
represents dependency relations between concepts. This represen-

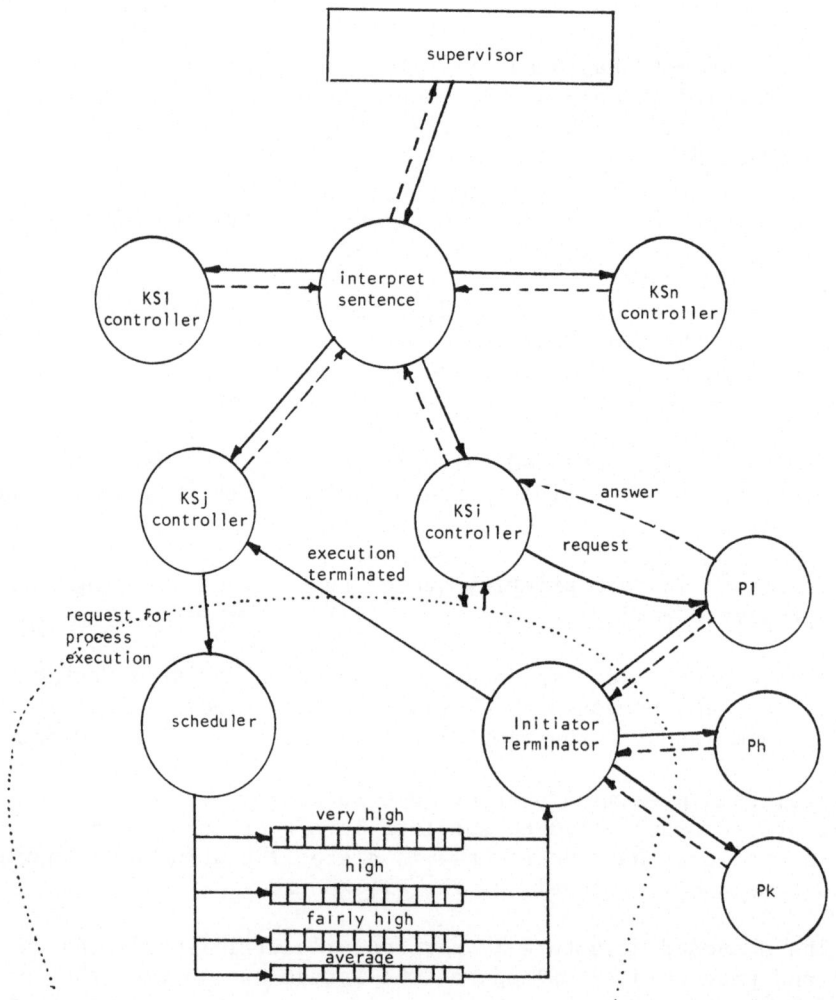

Fig. 8.13 Scheme of the "interpret-sentence" subsystem

tation is particularly useful for making inferences once the spoken sentence has been recognized.

A proper representation of the semantic knowledge has to be conceived when it has to be used for solving ambiguities and recovering from errors in the lattice of lexical hypotheses.

Such a representation should exhibit semantically acceptable relations between key words which can be hypothesized in the speech spectrogram.

It has been found that an efficient way to represent semantically acceptable relations between hypothesized or hypothesizable words is again with an Augmented Transition Network Grammar (ATNG), which will be referred to as Semantic ATNG in the following.

The main motivations are: 1) the hierarchy implicit in a semantic representation is well represented by the hierarchy of subnetworks of an ATNG; 2) sets of key words can be associated with concepts as arc conditions at every level of the hierarchy; 3) generation of symbols representing syntactic classes to be processed by the syntactic ATNG can be performed by actions associated with arcs. Finally, the semantic ATNG is conceived in such a way that it is straightforward to derive a set of production rules from it.

In fact, an arc of the semantic ATNG is described as follows:

(state (arc label/condition
       (action)
       (destination)))   (8.36)

Expression (8.36) can be translated into the following production rules:

    IF state active THEN try to move from it by verifying conditions;
    IF condition true THEN try to move on the arc;
    IF move successful THEN perform action and make destination active.

Conditions are represented by regular expressions of words which have to be hypothesized with enough evidence.

At the beginning, only the first state is active. Then the conditions associated with the arcs moving from it are tested. This may imply a search on the lexical hypotheses of a set of vistas for a model-driven word verification.

For all the arcs for which the conditions are met with enough evidence, instantiations are created for attempting to move on the

arcs.   Instantiations may be scheduled according to condition
evidences, complexity, and importance of the work which remains to
be done for obtaining a complete interpretation hypothesis about
the utterance.   Notice that the Semantic ATNG is organized in such
a way that it develops a hierarchy of semantic plans.   Subnetworks
are detailed subplans which are not executed if the conditions
associated with the invoking arc (representing a more general plan)
are not met with enough evidence.

An example of the semantic ATNG for a travel reservation and
information system is given in Table 8.1.   Conditions LWj are represented by regular expressions of lexical items, whose details are
omitted for the sake of simplicity.   Actions are indicated by a SET
primitive which prepares data structures for the syntactic network.

The type of the prepared data is represented by the name following SET.

The meaning of the state labels is quite evident from the abbreviations used.   VSPELL is the spelling of vowels, CSPELL is the
spelling of consonants.

TABLE 8.1

SEMANTIC NETWORK
(SP   (PUSH INFORMATION/LW1
                        (SET REQUEST OF INFORMATION)
                        (TO QP))
                        (PUSH RESERVATION/LW2
                        (TO QP))
                        (JMP/LW3
                        (SET CANCELLATION)
                        (TO Q1)))
(QP   (POP))
(Q1   (PUSH RESERVATION/LW2
                        (TO QP)))

INFORMATION   (PUSH COST-INF/LW4
                        (TO QP1))
                        (PUSH TIME-INF/LW5

```
                           (TO QP1))
                           (PUSH TRANSP-INF/LW6
                           (TO QP1))
                           (PUSH SEAT-INF/LW7
                           (TO QP1))
                           (PUSH RES-INQ/LW2   reservation inquiry
                           (TO QP1)))
(QP1                       (POP))

(RESERVATION               (PUSH OBJECT/LW8
                           (TO Q2)))
(Q2                        (PUSH TRAVEL/LW9
                           (TO Q3))
(Q3                        (PUSH NAME/LW10
                           (TO QP2)))
(QP2                       (POP))

(COST-INF                  (PUSH OBJECT/LW8
                           (TO Q4)))
(Q4                        (JMP/LW11
                           (SET TRANSPORTATION MEAN)
                           (TO Q5)))
(Q5                        (PUSH DEP-AR/(LW20 V LW21)
                           (TO QP3))
                           (JMP/LW12
                           (SET DAY-OF-WEEK)
                           (TO QP3))
                           (JMP/LW13
                           (SET DATE)
                           (TO QP3)))
(QP3                       (POP))

(TIME-INF                  (JMP/LW11
                           (SET TRANSPORTATION-MEAN)
                           (TO Q6)))
```

```
(Q6                (JMP/LW14
                   (SET TRANSP-D)
                   (TO QP4))
                   (PUSH DEP-AR/
                   (TO QP4)))
(QP4               (POP))
(SEAT-INF          (PUSH OBJECT/LW8
                   (TO Q7)))
(Q7                (PUSH TRAVEL/LW9
                   (TO QP5)))
(QP5               (POP))
(RES-INQ           (PUSH TRAVEL/LW9
                   (TO Q8)))
(Q8                (PUSH NAME/LW10)
                   (TO QP6)))
(QP6               (POP))
(OBJECT            (JMP/LW15
                   (SET QUANT)
                   (TO Q9)))
(Q9                (JMP/LW16
                   (SET PLACE)
                   (TO QP7)))
(QP7               (POP))
(TRAVEL            (JMP/LW11
                   (SET TRANSPORTATION-MEAN9
                   (TO Q11)))
(Q10               (JMP/LW14
                   (SET TRANSP-ID)
                   (TO QP8))
(Q11               (PUSH DATE/LW17
                   (TO Q10)))
                   (POP))
```

```
(QP8              (POP))

(NAME             (PUSH PSS-SP/T   word spelling controlled by the
                  (TO QP9)))       syllabic grammar

(QP9              (POP)
                  (PUSH NAME/T
                  (TO QP9)))

(QP9              (POP))

(PSS SP           (JMP/LW18
                  (SET VSPELL)
                  (TO Q12))
                  (JMP/LW19
                  (SET CSPELL
                  (TO Q13)))

(Q12              (JMP/LW19
                  (SET CSPELL
                  (TO Q13))
                  (JMP/LW18
                  (SET VSPELL)
                  (TO QP10)))

(Q13              (JMP/LW19
                  (SET CSPELL
                  (TO Q14))
                  (JMP/LW18
                  (SET VSPELL)
                  (TO QP11)))

(QP10             (POP))

(Q14              (JMP/LW18
                  (SET VSPELL)
                  (TO QP10))
                  (JMP/LW19
                  (SET CSPELL
                  (TO Q15))

(Q15              (JMP/LW18
```

```
                    (SET VSPELL)
                    (TO QP10)))

(DEP-AR             (JMP/LW20
                    (SET DEP)        departure
                    (TO Q16))
                    (JMP/LW21
                    (SET AR)         arrival
                    (TO QP17)))
(Q16                (JMP/LW22
                    (SET TOWN)
                    (TO QP11)))
(QP11               (POP)
                    (PUSH TIME/LW23
                    (TO QP12))
                    (JMP/LW21
                    (SET AR)
                    (TO Q17)))
(Q17                (JMP/LW22
                    (SET TOWN)
                    (TO QP12)))
(QP12               (POP)
                    (PUSH TIME/LW23
                    (TO QP13)))
(QP13               (POP))

(DATE               (JMP/LW24
                    (SET DAY)
                    (TO Q18))
                    (JMP/LW25
                    (SET MONTH)
                    (TO Q19)))
(Q18                (JMP/LW25
                    (SET MONTH)
                    (TO Q19)))
```

```
(Q19              (POP)
                  (PUSH TIME/LW23
                  (TO QP14)

(TIME             (JMP/LW26
                  (SET HOURS)
                  (TO Q20)))
(Q20              (POP)
                  (JMP/LW27
                  (SET MINUTES)
                  (TO QP14)))
(QP14             (POP))

(TRANSP-INF       (PUSH DEP-AR/LW20 V LW21)
                  (TO Q21))
                  (PUSH TRAVEL/LW9
                  (TO QP2)))
(Q21              (PUSH TRAVEL/LW9
                  (TO QP2))
                  (POP))
```

8.7.3   The syntactic knowledge

The syntactic knowledge is represented by another ATNG, a
fragmant of which is given in Table 8.2 as an example. In this
example it is assumed that a sentence SEN can be a sequence of a
Verb Phrase (VP) and a Noun Phrase (NP). Only the subnetwork NP is
given in the example, which is similar to one proposed by Bates
(1975). The actions on the arcs are: (SETR X Y), which replaces
the content of register X by the value of Y (an asterix, *, indicates
the current input); (ADDR XY), which adds the value of Y to the
contents of register X without destroying the old value; (GETF X),
which returns the value of the syntactic feature associated with
the current word; (ABORTIF x) which blocks the arc if the condition
x is not true.

TABLE 8.2

SYNTACTIC ATNG

(SEN	(PUSH VP/T	
	(TO Q1)))	
(Q1	(PUSH NP/T	
	(TO QP)))	
(QP	(POP))	
(NP	(CAT DET/T	
	(SETR DET*)	
	(TO Q2))	
	(JMP/T	
	(SETR DET NIL)	
	(TO Q2)))	
(Q2	(CAT ADJ/T	
	(ADDR ADJS *)	
	(ADDR NUA (GETF NUMBER))	**adjective number**
	(TO Q2))	
	(CAT N/T	
	(SETR N*)	
	(SETR NU (GETF NUMBER))	**noun number**
	(TO Q3)))	
(Q3	(PUSH PP/T	
	(ADDR PHODS *)	
	(TO Q3))	
	(POP) (ABORTIF (OR (NOT DETAGREE)(NOT	
	NADJAGREE)))	
	(POP))	
(PP	(CAT PREP/T	
	(SETR PREP*)	
	(TO Q4)))	
(Q4	(PUSH NP/T	
	(SETR NP*)	
	(TO Q5)))	
(Q5	(POP))	

The semantic component selects a list of word hypotheses in a way similar to that of the SPEECHLIS system (Nash-Webber, 1975), the main difference being that in the system described in this book, the list is built up by searching almost all the key words of the sentence through a hierarchy of processes.

This hierarchy is established by the semantic ATNG and key words are looked for by creating vistas on the lexicon.

Once a syntactic theory is complete, i.e., its words cover most of the time of the sentence, the words are ordered according to the time intervals in which they have been hypothesized and the uninterpreted gaps in the utterance are filled up from left to right, using the syntactic knowledge to make predictions.

Particular care is paid to updating the semantic ATNG in order to allow a large coverage of the input sentence by word hypotheses. This would require the syntactic ATNG to recognize only the so-called function words.

The semantic and the syntactic ATNG have mechanisms for taking into account the time intervals in which predictions or recognitions are made. These mechanisms are similar to those used by Stockman (1977).

Evidences of word hypotheses are combined by an algorithm similar to that used for the syllablés in the words. Loss values for the uninterpreted gaps are evaluated on the basis of probabilities of having words in given contexts and the possibilities that the actual system fails to detect them. The degree of agreement between Verb Phrases, Noun Phrases, and pitch contours is taken into account when the evidence of a sentence is evaluated. Evidences and loss values are combined together to obtain the total evidence of a sentence hypothesis.

For each acceptable sentence hypothesis covering the whole input utterance, a *procedural semantic* is applied to obtain a conceptual representation of the sentence for the user of the speech understanding system.

8.7.4 Pragmatics

The semantic knowledge source can be used by the pragmatic KS to carry on a dialog between the speaker and the system.

After a spoken sentence has been interpreted by the semantic component, it is possible that more ambiguous interpretations are obtained. In this case, the system may ask the speaker to solve its

ambiguities by pronouncing another sentence. There are other possibilities. The input sentence may have been interpreted unambiguously, but there may be items which do not fit the system knowledge
(for example a town may have been called by its previous name). In
this case, the system may ask the speaker to use other words to
express his request. Finally, after a sentence has been unambiguously
interpreted, the system may perform inferences to obtain some information which was implicit in the spoken sentence but was not said.
For example, the departure time and the departure place may allow
one to infer the place of arrival or the flight number. But, even
after performing inferences, it is possible that the information
supplied by the speaker is still incomplete. In this case the
system may ask the speaker to pronounce another sentence.

The pragmatic component controls the dialog between the user
and the system by a general algorithm which is given in the following.

The details of the procedures are omitted for the sake of
brevity. The requests of the system to the user are prepared by
procedures based on table-look-up algorithms, whose entries are
the procedure arguments. The details of the tables are task-
dependent and language dependent.

"Algorithm for controlling the dialog"

```
repeat
  repeat
    repeat
      enter a sentence;
      apply KSs to interpret the sentence;
      if there are ambiguous interpretations then
          ask for verification (ambiguities);
    until interpretation is unambiguous;
      if there are unknown items then
          ask to clarify (unknown items);
  until all detected items are known;
      if information supplied is incomplete then
          ask for more information (information available)
          knowledge sources)
until information supplied is complete.
```

8.8 Evaluating Language Complexity

The vocabulary size is not the best measure of language complexity. The reason is that complexity should be strictly related
to the difficulty of the understanding task. This difficulty mostly
depends upon the number of word candidates the system may have to
choose among the lexical items during the interpretation process.

Goodman (1976) has introduced the notion of *static branching factor* for a better characterization of the difficulties of the understanding task.

Let the language model be a finite-state automaton A:

$$A = (V, Q, q(o), F, \delta) \tag{8.37}$$

where:

V is the vocabulary of input words,
Q is a set of states,
q(o) is the initial state,
F is a set of final states,
δ is a mapping function from V x Q into Q.

Let $L(A) \subset V^*$ be the language accepted by the automaton A.

The *static branching factor* BFS(q) of a state q is defined as follows:

$$BFS(q) = |X|$$

where $|X|$ denotes the number of words in the set X which is composed of words $x \in V$ which appear as labels for arcs leaving the state q for another state different from the null state $\emptyset$.

The *static branching factor SBF* (A) for the automaton A is the average of the branching factors of the states of A i.e.:

$$SBF(A) = \frac{\displaystyle\sum_{q \in NFS(A)} BFS(q)}{|NFS(A)|} \tag{8.38}$$

where NFS(A) is the set of non-final states of A defined as follows:

$$NFS(A) = \{q \in Q / \exists\ x \in V\ and\ \delta(q,x) = \emptyset\} \tag{8.39}$$

$|NFS(A)|$ is the number of states in NFS(A).

The use of a static branching factor as a measure of language complexity may be criticized for the following reasons:

1) it depends on the automaton (there are many equivalent automata accepting a regular language);

2) it does not take into account the relative frequencies of occurrence of the sentences in the language;

3) it does not account for the phrase length.

For these reasons two new definitions of language complexity have been introduced: *entropy* and *perplexity* (Sondhi and Levinson, 1978).

Let L be a finite language over which a probability distribution $p(1)$ is defined $\forall 1 \in L$.

Let $L(k)$ be the set of phrases of length k (made of k words) and $|L(k)|$ be the number of such phrases. Let k be the maximum length of phrases in L.

The *entropy* $H(L)$ of a language L, given a probability distribution $p(1)$ over L, is given as follows:

$$H(L) = \frac{\sum_{\ell \in L} -p(1) \log_2 p(1)}{\sum_{\ell \in L} p(1)|1|} \qquad (8.40)$$

where $|1|$ is the number of words (length) of the phrase 1.

$H(L)$ is the average information rate of the language and is expressed in bit/word. The *perplexity* $PX(L)$ of the language L is defined as follows:

$$PX(L) = 2^{H(L)} \qquad (8.41)$$

The perplexity, also called dynamic branching factor, gives the average number of choices the system has to make when leaving a state of the automaton generating the language.

Different choices of $p(1)$ give different perplexity values. Among the possible distributions of $p(1)$, three types of distribution are particularly interesting. A distribution we shall call Pe, corresponding to all phrases having the same probability, a distribution Pm which maximizes the entropy, and a distribution Pt obtained from an experimental set of test phrases.

The three cases are represented by the following relations.

Case : all phrases equiprobable:

$$Pe(1) = \frac{1}{|L|}$$

$$H = \frac{\log_2 |L|}{\sum_{k=1}^{k} |L(k)|/|L|}$$

This is the distribution one assumes if there is no information available on the probability of phrases.

Case 2: Distribution corresponding to maximum entropy

According to Sondhi and Levinson (1978), the entropy H for this case is a solution of the equation:

$$\sum_{k=1}^{K} |L(k)| H^k = 1$$

Case 3: Distribution to be inferred from an experiment

A simple way to infer a probability distribution in this case is to consider equiprobable the phrases having the same length and to assume that the probability of every sublanguage L(k) is proportional to the number of phrases n(k) of length k and appearing in the sample set. In such a case one gets:

$$Pt(k) = \frac{n(k)}{n}$$

where n is the number of phrases in the sampling set.

Furthermore, one gets:

$$H = \frac{\sum_{k=1}^{K} Pt(k) \log_2 |L(k)|}{\sum_{k=1}^{K} Pt(k) k} \qquad (8.42)$$

Let us consider now how the quantities introduced so far can be used to evaluate the complexity of a recognition task.

Goodman (1976) proposes evaluating the complexity of a recognition task as related to the extension of the search space covered during a recognition experiment. Let $E(|l|)$ be the average length of a phrase of L.

On the average, the recognition of a sentence in L requires the generation of a Space State Representation whose depth is $E(|l|)$ and every node has PX sons. The number of leaves of such a tree is:

$$T = PX^{E(|l|)}$$

and the computational complexity CC can be defined as follows:

$$CC = \log_2 T = E(|1|) \, H(L) \qquad\qquad (8.43)$$

In the case of isolated words L = V. If the words are equi-probable, then: $T=|V|$, which is the vocabulary size. A language whose perplexity is 2 and has an average phrase length of 10 turns out to have the same complexity as a system of 1024 isolated words. As this seems rather strange, Quniton, (1980) has proposed another concept called *equivalent vocabulary*.

A basic assumption is made that the probability of recognizing a word w between v competitors is:

$$Pr(w) = c^{v-1} \qquad\qquad (8.44)$$

where c is a consonant.

If now the recognition of a phrase whose length is $E(|1|)$ implies the recognition of $E(|1|)$ words in a vocabulary of size PX, the probability Pr(L) of recognizing a phrase of the language L is given by:

$$Pr(L) = (c^{PX-1})^{E(w)} \qquad\qquad (8.45)$$

The above relation states that the recognition complexity of a language L is equivalent to the recognition complexity of an *equivalent vocabulary* (EV) whose size is:

$$|EV| = (PX-1)E(|1|)+1 \qquad\qquad (8.46)$$

8.9   Review of Recent Work on Task-Dependent Knowledge

8.9.1 Representation

The HEARSAY-II system (Erman, 1977) was designed on the basis of the hypothesize-and-test paradigm, using cooperating independent knowledge sources communicating through a global data structure where all types of hypotheses about the input sentence are written. The knowledge sources are self-activating, asynchronous, parallel processes; the contextual and support connections are explicitly specified; the system structure is modular and suitable for execution on a parallel processing system.

Syntax and semantics are separated in HEARSAY-II. Syntax is represented by a context-free grammar whose rules are in a template

normal form (no rules have more than two symbols in the right side),
(Hayes-Roth and Mostow, 1976).

A word-sequence hypothesizer generates initial word sequences,
starting from hypothesized syllables.

Consider an hypothesized word sequence

$$W = w_1 \ldots w_n$$

that can be derived by the productions:

$$S \rightarrow AT,$$

$$T \rightarrow w_1 V,$$

$$V \rightarrow UX,$$

$$U \Rightarrow w_2 \ldots w_n \qquad\qquad\qquad\qquad (8.47)$$

where $\Rightarrow$ represents a composite derivation. In such a case W is
considered an instance of S. A is said to be the closest left-
missing constituent and X the closest right-missing constituent.
The words that can be adjacent to $(w_1 \ldots w_n)$ are all rightmost
derivations of A or leftmost derivatives of X. Such words are
predictions subject to verification. A knowledge source for word
pair adjacency acceptance is activated after verification, in order
to decide whether the phrase may be extended by concatenating the
verified words.

A semantic template grammar is used for interpretation of
recognized word sequences.

The semantic grammar has distinct templates for each unique
type of semantic form.

Semantic interpretation is accomplished by matching word sequences and templates.

Another interesting system developed at Carnegie-Mellon
University (CMU) is named HARPY. It is a combination of the best
features of HEARSAY-I (Reddy, 1976) and the DRAGON system (J.K.
Baker, 1975) conceived on the basis of a Markov process. The
knowledge source of the HARPY system is a finite-state network
whose input symbols are phonetic transcriptions of the input
utterance. The most interesting feature of the HARPY system is
the use of the so-called focus model of search. It is a graph-
search technique in which all, except a beam of near-miss
alternatives around the best path, are pruned from the search tree

at each segmental decision point, thus constraining the exponential growth without requiring back-tracking.

The Stanford Speech Understanding system (Walker, 1978) has been designed with two major objectives: integration, the process of forming a unified system out of the collection of components, and control, the dynamic direction of the overall activity of the system during the processing of an input utterance. System integration gives a central role to the input-language definition, which is based on augmented phrase-structure (APS) rules. A rule consists of a phrase-structure declaration, which specifies the possible constituents of a phrase, and an augmentation, which is a procedure for computing attributes and factors. Attribute statements determine the properties of particular phrases constructed by the rule; they may compute values for attributes that relate to syntax, semantics, or discourse.

Factor statements make acceptability judgements on phrases and are scored by integers referenced symbolically by linguistic variables. Factors are used for traditional syntactic tests, such as agreement for person and number.

Together, the attribute and factor statements in the procedural part of the rules contain specifications for most of the potential interactions among system components.

The attributes and factors of a phrase are not allowed to depend on the context formed by other phrases that can combine with it to produce larger structures.

The grammar can be tuned to particular discourse situations and language users simply by adjusting factors that enhance or diminish the acceptability of particular interpretations, without the need to rewrite the language definition for each new domain.

The rule procedure is organized as nested conditional statements.

A definition compiler translates a language definition into a form for use by the executive that is the general controller of the system.

Semantics are represented by a network whose nodes represent physical objects, situations, events, sets, etc., and arcs represent binary relations between nodes.

Nodes and arcs are partitioned into spaces.

Semantic composition routines are called directly from the language-definition rules for relating the constituents of a phrase

to the network model. These routines build new network structures
to reflect the underlying meanings of acceptable phrases.

Paxton (1975, 1977) reports interesting comments about the use
of augmented phrase structures (APS) instead of augmented transition
networks (ATN). The main motivation for the use of APS is that
they avoid control commitments, by putting the augmentations into
a single procedure rather than spreading them over a network.

A large contribution to the representation of syntactic and
semantic knowledge for speech understanding has been provided by
Woods et al.(1976), at Bolt, Beranek and Newman (BBN) Inc., Cambridge,
Mass.

The evolution of their project led to a complete integration
of syntax and semantics through the design of a parser capable of
producing a syntactic tree of a sentence and its semantic interpretation.

The HWIM (Hear What I Mean) system follows SPEECHLIS, the
first system developed at BBN. In the older system, the parser
was driven by a modified ATN grammar, which permitted parsing to
start anywhere, not necessarily left-to-right. The semantic
component uses case frame tokens to check for the consistency of
completed syntactic constituents and the current semantic hypotheses.
In HWIM, the Syntax component embodies the syntactic, semantic, and
pragmatic knowledge sources. The parser is built around a pragmatic
grammar which accepts only those utterances that are grammatical,
meaningful, and appropriate to the pragmatic circumstance, given
the previous conversation.

In this grammar, the usual syntactic categories, like NP, VP,
PP are replaced by structures such as "meetings", "trips" and
"budget items". This specialization considerably increases the
predictive power of the grammar. The functions of the parser are:

- to judge the grammaticality of a given word sequence;
- to predict the possible extensions of a hypothesized word sequence;
- to build up a formal, semantic representation of the utterance.

Procedurally, the grammar structure is an ATN machine that is
requested by the control component to take any consecutive sequence
of word matches, called island, and to determine if it can be
parsed as an acceptable subsequence within the grammar. If so, the
syntactic component must be able to return to control a list of
words and categories that would form acceptable extensions from the
island at either end.

The parser must be able to start at any point and to work in
either direction. Predictions of adjacent words are based on the

entire context of the island. The parser treats each island as an
entity that is created either as a single word (i.e., a seed event)
or by adding a single word to an existing island (i.e., as a word
event).

The parser has four basic actions, namely:

- seed-event processing (creation of a new one-word island and
  representation of every allowed path through that word);
- word-event processing (addition of a new word to one end of an
  existing island, extension of those paths that are compatible
  with the word, and elimination of the paths that are not compatible
  with the new word);
- end-event processing, that takes place when an existing island
  has reached one end of the utterence;
- island-collision event processing (merging of two islands with
  a one-word gap between them).

A scoping mechanism is provided within the parser. It determines the set of states in the grammar having the property that a
context-sensitive action can be safely done if its execution in a
right-to-left parse is delayed until the parse has passed through
one of those states.

A joint study by BBN and Univac (Lea, 1976) took into account
prosodic information in order to detect syntactic boundaries.
Unfortunately the authors did not have enough time to evaluate the
effects of integrating the prosodic component into the system.

The influence of the prosodic component was to provide a
positive score increment associated with a parse path, if a prosodic
boundary occurs where predicted by the syntactic analysis, and a
negative score otherwise.

Other interesting work on prosody has been done in France
(Vaissière,1980),in the Soviet Union (Krinov et al., 1975), and at
the University of Washington (Holden and Strasbourger, 1976).

Just to give an idea of the areas of applications, Table 8.3.
reports examples of the TASK-DEPENDENT knowledge representation in
some systems. The complexity is represented by two letters; the
first one refers to the number of words (T: thousand, H: hundreds;
D: tens);   the second letter refers to the word branching factor
(H: >100, M is a branching factor between 20 and 100, L is for a
branching factor lower than 20).

Understanding speech under the control of a grammar is not
purely a parsing action because the input message to be interpreted
is often corrupted and ambiguous. If the grammar of admissible

sentences is simple and can be represented by a finite-state
automaton, a network can be conceived to represent all the possible
interpretations of the speech messages. The problem is that of
pruning the network in order to obtain, hopefully, a single path
that best matches the set of features extracted from the speech
waveform. Syntactic decoding and focus model of search solve the
problem for networks with low branching factors. In these cases,
the redundancy of the speech message compensates for the imprecision
of knowledge and feature extraction. For more complex and less
restricted protocols, approaches in the area of Artificial Intelligence have been proposed. They require complex structures for
knowledge representation, complex rules of inference, and sophisticated strategies for assigning priorities to the processes that
compete for the interpretation of the verbal message. These problems
will be discussed in the next sub-section.

8.9.2 Control strategies and scoring philosophies

Ambiguities and errors in the interpretation of the speech
waveform and the rather poor capacity for modeling the human
knowledge in speech understanding systems require the control of
such systems to be based on a strategy for the search for the best
path in a graph. The nodes in the graph are partial interpretations
of the utterance and the generation of the graph is controlled by
a grammar. The constraints on the search on such a graph depend on
peculiar aspects of the speech understanding mechanism.

Different solutions have been proposed and their evaluation is
still under investigation.

Some of the most used search techniques are the so-called
"best-first", "depth-first", "breadth-first". A critical review
of these methods and a discussion of their applicability to speech
understanding system controllers can be found in Nakagawa (1976).
Another view, followed by Lowerre (1976) consists in pruning the
search tree at each node, keeping only the best nodes at each
expansion, and processing the few remaining nodes in parallel.
Such a technique has been successfully applied in the HARPY, in
the LITHAN and, in a slightly different way in the CNET (Mercier
and Quinton, 1980) systems.

In the HARPY system, the majority of the knowledge is represented within a single network generated from the specifications
of the grammar of the language, the phonetic dictionary of the
lexical words, and the interword juncture rules. The network
consists of a set of states, where each state represents a phonetic
sound.

TABLE 8.3

System	Compl.	Type of Grammar	Semantic Representation	Task
HWIM BIGDICT	TH	Pragmatic ATN	Incorporated in grammar	Trip
HWIM MIDDICT	HM	Pragmatic ATN	Incorporated in grammar	Trip
SRI	HM	APS	Semantic network	Information about ships
HEARSAY I	DL	Procedural embedding	Acorn	Chess-Doctor-Desk calc, news, formants
DRAGON	DL	Markov network	Incorporated in grammar	Chess-Doctro-Desk calc, news, formants
HARPY	TL	Transition network	Incorporated in grammar	Article retrieval
HEARSAY II	TL	Template normal form	Semantic grammar	Article retrieval
IPIT-Moscow	HL	Markov network	Incorporated in grammar	Question-Answering
IBM	TH	Markov network	Incorporated in grammar	Laser patent
IBM	HL	Markov network	Incorporated in grammar	Rayleigh passage
LITHAN	HL	cfg restricted by predicates	Incorporated in grammar	Calendar-Fortran-Comp. network
KYUSHU	HL	Dependency grammar	Incorporated in grammar	Weather forecast
MUSASHINO	HL	List form	Train time tables	Question-Answering system
YAMANASHI	HL	Transition network	Train time tables	Fortran programs
	HL	Transition network	Train time tables	Landscape
TURIN	HL	ATN	Semantic network	Reservation and travel information system
BELL	HL	Transition network	—	Reservation and travel information system
CNET (KEAL)	DL	Transition network	—	Connected digits
	HL	cfg	—	Computer aided design
MYRTILLE	HL	Regular grammar	—	Metereology

Some heuristics are introduced into the HARPY system in order to reduce the computation time, avoiding the computation of all the probabilities required by a correct algorithm based on dynamic programming, as used in the DRAGON system (J.K. Baker, 1975).

The search is limited to considering only those states whose probability is within a threshold amount of the highest state probability at each time sample t.

One of the most interesting approaches to the control of the perceptual activity of a Speech Understanding System has been proposed by Woods et al. (1976). Central to this approach is an entity called a theory.

A theory represents a particular hypothesis about the interpretation of a part or the entire utterance.

Seed hypotheses for elementary constituents of an interpretation are initially emitted. From such partial theories, other hypotheses are derived by a mechanism of prediction and verification. Predictions are dealt with by two kinds of devices: monitors, which are processes waiting for certain entries in the data structure of the utterance interpretation, and proposals, which are elementary hypotheses that are to be evaluated against the input. When a monitor is triggered, an event is created calling for the evaluation of a proposal and the creation of a new theory if the proposed hypothesis is accepted.

In general, a number of events are competing for service by the processor at any moment. The events are maintained in a queue in order of priority, and the assignment of the priorities based on the scoring reached by the events is one of the major problems in designing system controls.

The assignment of priorities is also related to the problem of deciding when seeds should be formed and how the theories should grow. There are two main approaches to this problem: the first consists in making subsequent theories grow in both directions from the middle out, the other consists in making the theories grow from left to right.

The HWIM system contains about 25 flag variables that control different strategy options. Among them, the left-hybrid strategy as described by Wolf (1977) will be recalled here.

After an utterance has been acquired and the acoustic front-end processing has been done to produce a lattice of phonemic hypotheses, the lexical retrieval component is called to determine the words that match best at or near the left end of the utterance. Each of

these words is scored and put in an event queue as "seed" hypotheses.
A word verification component evaluates how well these words match
the signal and contributes, together with the scoring assigned by
the lexical retrieval component on the basis of phoneme scorings,
to the assignment of priorities to the word hypotheses.

The event of the queue with highest priority is presented to
the syntax component as a theory for evaluation. Syntax makes
proposals for all words and semantic categories that are grammatically possible at each end of this theory. A cycle consisting of
syntactic proposals, lexical retrievals, and verification of new
events is started and repeated until an event is processed that
spans the utterance and is acceptable to the grammar. The number
of theories allowed for interpreting an utterance is limited. When
the upper bound has been reached, the process of interpretation of
an utterance is stopped.

In HWIM, score of a theory is an estimate of its being correct,
given the information from the various knowledge sources. This
score is computed according to the following relation:

$$P(F_i/E_i) = P(E_i/F_i)P(F_i)/P(E_i)$$
(8.48)

where F_i is a theory and E_i is the evidence with which something
has been recognized.

When there are several types of evidences, Eq.   (8.48) is
factored as follows:

$$P(F_i/E_{j1}E_{j2}...E_{jn}) = P(E_{j1}/F_i)/P(E_{j1})$$

$$x(P(E_{j2}/F_i)/P(E_{j2}) \text{ x...x} P(E_{jn}/F_i)$$

$$/P(E_{jn}) \ P(E_i)$$

under the assumption that the different pieces of evidence are
independent. In order to base the system performances on its
ability to hear, rather than its ability to guess what will be said,
the a priori term $P(F_i)$ is considered equal for all utterances.

The terms $P(E_{jk}/F_i)/P(E_{jk})$ are learned by a process called
calibration.

The log of the probabilities obtained for a given theory are
called quality scores.

A new parameter, named shortfall score, is obtained as a difference between the particular quality score for a theory and an upper bound on possible quality scores for any theory covering the same portion of the utterance. Dividing the shortfall score by the duration of the "island" corresponding to the portion of the utterance under consideration, a shortfall density score is obtained.

Interesting theoretical properties of such types of scoring are given in Woods (1977).

An initial profile of maximum possible scoring is contructed during the initial scan for seed words.

Woods (1977) reports comparative results on 10 utterances using different scoring methods (quality, quality density, shortfall, shortfall density).

The best results (50% of correct interpretations) are obtained using shortfall densities.

Wolf (1977) reports results of 124 utterances by three speakers with a dictionary of about 1000 words. The rate of correct interpretation is 44% and the execution time is 1350 times real time.

The control of the SRI system centers on a system "Executive" applying the rules of the language definition, organizing hypotheses and results, and assigning priorities.

It builds a "parse net" to hold intermediate data. Two types of tasks interact to build the net: the predict task, which leads to predictions for words to be verified in the input waveform, and the word task, which gets words from the acoustic and uses them to construct new phrases. The predict task operates in a top-down manner, the word task operates in a bottom-up manner, and one task predicts the other recursively.

Whenever a new constituent is inserted into an incomplete phrase, any adjacent constituents that are missing can be predicted.

A prediction serves as an intermediary between two sets of incomplete phrases: consumer phrases that are all missing a constituent of the predicted category at the predicted location in the input, and producer phrases, all of which might supply the missing constituents.

Both tasks can operate bidirectionally through the input and are guided by a lookahead mechanism to avoid unnecessary operations.

Nodes in the parse set are either "phrases" or "predictions". Phrases correspond to words or composition rules. "Terminal" phrases contain a single word and are formed after acoustic tests. "Non-terminal" phrases are formed by applying the composition rules.

Priorities are set on the basis of linguistic and acceptability judgements. Priority setting is based on scores, provided by the acoustic mapper, and ratings, that depend on how well a phrase may be embedded in the other phrases to form a correct sentence.

These ratings are modified in dependence on the control strategy being used and the result is the priority of the task to be performed for the phrase.

Paxton (1977) carried out a set of interesting experiments based on the analysis of variance of system accuracy and run time. The analysis of variance is a statistical technique for computing the probability that the observed effects or interactions are really caused by experimental variables, rather than the result of random variation.

The following techniques were experimentally investigated for priority setting.

- Island driving: going in both directions from arbitrary starting points in the input utterance;
- Mapping all at once: testing all the words at once at a given location, retaining the ones reaching scores over a threshold;
- Context checks: taking into account the restrictions of the possible sentential contexts as part of setting priorities;
- Focus by inhibition: focusing the system on selected alternatives by inhibiting competition with candidates that are not likely to enter into the interpretation of the utterance, concentrating the attention on a particular set of potential interpretations.

The main results are that "context-checks" and "mapping-all-at-once" techniques had good effects while "focus by inhibition" and "island driving" had bad effects on performance when used for priority setting.

In the HEARSAY-II (Hayes-Roth and Lesser, 1977) each KS is data-directed, i.e., it monitors a blackboard for arrival of data matching its precondition pattern; whenever a precondition of a KS is matched, the KS is invoked and, after execution, a logic, associated with the KS, is evaluated to determine how to modify the database in the vicinity of the precondition pattern that triggered the invocation.

The data pattern matching the precondition is called stimulus frame (SF); the changes it makes are called response frame (RF).

The numerous potential activities of the KSs are scheduled by a controller whose strategy is based on a so-called "focus of attention".

Resource allocation and KS scheduling depend on where the attention of the control is focussed. Focus of attention results in the assignment of priorities to the various KS tasks. Priorities depend on the values of a function based on measures depending on the following principles.

(1) The competition principle: the best of several local alternatives should be performed first.

(2) The validity principle: KSs operating on the most valid data should be executed first.

(3) The significance principle: those KSs whose RFs are most important from an a priori point of view should be executed first.

(4) The efficiency principle: those KSs which perform most reliably and inexpensively should be executed first.

(5) The goal satisfaction principle: those KSs whose responses are most likely to satisfy processing goals should be executed first.

The major operational principle for focussing is to schedule for earliest execution the most desirable KS invocation according to the above-mentioned principles.

Each KS invocation is characterized by a number of attributes. Its SF has a credibility value estimating likelihood that the detected pattern of hypotheses and links is valid and satisfies the KSs precondition.

The objectives of the principles can be achieved if the desirability of a KS invocation is computed by an increasing function of the credibility of its SF, the estimated level, duration, and validity of RF hypotheses, and the estimate reliability of the KS.

Using these parameters, an expected RF value is computed.

The strategy of the controller is based on a comparison between the expected RF value and the state of the blackboard at time t. The state is indicated as $S(t)$ and is given by the maximum of the values of all hypotheses representing interpretations containing the point t.

The value $V(h)$ of an hypothesis is an increasing function of its level, duration, and validity (depending on the matching with the acoustic data).

S(t) can be considered as a degree of problem solution, and employed to decide whether a prospective action is likely to improve the current state of understanding.

If the estimated value of a RF hypothesis exceeds S(t) anywhere in the corresponding interval, the KS invocation should be considered very desirable.

TABLE 8.4

Strategy used by some speech understanding systems

System	Search strategy
HWIM	Hybrid best-first shortfall density
SRI	Context-check, mapping all-at-once
HEARSAY I	Best-first with backtracking
DRAGON	All paths in parallel with no back-tracking
HARPY	Best few in parallel with no back-tracking
IBM	Centisecond Markov model on CMU 1011 words AIXOS protocol
LITHAN	Pruning top-down
YAMANASHI Landscape	Rule-driven word prediction
MUSASHINO	Best-first with backtracking
NEC (40 sp.)	Dynamic Programming (connected words)
CNET	Best few in parallel
MYRTILLE	Dialogue-based understanding
UNIVERSITY OF TURIN	Best few in parallel
HEARSAY II	Focus of attention

Different types of competitions between KS tasks may be considered, depending on the extension of the hypotheses on which S(t)
and the estimated RF value are compared.

Experiments carried out by Hayes-Roth and Lesser are in favor
of word-specific competitions (77% of correct semantic interpretations and 71% correct word recognition on a corpus of 61 test
sentences). Table 8.4. shows the strategies used by some recent
systems. Among them, only the DRAGON and the HWIM system use
admissible strategies. Recognition performances depend on the task
complexity; systems like the NEC, HARPY and IBM have reached on a
single speaker and with very constrained protocols recognition rates
higher than 99%.

8.10 References

Books by Feigenbaum (1963) and Minsky (1968) are a good introduction to knowledge representation.

Books by Ernst and Newell (1969), Sacerdoti (1977), Nilsson
(1971), Chang and Lee (1973) are good references for problem solving.

Books by Newell et al. (1973), Reddy (1975), Dixon and Martin
(1979), Walker (1978), Lea (1980), Cole (1980) Fain (1977), Zagoruiko
(1976) give a good view of the state of the art in the design of
Speech Understanding Systems. Of great interest for knowledge representation and control strategies for speech understanding are the
final reports of the main contractors of the ARPA project (Walker,
1976, Woods et al.,1976, Reddy et al.,1977) and the Ph.D theses by
Erman (1974a), Paxton (1977), Lowerre (1976), Nakagawa (1976), and
Quinton (1980).

Advances in this area can be mostly found in the proceedings
of the International Joint Conference of Artificial Intelligence
(IJCAI), of the American Association of Artifical Intelligence (AAAI),
of the Conference of computational linguistics (COLING), ARSO (USSR),
and in the American Journal for Computational Linguistics, Artificial
Intelligence and Cognitive Sciences.

9. AUTOMATIC LEARNING OF FUZZY RELATIONS

9.1 Introduction

The problem considered in this chapter refers to those cases for which pattern descriptions cannot always be obtained in a precise way and, furthermore, in which the interpretation of a pattern is controlled by knowledge sources whose content may also be imprecise.

Such cases are usual in both speech and image understanding because the available systems used to recognize elementary items in the pattern under observation are imprecise and the knowledge representation inside an automatic information processing system has very often a degree of vagueness due to our inability to represent adequately all we can see or all we can hear.

Nevertheless, there is a great deal of redundancy in most of the patterns to be interpreted. Such redundancy can compensate the imprecision of the knowledge models and the pattern description system, allowing acoustic or visual patterns to be correctly interpreted. Very important for the achievement of such a task is the capability of correctly representing and processing the various degrees of vagueness implicit in the knowledge sources and used during the interpretation of a pattern.

This chapter proposes a method for automatic learning of the degrees of vagueness of a knowledge source that has to be used to interpret an imprecise pattern description.

It is assumed that the possibly partial interpretations of a pattern are to be obtained by the scheme of Fig. 9.1.

Let $f \epsilon F$ be the pattern to be interpreted. Preliminary operations, such as digitization, filtering, Fast Fourier Transform,
etc., lead to a pattern representation in the form of a matrix $A(f)$
of numbers.

Then a description process takes place, giving a description
$U(f)$ of the pattern. Such a description can be obtained by asking
questions relating to the numerical representation $A(f)$ of the
pattern and by combining the answers, that are linguistic variables,
to form a string. There is a possibility that more answers can be
obtained for the same question and that each answer has to be
assigned a degree of applicability to the pattern under analysis.

Following Zadeh (1975), the pattern descriptions $U_i(f)$, $\forall i \epsilon$
$(1, N_f)$, are phrases composed of fuzzy linguistic variables. Such
variables belong to an alphabet Σ and $U_i(f)$ is an element of Σ^*.
The index i will be omitted in the following for the sake of
simplicity.

Each description U obtained from the matrix $A(f)$ is assigned
a possibility value $\Pi_{A(f)}(U)$ obtained by composition of the
compatibility values of the fuzzy variables in U.

Pattern interpretation is seen as a naming relation in which
an interpretation $x \epsilon S_N$ is assigned to the pattern to be interpreted; S_N is the set of the possible names. Again, the naming
relation is a fuzzy relation because of the imprecise knowledge
we have about the correspondence between names and descriptions.
In order to profit from the information redundancies inside the
pattern and to infer correct global interpretations from the
interpretation of some pattern components properly, it is
necessary to evaluate Poss{f is x}, that is, the possibility that
the interpretation x applies to the pattern or the subpattern f.

Let $R_f(U)$ be the fuzzy restriction associated with the description U of f; let $R(x,V)$ be the fuzzy relation between a description
U of f and an interpretation x; the meaning of an interpretation x
of f is represented by the restriction $R_f(x)$, which is generally
obtained by the following general rule of inference:

$$R_f(x) = R_f(U) \circ R(x,U) \qquad\qquad (9.1)$$

Following Zadeh's theory of possibility (Zadeh, 1978), the following rule for computing Poss{f is x} is obtained from (9.1):

$$\text{Poss } \{f \text{ is } x\} = \underset{U\in\Sigma*}{\text{Sup}} \prod_{A(f)} {}^{(U)\wedge\mu}R(x,U) \qquad\qquad (9.2)$$

where $\wedge$ is the min operator and $\mu_{R(x,U)}$ is the compatibility of the name x with the description U.

The main motivation for using fuzzy relations to assign names to patterns of subpatterns is because names can be seen as partially overlapped languages, defined over the set U $\Sigma*$ of possible pattern descriptions.

Assume that there are only two names, x and y, defining two languages L_x and L_y over U. For those descriptions U for which:

$$U\in L_x \text{ and } U\notin L_x\cap L_y$$

there should hold $\mu_{R(x,U)} = 1$ and $\mu_{R(y,U)} = 0$; analogously, if:

$$U\in L_y \text{ and } U\notin L_x\cap L_y$$

there should hold $\mu_{R(x,U)} = 1$ and $\mu_{R(y,U)} = 0$, but, for those cases for which:

$$U\in L_x\cap L_y,$$

it is desirable to have $\mu_{R(x,U)}$ and $_{R(y,U)}$ different from zero and from one.

The distinction between possibility and probability, illustrated by Zadeh (1978) recently, can be applied to the above-mentioned naming problem, leading to the conclusion that the meaning of a naming is properly expressed by a possibility measure.

This chapter deals with the problem of learning the memberships of fuzzy naming relations from imprecise descriptions of the input patterns in order to maximize the number of cases that the right (known) name is assigned to a pattern with a degree of possibility higher than the other possible names. The learning algorithm is based on the frequencies of occurrence of the possibilities $\Pi_{A(f)}(U)$.

9.2   Formal Definition of the Problem and an Example of Application

9.2.1 Generalities

Assume for the sake of simplicity, that $A(f) = \{a_1(f), a_2(f),$ ..., $a_i(f),$ ... , $a_q(f)\}$ is a vector of attributes taken on a real-world phonomenon $f \epsilon F$. Let A be the universe on which the vector $A(f)$ may vary.

Let $\Sigma = \{u_1, u_2,...., u_j,...., u_m\}$ be an alphabet of linguistic variables describing $A(f)$. Let $U \subseteq \Sigma^*$ be the language of all the acceptable descriptions. Let $U \epsilon U$ be a string of this language describing a pattern or a subpattern f. The possibility $\Pi_{A(f)}(U)$ is a possibility distribution function induced by the proposition:

$$P = f \text{ is (described by) } U \tag{9.3}$$

For each symbol u_j of the string U, a fuzzy restriction:

$$R_j(A(f)) = u_j \tag{9.4}$$

can be considered. The restriction (9.4) represents the fact that u_j is a fuzzy linguistic variable whose meaning is ill-defined, as in the following statements: "a_i is high", "a_i and a_k are very different". The description U, being composed of fuzzy linguistic variables, is itself a composite fuzzy restriction; $A(f)$ is an implied composite attribute of f which takes values on the vector space A.

Due to (9.4.) the possibility distribution:

$$\Pi_{A(f)}(u_j) = \mu_{u_j}(A(f)) \tag{9.5}$$

expresses the degree of worthiness of U_j given the vector $A(f)$ of attributes. The string U is a concatenation of the variables u_1, $u_2,....,u_m$; thus its possibility distribution over A is given by:

$$\Pi_{A(f)} (u_1 \cdots u_m) = \Pi_{A(f)} (u_1) \wedge \cdots \wedge \Pi_{A(f)} (u_m) \qquad (9.6)$$

In practice, given a phenomenon f, a set of attributes $A(f)$ is evaluated. One or more linguistic variables u_i can describe $A(f)$ and several strings can be constructed using these variables. Each string U is assigned a degree of possibility $\Pi_{A(f)} (U)$ that can be evaluated by (9.6).

In order to make the scheme of Fig. 9.1 applicable to real problems, it is necessary to know the following functions:

$$\Pi_{A(f)} (u_j) \quad \forall j \in [1,m] \dagger$$

for every u_j that may describe $A(f)$ and for all the possible values of $A(f)$ and $\mu_{R(x,U)}$, $\forall x \in S_N$, $\forall U \in U \subseteq \Sigma^*$.

The possibilities $\Pi_{A(f)} (u_j)$ can be established from histograms on the basis of experience or in a purely subjective way, as in the following sample.

Example 9.2.1

Let f represent a person of a group, $A(f) = a_1(f)$ is the age of f, and u_1 is the adjective 'young', a fuzzy subset of $I = [0,100]$, characterized by the membership function proposed by Zadeh (1975):

$$\mu_{young} = 1 - S(a_1; \alpha, \beta, \gamma)$$

with:

$$S = 0 \qquad \qquad \text{for } a_1 < \alpha$$

$$S = 2 \left(\frac{a_1 - \alpha}{\gamma - \alpha} \right)^2 \qquad \text{for } \alpha \leq a_1 \leq \beta$$

---

† We will denote, in the following, by $[z_1, z_2]$ the closed interval $z_1 \leq z \leq z_2$ and by (z_1, z_2) the open interval $z_1 < z < z_2$.

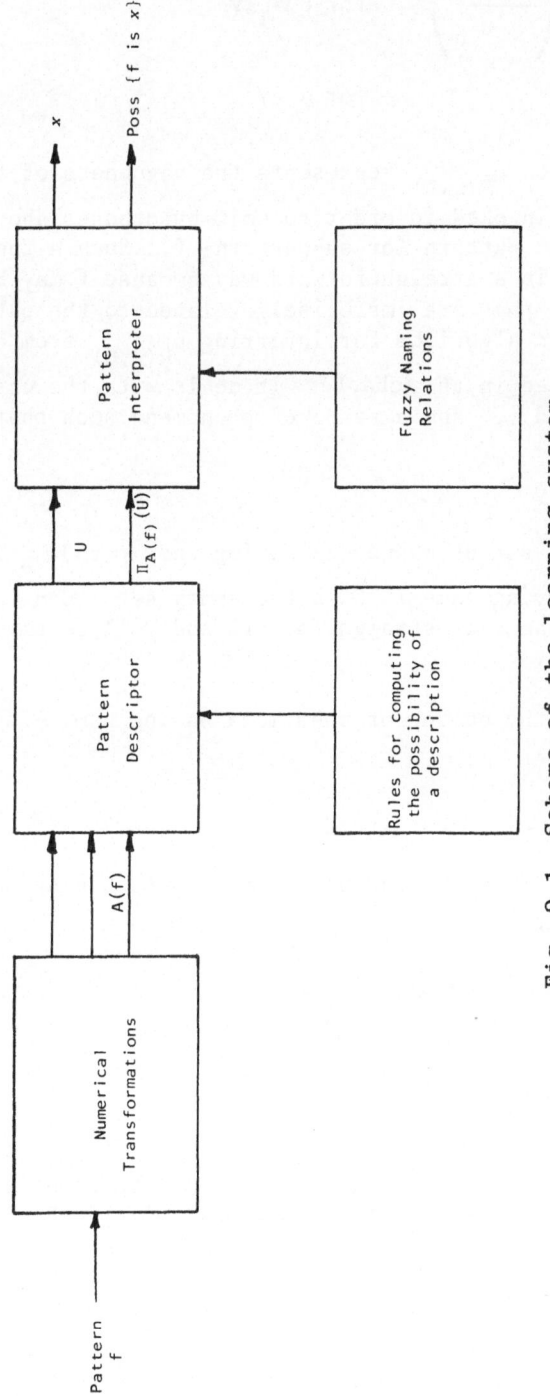

Fig. 9.1   Scheme of the learning system

$$S = 1 - 2 \left(\frac{a_1 - \alpha}{\gamma - \alpha} \right)^2 \qquad \text{for } \beta < a_1 \leq \gamma$$

$$S = 1 \qquad\qquad\qquad\qquad\qquad \text{for } \alpha \geq \gamma \qquad\qquad\qquad\qquad (9.7)$$

The function $\mu_{R(x,U)}$ represents the vagueness of the knowledge source that is invoked in order to emit hypotheses about the interpretation of the pattern (or subpattern) f. Such a function cannot be established in a straightforward way because f may be described by many strings that are imprecisely related to the pattern to be interpreted. An algorithm for inferring $\mu_{R(x,U)}$ from experiments will be presented in this chapter. It deals with the case of two names, say x and y, and a set F of phenomena such that:

$$F = F_x \cup F_y \qquad F_x \cap F_y = \emptyset$$

where F_x is the set of phenomena having name x and F_y is the set of phenomena having name y; $\emptyset$ is the empty set. Generalization to more than two names is straightforward and will be omittted for the sake of brevity.

Let E_x be the number of phenomena having name x and for which (9.2) gives:

$$\text{Poss } \{f \text{ is } y\} \geq \text{Poss } \{f \text{ is } x\}$$

Conversely, let E_y be the number of phenomena having name y and for which (9.2) gives:

$$\text{Poss } \{f \text{ is } x\} \geq \text{Poss } \{f \text{ is } y\}$$

Let

$$E = E_x + E_y \qquad\qquad\qquad\qquad\qquad\qquad\qquad (9.8)$$

be the total number of the cases for which the wrong name is assigned to a phenomenon with higher possibility than the true name.

The algorithm developed in the following has the purpose of finding a set of functions $\mu_{R(x,U)}$ and $\mu_{R(y,U)}$ for which the value of E; defined by (9.8), is minimum. Such a criterion allows one to infer naming relations allowing the right name to be assigned by (9.1) to a phenomenon f with the highest plausibility value the highest number of times.

The proposed algorithm allows one to learn the naming relations by steps, that is with the possibility of modifying the relations learned with some sets of input samples, given a new input sample.

9.2.2 An example of application

The need for the theory presented in this chapter arose during investigations on speech understanding systems. One of the many cases where fuzzy naming relations are of interest is the assignment of phonetic features.

As it has been described in Chapter 4, if a consonant is 'nonsonorant', broad frequency intervals of energy concentrations are used for further classification.

After the intervals corresponding to vowels and consonants of a spoken sentence are delimited, the features introduced for the consonants in Chapter 3 are renamed as follows:

a_1 = R,

a_2 = minimum depth of the dip of R in the consonant,

a_3 = absolute value of the minimum of S in the dip of S during the consonant,

a_4 = duration of the consonant,

a_5 = minimum depth of the dip of S,

a_6 = maximum depth of the dip of R.

For every consonant (subpattern f), the six attributes a_i ($i \in$ [1,6]) are evaluated. For each attribute a_i two linguistic variables h_i (high a_i) and l_i (low a_i) are defined as fuzzy subsets of the interval on which a_i may vary.

The two names, sonorant (x) and nonsonorant (y) are defined over a set of strings $U \in \Sigma^*$, where:

$$\Sigma = \{h_1, l_1, \ldots, h_i, l_i, \ldots, h_6, l_6\}$$

On the basis of experience and in accordance with experimental phonetics it is possible to consider U as the union of two subsets $U_x \subseteq \Sigma_x^*$ and $U_y \subseteq \Sigma_y^*$, where Σ_x is the set of symbols corresponding to properties of sonorant consonants and Σ_y is the set of symbols corresponding to properties of nonsonorant sounds.

For the example considered in this section, the following definitions are acceptable:

$$\Sigma_x = \{h_1 \; l_2 \; l_4 \; l_5 \; l_6\}$$

$$\Sigma_y = \{l_1, h_2, l_3, h_4, h_5, h_6\}.$$

Learning consists in determining the memberships associated to the naming relations defining x over U_x and y over U_y, given a set of examples of sonorant and nonsonorant consonants. For each sample f, the membership functions $\mu_{h_i}(f)$ and $\mu_{l_i}(f)$ are obtained from fuzzy restrictions; it is assumed that:

$$\mu_{h_i}(f) = 1 - \mu_{l_i}(f).$$

An experiment has been carried out with 400 consonants extracted from sentences pronounced by 4 speakers. Using non-fuzzy relations and a definition of error according to (9.8), an error rate of about 10% was obtained; this error reduces to 5% with fuzzy naming relations and did not show significant variation in other experiments, consisting of sets of different sentences enunciated by different speakers.

9.3   A Simple Preliminary Learning Case

A simple preliminary case will be considered in this section to show how the problem of learning fuzzy naming relations has been approached.

Let F be a set of phenomena (patterns or subpatterns) partition-ed   into two subsets, namely F_x and F_y, i.e.:

$$F = F_x \cup F_y \text{ and } F_x \cap F_y = \emptyset$$

Let $A(f) = \{a_1(f), a_2(f)\}$ be the set of attributes taken on every phenomenon $f \in F$.

Let $\Sigma = \{u_1, u_2\}$ be the alphabet of the symbols from which the descriptions of f are made. Let $\Sigma_x = \{u_1\}$ and $\Sigma_y = \{u_2\}$.

We are interested in the possibilities of the naming relations, assigning names x and y to a phenomenon $f \in F$, that is:

$$\text{Poss } \{x \text{ is } f\} = \mu_{R(x,u_1)} \land \Pi_{A(f)}(u_1) \qquad (9.9)$$

$$\text{Poss } \{y \text{ is } f\} = \mu_{R(y,u_2)} \land \Pi_{A(f)}(u_2)$$

The problem of abstraction consists in inferring $\mu_{R(x,u_1)}$ and $\mu_{R(y,u_2)}$ in order to minimize the total error defined by (9.8). For the sake of simplicity, the right-hand possibilities are renamed as follows:

$$\mu_{R(x,u_1)} = x$$

$$\mu_{R(y,u_2)} = y$$

$$\Pi_{A(f)}(u_1) = h$$

$$\Pi_{A(f)}(u_2) = k \qquad (9.10)$$

Let

$$Q_N = \{f_1, f_2, \ldots, f_j, \ldots, f_N\}$$

be the training set of experiments; Q_N be partitioned into four sets Q'_x, Q''_x, Q'_y, Q''_y according to the following definitions:

$$f_j \in Q'_x \qquad \text{iff} \qquad (f_j \in F_x) \land (h \geq k)$$

$$f_j \in Q''_x \qquad \text{iff} \qquad (f_j \in F_x) \land (h < k)$$

$$f_j \in Q'_y \qquad \text{iff} \qquad (f_j \in F_y) \land (k \geq h)$$

$$f_j \in Q''_y \qquad \text{iff} \qquad (f_j \in F_y) \land (k < h) \qquad (9.11)$$

Let N'_x, N''_x, N'_y, N''_y be respectively the cardinalities of Q'_x, Q''_x, Q'_y, Q''_y, respecting the relation:

$$N = N'_x + N''_x + N'_y + N''_y$$

Let h_j and k_j be the values assumed by h and k for the phenomenon f_j; for the sake of simplicity the right-hand possibilities of (9.9) are renamed as follows: ·

$$\text{Poss } \{f_j \text{ is } x\} = \Pi_j^{(x)}$$

$$\text{Poss } \{f_j \text{ is } y\} = \Pi_j^{(y)} \tag{9.12}$$

The value of $\Pi_j^{(x)}$ depends on x and h_j, the value of $\Pi_j^{(y)}$ depends on y and k_j. Assuming that x and y can both vary in the interval $[0,1]$, an assignment of memberships (x,y) may vary in a square Q of the x-y plane. Given a phenomenon f_j, Q may be subdivided into two domains, Q_j^t and Q_j^f. Q_j^t is the domain representing the assignments of memberships for which the possibility of the right hypothesis is higher than the possibility of the wrong hypothesis; Q_j^f is the domain representing the assignments of memberships for which the possibility of the wrong hypothesis is higher than the possibility of the right hypothesis. Figure 9.2 shows the domains Q_j^t and Q_j^f for the four cases defined by (9.11).

It is interesting to consider how the Q_j^t and Q_j^f, $\forall j \in [1,N]$, may be used to subdivide Q into error domains labelled $Q_k(Q_N)$, such that $Q_k(Q_N)$ contains the set of assignments (x,y) for which k phenomena of Q_N receive a wrong name with higher possibility.

Subdividing the set Q_N into the four subsets Q'_x, Q''_x, Q'_y, Q''_y according to (9.11), the phenomena f_j, belonging to each subset, can be ordered according to decreasing values of k_j for Q'_x increasing values of h_j for Q''_x, decreasing values of h_j for Q'_y and increasing values of k_j for Q''_y. For the sake of simplicty, the four ordered sequences will be obtained by renaming the k_j and h_j as follows:

$$f \in Q_x \quad \begin{cases} k_j \to k_p \geq k_{p+1} & (p \in [0,N'_x], \; k_0 = 1, \; k_{N'_x+1} = 0) \\ h_j \to h_q \leq h_{q+1} & (q \in [0,N''_x], \; h_0 = 0, \; h_{N''_y+1} = 1) \end{cases}$$

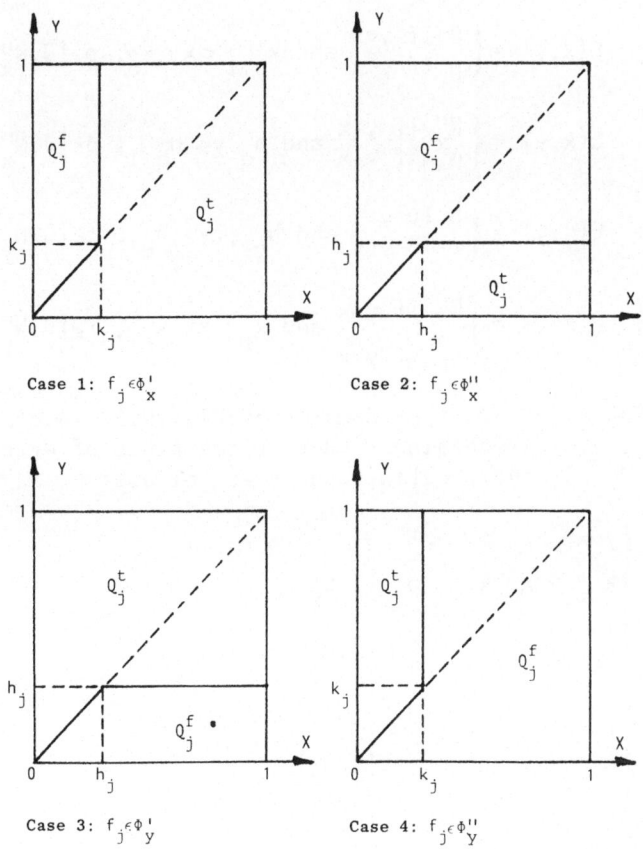

Fig. 9.2 Domains Q_j^t and Q_j^f for all possible assignments (x,y) in the x-y plane.

$$f \epsilon Q_y \begin{cases} k_j \to k_r \leq k_{r+1} & (r \epsilon [0, N_y''], \quad k_0 = 0, \quad k_{N''+1} = 1) \\[3mm] h_j \to h_t \geq h_{t+1} & (t \epsilon [0, N_y'], \quad h_0 = 1, \quad h_{N'+1} = 0) \end{cases}$$

An example of the domains $Q_k (Q_N)$ is represented in Fig. 9.3. for the four cases listed in the following:

Case 1: $E(x,y) = \begin{cases} 0 & \text{if } y<x \\ p & \text{if } y \geq x \end{cases}$ and $k_{p+1} < x \leq k_p$, $p \epsilon [0, N_x'']$

Case 2: $E(x,y) = \begin{cases} N_x'' & \text{if } y \geq x \\ q & \text{if } y<x \end{cases}$ and $h_q \leq y < h_{q+1}$, $q \epsilon [0, N_x'']$

Case 3: $E(x,y) = \begin{cases} 0 & \text{if } y>x \\ t & \text{if } y \leq x \end{cases}$ and $h_{t+1} < y \leq h_t$, $t \epsilon [0, N_y']$

Case 4: $E(x,y) = \begin{cases} N_y'' & \text{if } y \leq x \\ r & \text{if } y>x \end{cases}$ and $k_r \leq x < k_{r+1}$, $r \epsilon [0, N_y'']$

Combining the four error domains of Fig. 9.3 into a single diagram, Fig. 9.4 is obtained. Here three types of error domains can be considered; their definition is as follows:

$$\Lambda_{p,r} = \begin{cases} y>x \\ k_{p+1} < x \leq k_p & p \epsilon [0, N_x'] \\ k_r \leq x < k_{r+1} & r \epsilon [0, N_y''] \end{cases}$$

$$\Theta_{t,q} = \begin{cases} y<x \\ h_{t+1} < y \leq h_t & t \epsilon [0, N_y'] \\ h_q \leq y < h_{q+1} & q \epsilon [0, N_x''] \end{cases}$$

$$\Omega_{p,t} = \begin{cases} y=x \\ k_{p+1} < x \leq k_p & p \epsilon [0, N_x'] \\ h_{t+1} < y \leq h_t & t \epsilon [0, N_y'] \end{cases}$$

The number of errors $E_\Lambda(p,r)$, $E_\Theta(t,q)$ and $E_\Omega(p,t)$, corresponding respectively to the error domains $\Lambda_{p,r}$, $\Theta_{t,q}$ and $\Omega_{p,t}$, are expressed as follows:

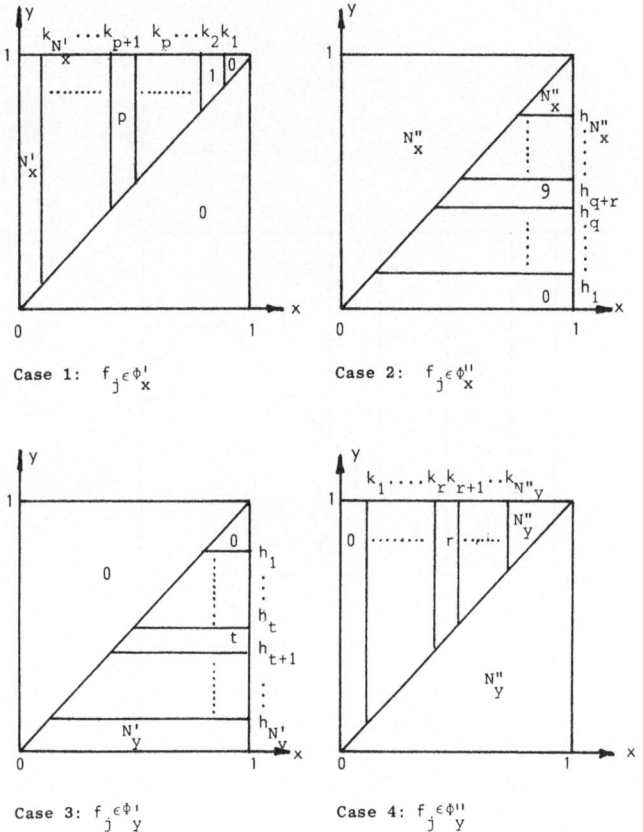

Fig. 9.3   Error distributions for all possible assignments $(x\text{-}y)$
in the $x\text{-}y$ plane, for $f \in Q'_x$, Q''_x, Q'_y, and Q''_y respectively

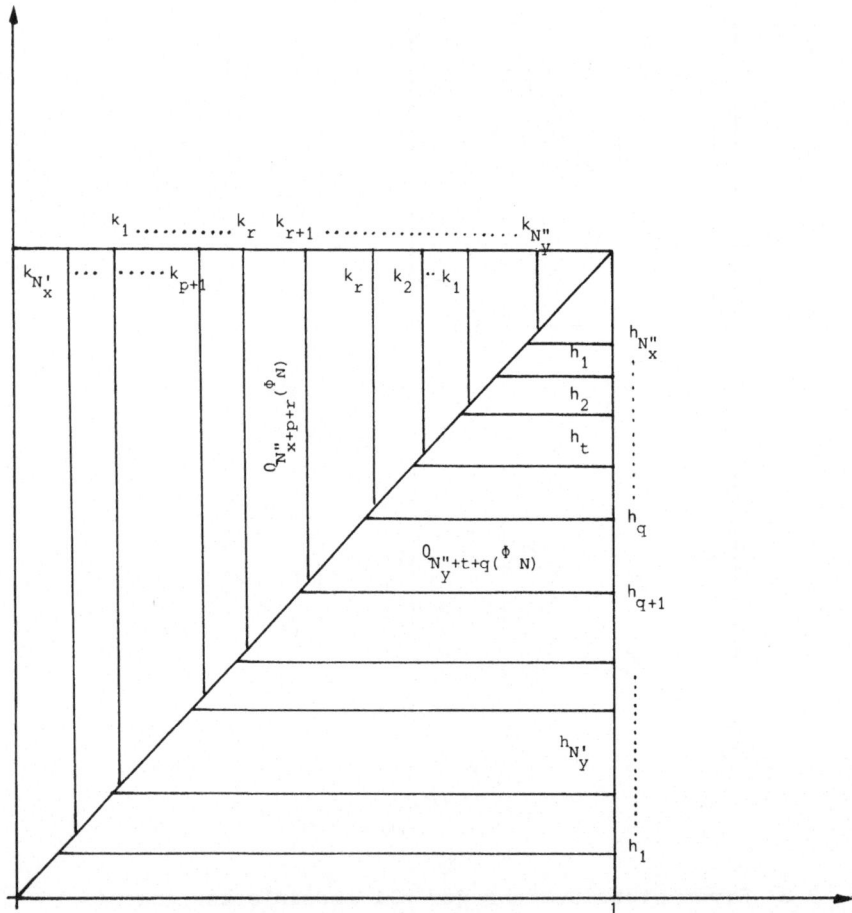

Fig. 9.4 Error domains $Q_k(Q_N)$ **for the whole sample** Q_N.

$$E_\Lambda(p,r) = N''_x + p + r \qquad p\epsilon[0,N''_x], \ r\epsilon[0,N''_y]$$

$$E_\Theta(t,q) = N''_y + t + q \qquad t\epsilon[0,N'_y], \ q\epsilon[0,N''_x]$$

$$E_\Omega(p,t) = N''_x + N''_y + p + t \quad p\epsilon[0,N'_x], \ t\epsilon[0,N'_y]$$

Notice that in particular cases some error domains may not exist and moreover, as $r \le N''_y$ and $q \le N''_x$, the following relations hold:

$$E_\Lambda(p,r) \le E_\Omega(p,t) \ \text{ and } \ E_\Theta(t,q) \le E_\Omega(p,t)$$

We can then avoid further consideration of the domains $\Omega_{p,t}$.

The problem of assigning the memberships of the naming relations in order to minimize the total error E defined by (9.8) requires the values of k_p, h_q, h_t, and k_r to be known for some training set of phenomena belonging to F_x and F_y respectively.

Without any loss in generality it may be assumed that the above-mentioned possibilities have joint probability densities represented by $\gamma(h_q, k_p)$ and $\delta(h_t, k_r)$. We can also define their marginal probability densities $f(k_p)$, $g(k_r)$, $p(h_t)$, $d(h_q)$ and the corresponding cumulative functions $F(k_p)$, $G(k_r)$, $P(h_t)$, and $D(h_q)$.

The algorithm continues with the computation of the probability of having a certain number of errors in some areas α and β of the square Q in the x-y plane. Then the average number of errors in such areas, made over all the possible samples, is derived and those areas in which the average error defined by (9.8) is minimum are found. The details of this computation can be found in De Mori and Saitta (1980), together with an algorithm for extending the conditions of minimum error to those cases for which frequency distributions of k_p, h_q, h_t, and k_r rather than probability densities are available in discrete domains. The development of this research has shown some conditions for the existence of a solution to the learning problem. The relations obtained have been further generalized to the case of finite languages for which an algorithm has been presented for inferring by experiments the areas of minimum error.

10. TOWARDS A PARALLEL SYSTEM

10.1   A New Model for Lexical Access

An efficient knowledge use is made possible by a task decomposition which produces an iterative array of computational activities. This is another novelty introduced here which extends to complex tasks, such as perceptual models, the advantages of using iterative computational structures. These advantages have been widely appreciated, for example, in the design of cellular iterative arrays for performing arithmetic operations in modern computers.

Figure 10.1 shows an example of a three-dimensional iterative network of computational activities. Each computational activity is performed by a set of concurrent or cooperating processes. The horizontal dimension of the network corresponds to time, while the vertical dimension corresponds to levels of abstraction; along the vertical dimension, there are circles representing processes for the generation of sets of lexical hypotheses starting at different time instants. These time instants can only correspond to the beginning of PSS hypotheses.

The activities $A1, A2, \ldots, An, \ldots, AN (1 \leq n \leq N)$ are auditory activities. A collection of computational activities called "Experts" transforms the input signal into a description of acoustic cues. When the acoustic cues of a pseudo-syllabic segment are found, messages containing descriptions of acoustic cues are sent to upper perceptual levels. The first activity A1 is accomplished

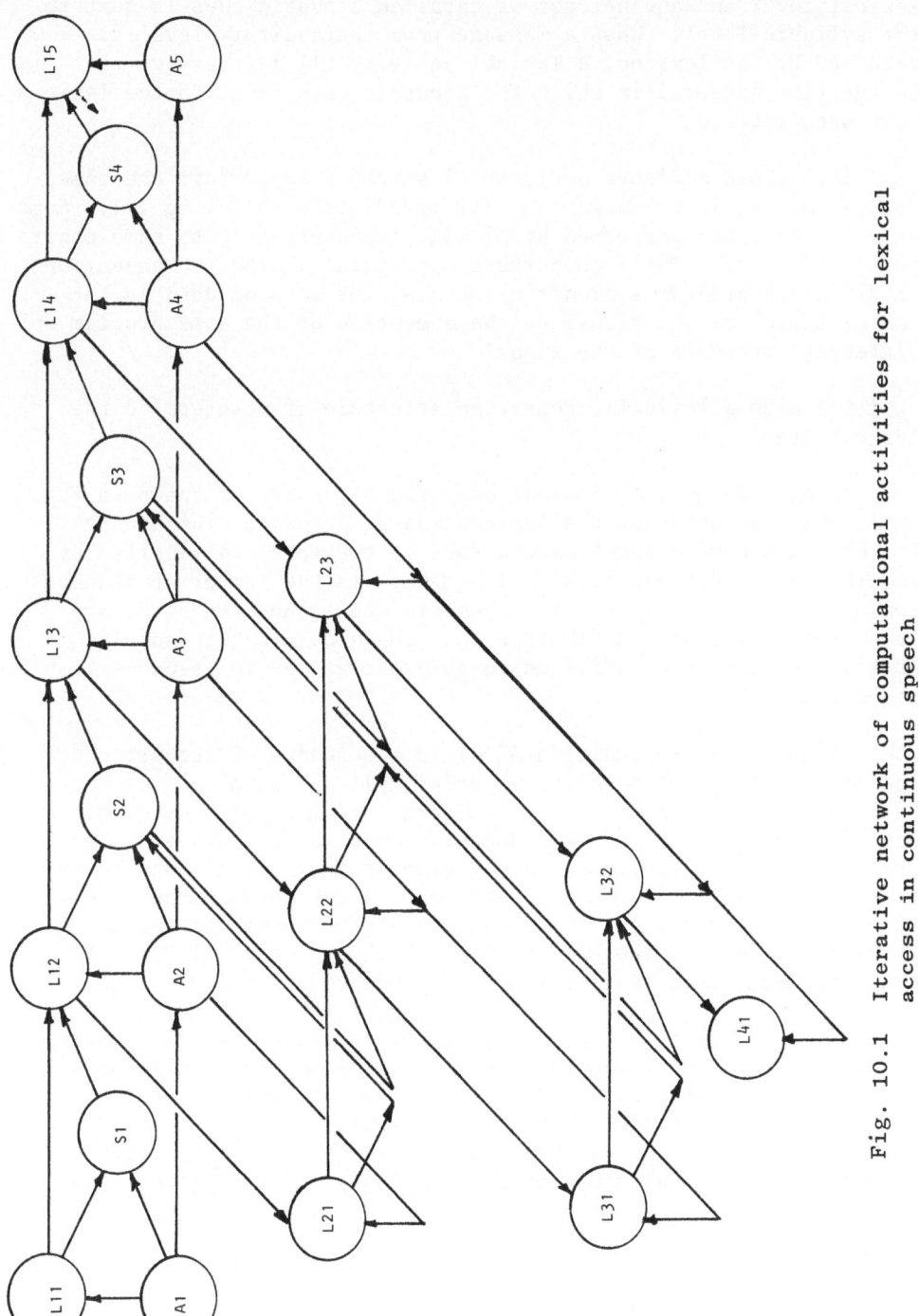

Fig. 10.1 Iterative network of computational activities for lexical access in continuous speech

after a set of acoustic cues is obtained and transmitted to the
lexical level and another set of detailed acoustic cues is sent to
the syllabic level. When a message from the auditory level is
received by the lexicon, a lexical activity L11 is created. Let TI1
be the time interval in which the acoustic cues, sent to the lexicon,
have been detected.

The second auditory activity A2 attempts to perform the same
task as A1 but in a time interval TI2 immediately following TI1. Some
of the operations performed by A2 can be concurrent with some others
performed by A1. These concurrent operations can be the execution
of different program segments using disjoint sets of data in the
same interval of the signal or the execution of the same program on
different intervals of the signal.

L11 uses a lexical-access tree structure of pointers to the
lexical items.

An example of a lexical-access tree structure is shown in Fig.
10.2. All the words of the lexicon can be accessed from the root.
The branches coming out from the root of the tree are labelled by
acoustic preconditions $AC(k)$, $(1 \leq k \leq k)$; k is the number of branches
coming out from the root. The acoustic cues generated by A1 are
such that they cannot match more than one acoustic precondition
$AC(k)$. Acoustic preconditions roughly correspond to pseudo-syllabic
segments.

An acoustic precondition $AC(k)$ is a sequence of acoustic cues
$(ac(k1), ac(k2),..., ac(ki),..., ac(kl(k)))$ and each cue may have
associated conditions on its parameters. $l(k)$ is the number of
acoustic cues for $AC(k)$. The acoustic cues of $AC(k)$ have to be
detected in the indicated sequence even if they are allowed to be
not contiguous in time. The duration of a gap between the detection
of two successive acoustic cues in the signal cannot exceed a
threshold $tg(k)$. Values of the threshold gaps $tg(k)$ $(1 \leq k \leq k)$
are determined by a small set of word-independent rules.

Let $AC(k)$ be the acoustic precondition which matches the string
of acoustic cues A1AC generated by A1; $AC(k)$ points to all the words
of the lexicon for which $AC(k)$ solves a sub-problem corresponding
to a "stimulus". Notice that here a stimulus does not come from a
blackboard of hypotheses, but is a message sent by an auditory
activity to a lexical computational activity. Furthermore, this
message describes, unambiguously, a segment of speech data.

Acoustic preconditions may correspond to phonetic or prosodic
features.

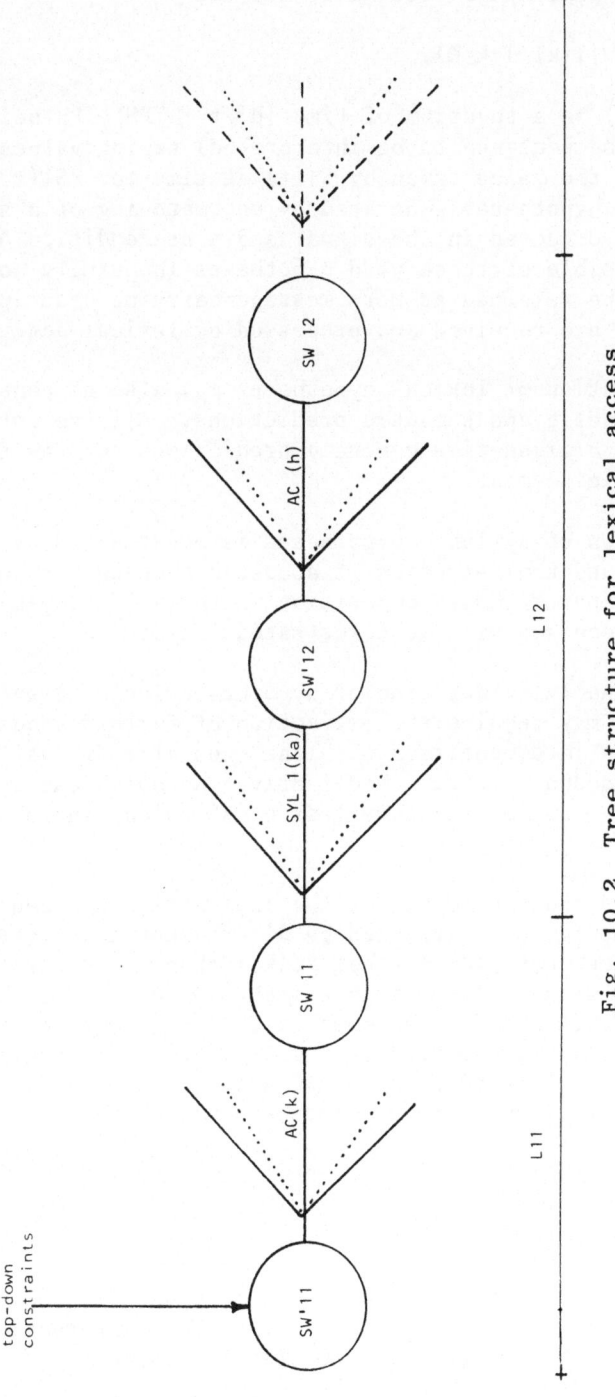

Fig. 10.2 Tree structure for lexical access

In Fig. 10.2, each branch labelled by an acoustic precondition AC(k) ends in a set of word candidates. Let SW11(k) be the set following the branch labelled AC(k) and let

SP1 = {SW11(k)/1≤k≤K}.

Let S1(t) be a function of time (0≤ t ≤TPH),TPH being the duration of the sentence to be interpreted) taking values on SP1. Let S1(ta) be the value taken by S1(t) at time ta. S1(ta) indicates a set of word hypotheses generated by the matching of a sequence of acoustic cues detected in the signal and a precondition AC(k). Only the most plausible of these word hypotheses (hopefully no more than one) have to be retained as more messages carrying descriptions of acoustic cues are received and processed by lexical activities.

The selection of lexical hypotheses may also be constrained by top-down syntactic and semantic predictions. All the active word hypotheses at a given time generate predictions for the following pseudo-syllabic segment.

Generation of syllabic hypotheses is constrained by lexical expectations and the detection of acoustic cues and is performed by a "syllabic expert" whose composition in terms of cooperating and concurrent processes will be illustrated later.

Exceptionally, generation of hypotheses inside a syllabic segment (PSS) may require the extraction of further acoustic cues which won't be in competition with the ones already available but will just be added to them. Model-driven acoustic cue extraction would make the control strategy less parsimonious and should be avoided.

Fig. 10.3 shows a detail of the interaction between L1, A1 and S1. The acoustic cues extracted by A1 are sent as a message A1AC to L11. Assume that A1AC matches AC(k); then L11 generates a set of word hypotheses SW11 compatible with A1AC and top-down constraints. Based on SW11, L11 sends a request to S1 for generating hypotheses in the set SYL11(k), composed of the first syllabic segments of the words in SW11. S1 receives a set of acoustic cues A1SC from A1 and generates a set of syllabic hypotheses SYL11 with

SYL11 = SYL11(k).

The set SYL11 is sent to L12.

The request for the extraction of more acoustic cues is indicated by a dashed arrow coming out of S1. Interactions with levels superior to the leixcon are also indicated.

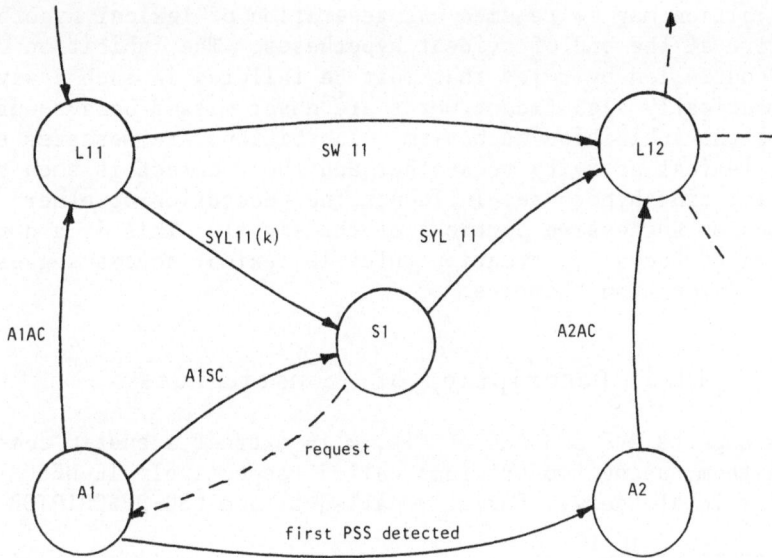

Fig. 10.3 Messages exchanged between computational activities

After having received the hypotheses about those syllables with which the words of SW11(k) may begin, L12 evaluates the evidences of partial word hypotheses and decides which of them can be retained.

L12 keeps active a set of word hypotheses SW'12 such that:

SW'12 = SW11.

The basic scheme described so far is repeated iteratively. A2 generates acoustic cues, based on which SW'12 can be further reduced to SW12; L12 will then send a message to S2 for the verification of the second syllables of the active word hypotheses. These successive refinements of lexical hypotheses can be reduced to a simple consistency evaluation when a single evident word hypothesis remains that univoquely allows its continuation to be predicted.

A new lexical activity L21 is created by A2 in parallel with L12. L21 accounts for words which may start with a syllable in the time interval TI2. Inhibition messages, represented by arrows in the third dimension in Fig. 10.1, may inhibit the generation of a new set of lexical hypotheses or the continuation of the verification of the hypotheses in already generated sets. Notice that a single auditory and a single syllabic activity are performed for many lexical activities in the same time interval even if lexical activities constrain the generation of syllable hypotheses. This is another peculiar characteristic of the model proposed here.

Inhibition may be resumed and generation of lexical hypotheses may restart at the end of evident hypotheses. The inhibition mechanism is controlled by rules that must be tailored in such a way that semantically significant words are never missed because of faults of the inhibition mechanism. Inhibitions are messages sent from one lexical activity to another and their effect is such that already existing hypotheses influence the generation of other hypotheses on successive portions of the signal. This is a novel conception of focus of attention rules in lexical access suggested by speech perception theories.

10.2 Description of Acoustic Cues

The experts AVE and AEC of Fig. 4.16 extract acoustic cues and describe them, using two programs called respectively TE-DESCRIPTOR (TE is for Total Energy, formerly called ·S) and GSG-DESCRIPTOR.

TE-DESCRIPTOR has the task of describing the time evolution of the total energy of the signal (TE) in terms of peaks and valleys. At the same time, AEPDST goes on, transforms another portion of the signal and sends a message to TE-DESCRIPTOR. This operation is repeated until a sentence end-point is detected.

The LTM of TE-DESCRIPTOR contains a grammar GTEDES that controls a coding of TE in terms of peaks and valleys. This grammar and its use were described in Section 3.3.

The terminal alphabet VT of GTEDES is composed of symbols representing peaks and valleys:

VT = {PEAK,DIP}

The nonterminal alphabet consists of the starting symbol SENTENCE and the term ZETA. The rewriting rules are:

SENTENCE := ZETA

           := ZETA SENTENCE

ZETA   := DIP PEAK

         := PEAK

The times of beginning and end are attributes associated to the detected terminal symbols. The maximum value of the peak and ther minimum value of the dip are also associated to PEAK and DIP respectively.

Descriptions of the signal energy (TE) are sent to "GSF-DESCRIPTOR", which provides the acoustic cues for segmentation.

The LTM of GSF-DESCRIPTOR contains an integration of the structural and the procedural knowledge for obtaining a description of the gross spectral features represented by the time evolution of the parameters defined in Section 3.3.

TE : the total energy (formerly S)

E12 : the energy in the 3-5 KHz frequency band (formerly A)

R12 : the ratio of the energies in the frequency bands
     B1 = 200 - 900 Hz, B2 = 5000 - 10000 Hz (formerly R)

The knowledge of GSF-DESCRIPTOR is a hierarchical network of plans represented by a grammar of frames. The network of plans represents a control strategy according to which knowledge is applied for extracting acoustic cues from spectral information.

A frame is an information structure made of a frame-name and a number of slots. A slot is the holder of information concerning a particular item called "slot filler" (Minsky 1974). Slot-fillers may be descriptions of events, relations or results or procedures. Attempts to fill the slots are made during a frame instantiation. A frame instantiation can be started by a simple reasoning program of an expert after having received a message. Once the instantiation of a frame is started, the expert attempts to fill the slots sequentially according to their order.

An event filling a slot may be the instantiation of another frame. Eventually default conditions may also appear in frame slots to specify what the system can assume if its expectations do not match the data.

Slot-fillers may be invocations of procedures for extracting some information from the data. In such a case, a slot can be filled only if extraction has been successfully completed. In particular, a procedure P-READ (Parameters) is introduced, which allows information from the outside of the frame to be loaded into a specified slot. Parameters to be read are assigned during the instantiation of a frame. Another procedure P-APPEND (Slot-List) has the effect that only the contents of the slots in a slot-list are returned as result of the frame instantiation. A default condition of the P-APPEND returns all the slots of the frame. A slot may contain a relation involving predicates; relations may imply the evaluation of functions defined by semantic attachments.

When a frame is invoked for interpreting a set of data, an instantiation of it is created in the Short-Term-Memory. When all the slots of a frame instantiation are filled, the frame instantiation is complete.

The execution of a procedure can be invoked by the attempt to fill a frame slot. The parameters of the procedure can be taken only from slots already filled in the same instantiation. The result of the procedure can be the filler of the slot.

Many procedural rules used here are derived from hueristic knowledge related to speech perception, spectrogram reading, speech production, and the experience gained in designing rule-based speech recognition systems. Frame structures are precisely defined by the rules of another grammar. Table 10.1 shows the rules of this frame-structure grammar. The exponent k>1 of an expression means that the expression can be rewritten any number of times greater than or equal to 2. The asterix means that the expression can be absent, present, or repeated any number of times.

The frame-structure grammar defines a language for representing LTM knowledge. The semantics of this language have been introduced before when it was stated how frames are instantiated and slots have to be filled. The following considerations complete the description of the semantics of this language. Such a description is informal, for the sake of brevity.

Attempting to fill frame slots is the execution of a network of plants. This network has the properties of the NOAH System proposed by Sacerdoti (1977). The main advantage of this model is that it specifies which plans can be executed in parallel and the goal is achieved with no backtracking. Avoiding backtracking is a way of implementing a "parsimonious" strategy of speech perception. The correspondence between a frame structure and a network of plans is established by the following rules:

Rule 1. Frame invocation corresponds to the expansion of a plan into more detailed sub-plans, each sub-plan being related to the attempt of filling a slot of the invoked frame.

Rule 2. A sequence of slots in a frame is filled by a sequence of plans.

Rule 3. When a slot can be filled by a conjunction or a disjunction of frame invocations, all the invocations are performed in parallel by plans. There is one plan for each invocation.

Rule 4. Conditional statements correspond to pre-conditions that have to be verified before expanding a plan into a more detailed network of sub-plans. Due to the particular tasks performed by these plans there is no need to perform revisions like those indicated

Table 10.1

Rewriting rules of the frame-structure grammar

`<FRAME>`	`:= (<NAME> <SLOT-LIST>)`
`<SLOT-LIST>`	`:= (<SLOT> (<SLOT>)*)`
`<SLOT>`	`:= (<NAME> [(<DESCRIPTIONS>)])`
`<DESCRIPTION>`	`:= ( described-as <NAME>)`
	`:= (<CONNECTIVE> <DESCRIPTION>`$^{k>1}$`)`
	`:= (not <DESCRIPTION>)`
	`:= (filled-by <NAME>)`
	`:= <CONDITIONAL>`
	`:= (result-of <PROC>)`
`<CONDITIONAL>`	`:= (when <NAME>`
	`     <DESCRIPTION> <DESCRIPTION>`
	`     [(else <DESCRIPTION>)])`
	`:= (when <PREDICATE EXPRESSION>`
	`     <DESCRIPTION>`
	`     [(else <DESCRIPTION>)])`
	`:= (unless <DESCRIPTION> <DESCRIPTION>)`
`<CONNECTIVE>`	`:= or`
	`:= and`
	`:= xor`
`<PREDICATE EXPRESSION>`	`:= <PREDICATE>`
	`:= (not <PREDICATE>)`
	`:= (<CONNECTIVE> <PREDICATE>`$^{k>1}$`)`
`<PROC>`	`:= F-<function>`
	`:= P-<procedure>`

by Sacerdoti as CRITICS.

Rule 5. A default condition corresponds to a plan
that is executed after the unsuccessful execution of
the plans which had to fill the same slot as the
default statement.

Brackets in Table 10.1 contain optional items which can be
repeated any number of times. Parentheses and words in lower case
letters are keywords of the frame description language.

Table 10.2 contains the grammar and the description of the
frames stored in the LTM of GSF-DESCRIPTOR. Predicates are indicated
in capital letters by words ending with -P and are defined by
semantic attachments which will be described informally for the sake
of brevity. Functions are indicated by names starting with F-.
Procedures are indicated by names starting with P-.

Frame instantiations must be in accordance with the rules
indicated at the beginning of Table 10.2; the rules represent a
basic iteration consisting in filling the slots of a total energy
peak description and a total energy dip description. This iteration
is a plan sequence repeated until a sentence end point is found.

Example 10.2.1

Fig. 10.4 shows the network of plans corresponding to the
grammar of Table 10.2 and a detailed expansion of the plan for
filling the slots of an instantiation of PKTE. The symbols used
are the same as in Sacerdoti (1977). The plan labelled PKTE is
expanded into a sequence of some plans, one for each slot. The plan
PCONT is expanded into a disjunction of more detailed plans.

The hierarchical structure of the planning system is expressed
by the fact that a plan can be expanded into a more detailed network
of plans, which can be further expanded into a more detailed representation, and so on.

The expansion of a plan into the execution of more detailed
plans can be conditioned. Conditions are expressed by conditional
statements.

10.3 The Knowledge of the Descriptor
of the Global Spectral Features

This section is devoted to the description of Table 10.2, which
contains many characteristic aspects of the planning system and its
frame-based representation. The network of plans described by the

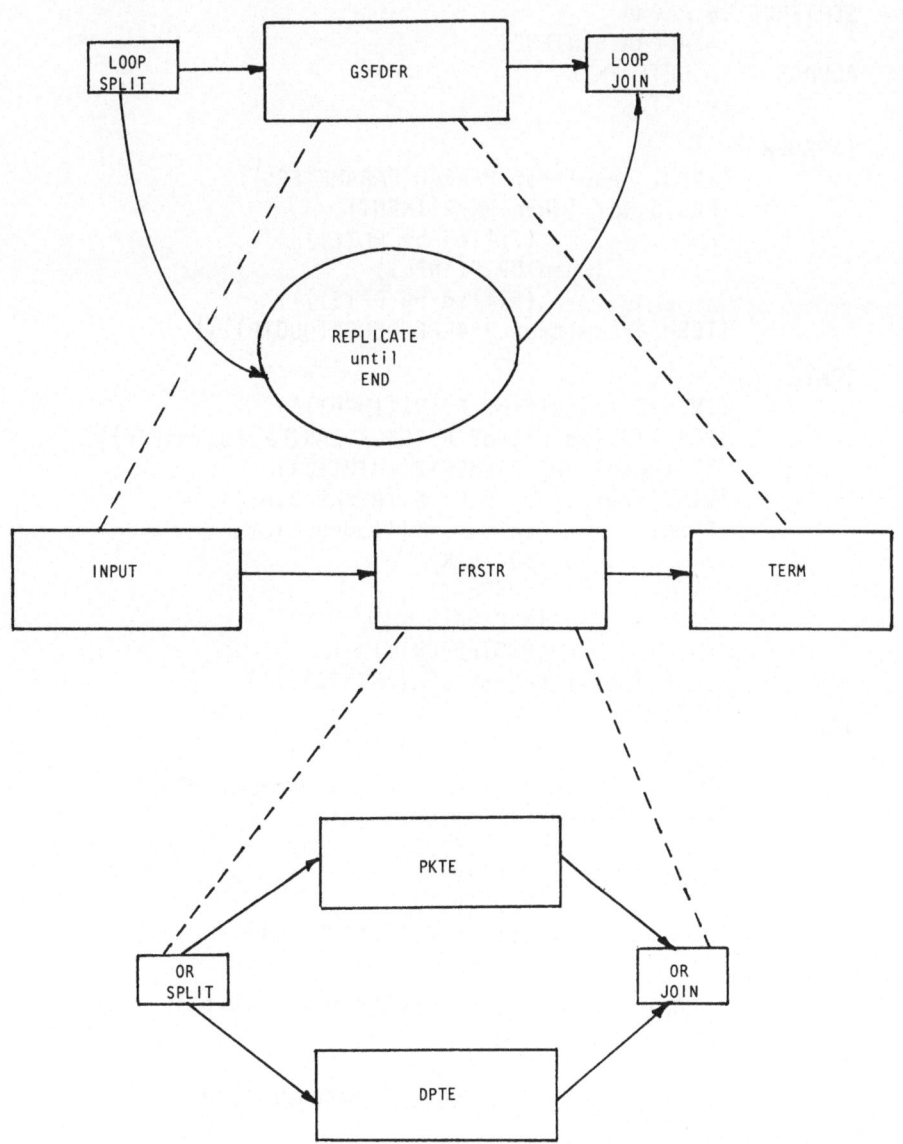

Fig. 10.4 Example of a network of plans

Table 10.2

The LTM of GSF-DESCRIPTOR

```
SENTENCE := ALPHA
         := ALPHA SENTENCE
ALPHAS   := DPTE PKTE
         := PKTE

(GSFDFR
          (INPUT result-of P-READ(PARAMETERS))
          (FRSTR (or (when PK-P(INPUT)
                        (filled-by PKTE))
                    (when DP-P(INPUT)
                        (filled-by DPTE))))
          (TERM (result-of P-APPENDESCR(QUOTP))))

(PKTE
          (INTPTE (result-of F-INT(INPUT)))
          (PEAKE12 (result-of F-DESCRPEAK(Fa,Fb,INTPTE)))
          (HR (result-of F-FHGR12(INTPTE)))
          (VINT (result-of F-CVINT(PEAKE12,HR)))
          (PCONT      (unless (filled-by (or
                      (VOCPEAK)
                      (SONPEAK)
                      (NSPEAK)
                      (BRSTPEACK)))
                (described-as UPK(INTPTE)))))

(DPTE
          (CDINT (result-of F-INT(INPUT)))
          (DIPPAR      (filled-by P-EXTDPPAR(CDINT)))
          (DCONT (unless (filled by (or
                      (SONDIP)
                      (SILDIP)
                      (BUZDIP)))
                (described-as VPC(CDINT)))))

(VOCPEAK
          (WCONT (when (and (HDURPKTE-P)(HPR12-P))
                (filled-by (or
                      (VOCCUESET)
                      ((LEFTVOW))(CONSVOW))))))

(VOCCUESET
          (LOWR (result-of F-FLOWR(INTPTE)))
          (TRINT (result-of F-TRNF(INTPTE)))
          (VWINT (result-of F-(VCINT)))
          (HGR (result-of F-CONSHR(INTPTE VWINT)))
          (VCONT (filled-by (or
                      (VOW)
```

```
                              (CONSVOW)
                              (VOWCONS)))))
     (VOW
             (when (and (HGVINT-P)(AEG-P(VWINT,INTPTE)))
             (described-as VOC(VWINT))))

     (CONSVOW
             (or    (when SEQ-P(LOWR,VWINT))
                           (described-as (NS(LOWR) VOC (VWINT)))
                    (when SEQ-P(PRNINT,VWINT))
                           (described as (TRNS(TRNINT) VOC(VWINT)))
                    (when SEQ-P(HGR,VWINT)
                           (described-as (SN(HGR) VOC(VWINT)))))

     (VOWCONS
             (or    (when SEQ-P(VCINT,LOWR)
                           (described-as (VOC(VWINT) NS(LOWR))))
                    (when SEQ-P(VCINT,HGR)
                           (described-as (VOC(VWINT) SN(HGR)))))

     (LEFTVOW
             (VLINT(when LEFTSIDE-P(VINT)
                           (result-of F-LINT)))
             (VOWDES       (described-as VOC(VLINT)))))

     (NSPEAK
             (when (and (not HDUR PKTE-P)(not HPR12))
                           (described-as NSPK(INTPTE))))

     (BRSTPEAK
             (when (and (not HDUR PKTE-P)(not HPR12))
                           (described-as BRSTPK(INTPTE)))

     (SONPEAK
             (when (and (not HDUR PKTE-P)(HPR12-P))
                           (described-as SNPK(INTPTE))))

     (SONDIP
             (MUSON(result-of F-APPLY(DIPPAR,SON)))
             (SNDS (when HMUSON-P
                    (described-as SNDIP(CDINT MUSON)))))

     (SILDIP
             (MUSIL(result-of F-APPLY(DIPPAR,SIL)))
             (SLDS (when HMUSIL-P
                    (described-as SLDIP(CDINT,MUSIL)))))

     (BUZDIP
             (MUBS (result-of F-APPLY(DIPPAR,BUZ)))
             BZDES (when HMUBZ-P
                    (described-as BZDIP(CDINT,MUBZ)))))
```

frame system reported in Table 10.2 contains a large variety of the frame structures which can be generated by the rules of Table 10.1. For the sake of simplicity the details of the semantic attachments, defining the procedures invoked for filling slots in Table 10.2, are not given. Rather, a description of the procedure behavious is given formally, together with the description of the frames.

The basic grammar for describing the signal energy is a regular one. It establishes that a sentence is a sequence of combinations of dip and peak descriptions. TE-DES sends to the GSF-DESCRIPTOR peak and dip descriptions through a queue named QUINP. A Pascal-like version of the reasoning program of the GSF-DESCRIPTOR is given below. The parameters passed in an instantiation are obtained from the read message by the procedure:

COMPUTE(PARAMETERS,MESSAGE).

```
"gsf-descriptor"
begin
      END := false;
      repeat
                RECEIVE (MESSAGE,QUINP)
                if MESSAGE <> ENDSENTENCE
                    begin
                    COMPUTE(PARAMETERS,MESSAGE);
                    QUOUT = INSTANTIATE (GSFDR,PARAMETERS)
                    end
                else END := true;
      until END
   end
```

Whenever the frame GSFDFR is instantiated, a process for filling the frame slots is created and a node in the output queue QUOUT is created.

The node in the output queue has a position in the queue which depends on the time intervals of the description that caused the instantiation. Descriptions which fill slots after the "described-as" keyword are appended to the node in QUOUT created at the beginning of the frame instantiation. This is performed by the procedure P-APPENDESCR whose termination causes the end of an instantiation of GSFDFR by putting a TERM mark in the corresponding slot. After termination the space of the Short-Term-Memory occupied by the frame instantiation is cleared.

Whenever a peak is received by the GSF-DESCRIPTOR, an instantiation of the frame PKTE is created into the STM of GSF-DESCRIPTOR
by the attempt to fill the slot FRSTR of GSFDRF. The execution of
the corresponding plan is then initiated.

This process attempts to fill sequentially the slots of PKTE.
INTPTE is filled by the result of the application of the function
F-INT on the argument INPUT. This function gives the time of
beginning, the time of ending and the duration of the peak of total
energy described in INPUT. INTPTE is written into the STM after
PEAKTE in the instantiation of PKTE.

The next slot of PKTE is filled by the result of the function
F-DESCRPEAK which describes the peak of the energy in the frequency
band between frequencies Fa and Fb and in the time interval written
in INTPTE.

Successively, the function F-FHGR12 is executed. It gives the
description of the time intervals inside INTPTE in which the ratio
R12 between energies in the bands B1=(300-900 Hz) and B2=(5-10 KHz)
is high (higher than a threshold TH1). This description fills the
slot corresponding to the name HR.

The function F-CVINT computes the time intervals inside the
peaks in PEAKE12 in which R12 is high. Its result fills the slot
VINT.

The last plan of the sequence attempts to fill the slot PCONT.
This slot can be filled by a disjunction of frame instantiations
called VOCPEAK, SONPEAK, NSPEAK, BRSTPEAK. Each invoked frame corresponds to a hierarchy of more detailed plans which are executed
for attempting to fill the frame slots.

If none of the frame instantiations can be completed, a default
condition is assumed, consisting in filling PCONT with the description UPK(INTPTE). UPK is the description of an uncertain peak
detected in the time interval INTPTE. A similar network of plans
is used for attempting to fill the slots of DPTE.

The procedure P-EXTDPPAR(CDINT) extracts the parameters of the
dip in the interval CDINT useful for a more detailed description of
the dip. UDP(CDINT) is a description of an uncertain dip and is
generated as a default, if no other more detailed description can
be generated.

The execution of more detailed plans for filling the slots of
VOCPEAK is conditioned by the verification of the truth of the two
predicates HDURPKTE-P and HPR12-P. HDURPKTE-P is true if the
duration of the signal energy peak is high, i.e., higher than a

threshold. HPR12-P is true if there is at least one peak of R12 in
INTPTE whose maximum value is higher than another threshold.

VOCCUESET has slots which are filled by the extraction of acoustic cues which usually appear in a total energy peak containing at
least one vowel.

F-TRNF(INTPTE) extracts an interval at the beginning of a peak
where cues of a consonantal transient, typical for example of plosive
sounds, has been found.

F-INT(VCINT) looks for the description of vocalic cues in VINT.
The time interval in which these cues have been found fills the slot
VVINT. F-FLOWR extracts the intervals in which R12 is low, F-CONSHR
extracts the consonantal interval in which R12 is high. The default
value for both the functions is zero.

The predicate HGVINT-P is true when the maximum energy in the
band between Fa and Fb is high in the time interval VINT. AEQ-
P(VWINT,INTPTE) is true when the two time intervals VWINT and INTPTE
are almost coincident. If the above predicates are both true, the
peak is described as a vocalic one and the description is VOC(VWINT).

The CONSVOW frame is filled by a sequence of vocalic and consonantal cues.

SEQ-P(A,B) is a predicate that is true when B follows A in time.

The consonantal portion of the peak can be described by SN(HGR),
if the acoustic cues are of sonorant type or NS(LOWR) if they are
of nonsonorant type.

LEFTSIDE-P(VINT) is true when there are two vocalic segments in
a peak and one starts at its left side. In such a case the time
interval of the left vowel is detected by the function F-LINT(VINT).

Consonantal peaks of signal energy can be described as NSPK
(fricative-type peaks) or BRSTPK (burst-type peaks) or SNPK (sonorant
peaks).

The dips of the signal energy can be described as sonorant
dips (SNDIP), silence dips (SLDIP) and buzz-bar type dips (BZDIP).
In order to obtain these descriptions, the function F-APPLY is
executed on the dip parameters DIPPAR for extracting degrees of
evidence MUSON, MUSIL, and MUBZ. When these degrees of evidence are
high enough (above a fixed threshold) the corresponding predicates
are true, allowing the corresponding description to fill a slot. If
two or more descriptions for the same dip fill a slot, they are
combined into a single one by P-APPENDDESCR. For example, SNDIP
(CDINT,MUSON) and SLDIP(CDINT,MUSIL are combined into SNDLIP(CDINT,
MUSON,MUSIL).

Sonorant dips can be further subdivided into high, middle, short, very short and long dips. SILDIPS are further subdivided into VERY LONG DIPS and LONG DEEP and SHORT DEEP dips. These extensions may allow ambiguities to be avoided. The details for the assignment of these new descriptions are omitted here for the sake of brevity.

Fig. 10.5 shows the time evolutions of the gross spectral features TE, E12, R12 of the piece of sentence /elega/ extracted from continuous speech.

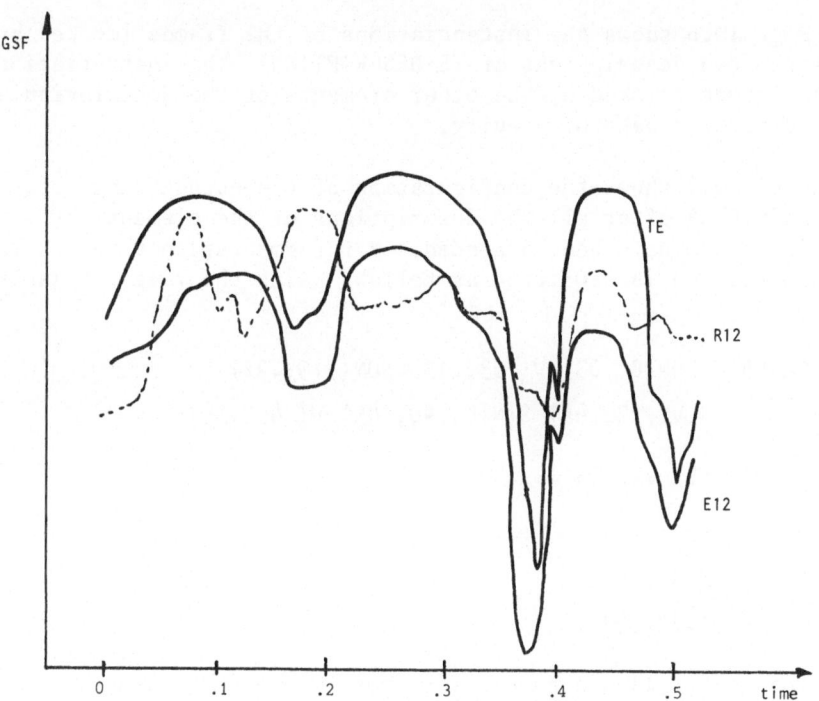

Fig. 10.5 Example of time evolution of gross spectral features

The expert TE-DESCRIPTOR sends the following description to the GSF-DESCRIPTOR:

TE-DESCRIPTOR: PEAK(0,.19,M1) DIP(.19,.22,m1)

　　　　　　　　　PEAK(.22,.40,M2) DIP(.40,.42,m2)

PEAK(.42,.45,M3) PEAK(.45,.59,M4)

The values of the maxima Mi (1≤i≤4) of the peaks and the minima mj (1≤j≤2) of the dips are not given because they are not relevant to this example.

The reasoning program of GSF-DESCRIPTOR creates an instantiation of GSFDFR and a node in the output queue for every item of TE-DESCRIPTION.

All the instantiations of GSFDFR which have been created are written in the STM of GSF-DESCRIPTOR. For each instantiation, a process is created that attempts to fill the slots of the invoked frames. All these processes are executed concurrently.

Fig. 10.6 shows the instantiations of the frames invoked by the first two descriptions of TE-DESCRIPTION. The instantiations of the frames invoked by the other elements of the description are omitted for the sake of brevity.

Fig. 10.7 shows the configuration of the output queue of GSF-DESCRIPTOR after all the descriptions of the elements of TE-DESCRIPTION have been appended. For segmentation purposes the GSF description is rewritten as follows using the rules of Table. 10.3:

DESCR1:   UN(0,.03) V(.03,.19) SON(.19,.22)

V(.22,.40) SINIL(.40,.45) V(.49,.59)

which gives three PSSs:

PSS1:   {0,.19}

{.03,.40}

{.22,.59}.

For each syllabic hypothesis, more phonetic hypotheses can be generated using the PRR representations and the rules given in Chapters 5-6. This generation can be invoked or constrained by lexical predictions. Phonetic hypotheses can be generated in parallel.

10.4 Conclusions

The purpose of this book was to describe the results of years of research primarily devoted to the analysis of basic problems related to the conception of perceptual models.

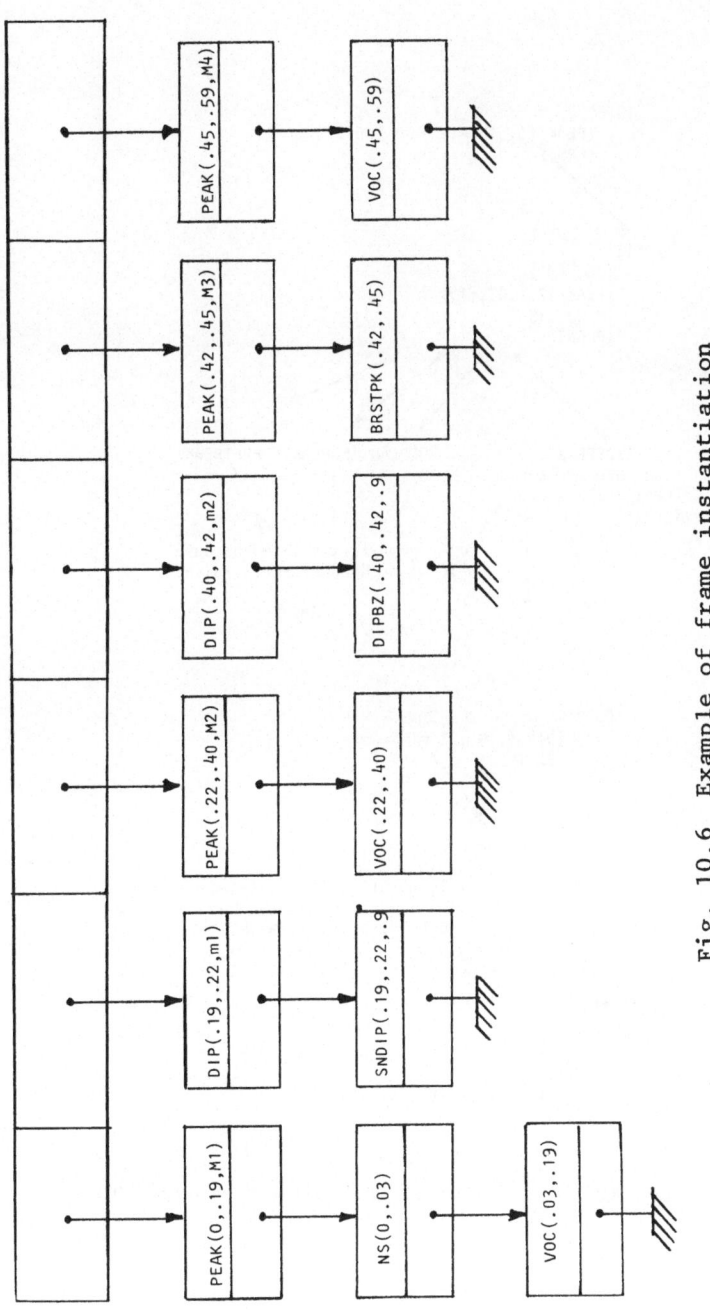

Fig. 10.6 Example of frame instantiation

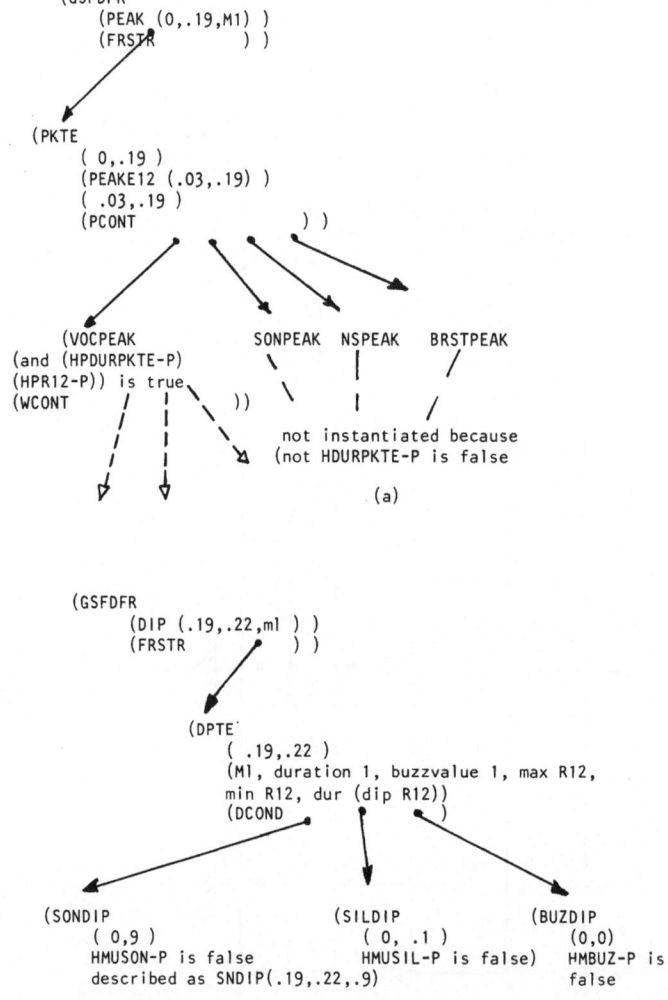

Fig. 10.7 Example of output queue configuration

```
(VOCCUESET
    ( 0,.03 )
    ( .03,.19 )
    ( 0 , 0 )
    (CINT      ) )
```

```
(CONSVOW
    SEQ-P (LOWR, VWINT is true
    described as NS (0,.03) VOC(.03,.19 )    (c)
```

There is no room in INTPTE for another instantiation of VOCCUESET and there
is no left vowel. Thus WCONT cannot be filled with other instantiations.

Fig. 10.7 continued

The availability of flexible models with learning capabilities
can be of great help for advancing investigation in the broad area
of the speech science.

These models are rich with suggestion for improving the design
of practical systems.

The conception of a model for speech understanding is research
work at the furthest frontiers of Computer and Information Sciences.

The algorithms proposed in this book often have a high degree
of parallelism involving sophisticated computational models for
which system architectures with distributed intelligence are required.

Most of the content of the book is centered around items of
Artificial Intelligence, such as knowledge representation and
strategies for its use.

It is characteristic of speech and image understanding that
knowledge is often imprecise and the description of the word under
observation can be vague.

A large part of the research described in this book has been
devoted to the inference of rules for relating descriptions of
acoustic cues and syllables of a Language. A point of view has

been taken for which the structural aspect of the rules is more important than the statistics of the acoustic parameters. A complete set of rules for hypothesizing phonetic features and phonemes in every syllabic context of the Italian Language has been presented.

Research is in progress in other countries for establishing similar rules for other languages, towards the goal of making a task-independent, speaker-independent system for generating phoneme hypotheses on continuous speech.

Finding simpler, more general, and effective rules is an open area for research in the next ten years.

Vagueness and imprecision have been taken into account by introducing fuzzy rules with fuzzy linguistic variables.

The excellent results in phoneme hypothesization reported in the book give a strong a posteriori motivation for using a fuzzy algebra achieving simplicity and flexibility in system design. The control strategies have been based on the statistic of fuzziness.

Statistics have been used for making decisions and for inferring rules whenever a frequency of occurrence was involved, especially for minimizing error rates.

The algorithms and the rules proposed here can be simplified by forcing the system to generate the best scored hypothesis at each step.

This simplification can be introduced in the design of low-cost speech recognition systems in which the input signal is described by a string of symbols. This string can be compared with prototypes using the dynamic programming algorithms described in Chapters 2 and 8.

In the more sophisticated project described in this book, the structural knowledge as well as the rules for composing evidences are embedded in a cascade of Augmented Transition Network Grammars, using the same formalism for representing knowledge at various levels and for composing evidences.

A further step forward could be to represent the various components of the system as experts of a society. The task of these experts would be to cooperate for understanding the spoken message.

The knowledge of each expert could have a complex structure in which an ATNG is an important component.

The reasoner, i.e., the expert who coordinates the activation of the various knowledge sources, realizes a complex plan. This plan is nonlinearly decomposed into a hierarchy of subplans some of which can be parsers, problem solvers or production systems. Examples of planning have been proposed in the book. Certainly the design of an efficient planning for a speech understanding system is another open subject for research.

Although some subplans can be developed by executing one of the classical graph search algorithms, a major problem is that of switching the focus of attention from one interpretation task to another. This is a typical control problem that has been approached in a framework of approximate reasoning. Conceiving plan hierarchies and control strategies for focusing the attention on an interpretation task are also great open problems to be further investigated.

TABLE 10.3

Relabelling of GSF-DESCRIPTIONS for segmentation

GSF-DESCRIPTIONS	Relabelling
VOC	V
SN	SON
SNPK	SON
SNDIP	SON
UPK	*
UPD	UN
NS	UN**
SNPK	UN
BRSTPK	UN**
SLDIP	SON,UN
BZDIP	SINIL***

* UPK is not relabelled because it is not used for segmentation.

** if not involved in SINIL, otherwise it is ignored.

*** BZDIP is relabelled as SINIL if precedes the contexts (VOC) or (BRSTPK) (VOC).

BIBLIOGRAPHY

1. Abercombie, D., (1967), *Elements of General Phonetics*. Aldine, Chicago.

2. Abramson, N., (1963), *Information Theory and Coding*. McGraw Hill, New York.

3. Ackenhusen, J.G. and Rabiner, L.R., (1981), Microprocessor implementation of an LPC based word recognizer. *Proc. ICASSP*, Atlanta, pp. 746-749.

4. Ackroyd, M.H., (1980), Isolated word recognition using the weighted Levenshtein distance. *IEEE Transactions*, vol. ASSP-28, no. 2, pp. 243-244.

5. Aho, A.V. and Ullman, J., (1972), *The Theory of Parsing, Translation and Compiling*. Prentice Hall, Englewood Cliffs, New Jersey.

6. Aldefeld, B. Rabiner, L.R. Rosenberg, A.E. and Wilpon J.G., (1980), Automated directory listing retrieval system based on isolated word recognition. *Proc. IEEE*, vol. 68, pp. 1364-1379.

7. Allen, J., (1973), Speech synthesis from unrestricted text. in: *Speech Synthesis*, ed. by J.L. Flanagan and L.R. Rabiner (1973).

8. Allen, J.B. and Rabiner, L.R., (1977), A unified approach to short-time Fourier analysis and synthesis. *Proc. IEEE*, vol. 65, no. 11, pp. 1558-1564.

9. Atal, B.S. and Hanauer, S.L., (1971), Speech analysis and synthesis by linear prediction of the speech wave. *J. A. S. A.*, vol. 50, no. 2, pp. 637-655.

10. Atal, B.S., (1975), Towards determinating articulator position from the speech signal. in: *Speech Communication*, ed. by G. Fant.

11. Atal, B.S. and Rabiner, L.R., (1976), A pattern recognition approach to voiced-unvoiced-silence classification with applications to speech recognition. *IEEE Transactions*, vol. ASSP-24, no. 3, pp. 201-212.

12. Atal, B.S. and Schroeder, M.R., (1978), Linear prediction analysis of speech based on a pole zero representation. *J. A. S. A.*, vol. 64, no. 5, pp. 1310-1318.

13. **Bahl, L.R.** and **Jelinek, F.,** (1975), Decoding for channels
     with insertions, deletions and substitutions with applications
     to speech recognition. *IEEE Transactions,* vol. **IT-21,** no. 4,
     pp. 401-411.

14. **Bahl, L.R., Baker, J.K., Cohen, P.S., Jelinek, F., Lewis, B.L.**
     and **Mercer, R.L.,** (1978), Recognition of a continuously read
     natural corpus. *Proc. ICASSP,* Tulsa, OKLA, pp. 422-425.

15. **Bahl, L.R., Jelinek, F.** and **Mercer, R.L.,** (1980), Continuous
     Speech Recognition. *Statistical Methods. Lecture notes,*
     CISM, Udine, Italy, 1980.

16. **Baker, J.K.,** (1975), The DRAGON system: an overview. *IEEE*
     Transactions, vol. **ASSP-23,** pp. 24-29.

17. **Baker, J.M.,** (1975), *A new time-domain analysis of human*
     speech and other complex waveforms. Ph. D. dissertation,
     Carnegie-Mellon University, Pittsburg, PA.

18. **Bakis, R.,** (1977), Continuous speech recognition via centi-
     second acoustic states. *91st meeting of the Acoustical*
     Society of America.

19. **Baldwin, J.F.,** (1979), A new approach to approximate reasoning
     using a fuzzy logic. *Fuzzy Sets and Systems,* vol. **2,** pp.
     309-325.

20. **Baldwin, J.F.** and **Pilsworth, B.W.,** (1980), Axiomatic approach
     to implication for approximate reasoning with fuzzy logic.
     Fuzzy Sets and Systems, vol. **3,** pp. 193-219.

21. **Barasova, T.A., Rudny, B.N., Trunin-Donskoj, V.N.,** (1972),
     On automatic segmentation of the speech signal. in: *Rechevoie*
     upravlenie (Machine control by speech), Academia Nauk, Moscow,
     pp. 17-28 (in Russian).

22. **Bates, M.,** (1975), The use of Syntax in a speech understanding
     system. *IEEE Transactions,* vol. **ASSP-23,** pp. 112-117.

23. **Baudry, M.** and **Dupeyrat, B.,** (1978), Utilisation de méthodes
     syntaxiques et de filtrage logique en reconnaissance de la
     parole. *Proc. AFCET/IRIA Conference on Pattern Recognition,*
     Chatenay-Malabry.

24. **Bawden, A., Greenblatt, R., Holloway, J., Knight, T., Moon, D.**
     and **Weinreb, D.,** (1977), *LISP Machine progress report.* AI
     Memo no. 444, M.I.T., Artificial Intelligence Laboratory,
     Cambridge, MA.

25.   **Becker, R.** and **Poza, F.**, (1975), Acoustic processing in the SRI speech understanding system. *IEEE Transaction,* vol. ASSP-23, pp. 416-426.

26.   **Bellmann, R.** and **Dreyfus, S.E.**, (1962), *Applied Dynamic Programming.* Princeton University Press.

27.   **Bellman, R.E.** and **Zadeh, L.A.**, (1977), Local and fuzzy logics. in: *Modern Uses of Multiple-valued logic,* ed. by J.M. Dunn and G. Epstein, Reidel, Amsterdam.

28.   **Bedzek, J.**, (1981), *Pattern Recognition with Fuzzy Objective Function Algorithms.* Plenum Press, New York.

29.   **Blumstein, S.E.**, **Stevens, K.N.** and **Nigro, G.N.**, (1977), Property detectors for burst and transitions in speech perception. *J.A.S.A.,* vol. **61**, pp. 1301-1313.

30.   **Blumstein, S.E.** and **Stevens, K.N.**, (1979), Acoustic invariance in speech production: evidence from measurements of the spectral characteristics of the stop consonants. *J.A.S.A.,* vol. **66**, no. 4, pp. 1001-1017.

31.   **Bobrow, D.G.** and **Collins, A.**, (1975), *Representation and Understanding.* Academic Press, New York.

32.   **Bobrow, D.G.** and **Winograd, T.**, (1977), An overview of KRL, a Knowledge Representation Language. *Cognitive Science,* vol. **1**, pp. 3-46.

33.   **Bond, Z.S.** and **Garnes, S.**, (1980), Misperception of fluent speech. in: *Perception and Production of Fluent Speech,* ed. by R. A. Cole, Arlbaum Associates, New York.

34.   **Bondarko, L.V.**, (1969), The syllable structure of speech and distinctive features of phonemes. *Phonetica,* vol. **20**, pp. 1-40.

35.   **Brady, P.T.**, **House, A.S.** and **Stevens, K.N.**, (1961), Perception of sounds characterized by a rapidly changing resonance frequency. *J.A.S.A.,* vol. **33**, pp. 1357-1362.

36.   **Bridle, J.S.**, (1973), An efficient elastic template method for detecting given words in continuous speech. *Proc. British Acoustical Society Meeting,* London, paper 73SHC3.

37.   **Brinch Hansen, P.**, (1973), *Operating System Principles.* Prentice Hall, New York.

38.   **Broad, D.J.** and **Fertig, R.H.**, (1970), Formant-frequencies trajectories in selected CVC-syllable nuclei. *J.A.S.A.,* vol. **47**, no. 6, pp. 1572-1582.

39.   **Broad, D.R.** and **Soup, J.E.**, (1975), Concepts for acoustic phonetic recognition. in: *Speech Recognition,* ed. by D.R. Reddy, Academic Press, New York.

40.   **Bruce, B.C.**, (1973), A model of temporal references and its application in a question answering program. *Artificial Intelligence,* vol. **3**, pp. 1-26.

41.   **Butterworth, B.**, ed. (1979), *Language production.* Academic Press, New York.

42.   **Caelen, J.**, (1979), *Un modèle d'oreille. Analyse de la parole continue. Reconaissance phonémique.* Thèse d'état, Université de Toulouse, France.

43.   **Chang, C.L.** and **Lee, R.C.T.**, (1973), *Symbolic Logic and Mechanical Theorem Proving.* Academic Press, New York.

44.   **Charniak, E.**, (1976), Inference and Knowledge. in: *Computational Semantics,* ed. by E. Charniak and Y. Wilks, North-Holland Publ. Co..

45.   **Chiba, T.** and **Kajiyama, M.**, (1941), *The vowel, its Nature and Structure* (in Japanese). Tokyo-Kaiseikan.

46.   **Chiba, S.**, **Watari, M.** and **Watanabe, T.**, (1978), A speaker-independent word recognition system. *Proc. 4th IJCPR,* Kyoto, Japan, pp. 995-999.

47.   **Chistovich, L.A.**, **Fyodorova, N.A.**, **Lissenko, D.M.** and **Zhokova, M.G.**, (1975), Auditory segmentation of acoustic flow and its possible role in speech processing. in: *Auditory Analysis and the Perception of Speech,* ed. by Fant and Tatham, Academic Press, London, New York, pp. 221-231.

48.   **Chitavichius, A.**, ed. (1977), *Analysis and Recognition of Speech Signals.* Inst. of Mathematics and Cybernetics, USSR Academy of Sciences, Kaunas, Lithuania (in Russian).

49.   **Chomsky, N.** and **Halle, M.**, (1968), *The Sound of Pattern of English.* Harper and Row, New York.

50.   **Chorayan, D.G.**, (1979), *Fuzzy Algorithms in Thinking Processes.* Rostov University Press, Rostov, USSR (in Russian).

51.   Christiansen, R.V. and Rushforth, C.H., (1977), Detecting and
      locating key words in continuous speech using linear predic-
      ting coding. *IEEE Transactions,* vol. **ASSP-25**, no. 5, pp.
      361-367.

52.   Cohen, A. and Noteboom, S.G., eds. (1975), *Structure and
      Process in Speech Perception.* Springer-Verlag, New York.

53.   Cohen, P.S. and Mercer, R.L., (1975), The phonological compo-
      nent of an automatic speech recognition system. in: *Speech
      Recognition,* ed. by D.R. Reddy, Academic Press.

54.   Cole, R.A., ed. (1980), *Perception and Production of Fluent
      Speech.* Arlbaum Associates, New York.

55.   Cole, R.A. and Jakimik, J., (1980), How are syllables used
      to recognize words. *J.A.S.A.,* vol. **67**, no. 3, pp. 965-970.

56.   Cole, R.A., Jakimik, J. and Cooper, W.E., (1980), Segmenting
      speech into words. *J.A.S.A.,* vol. **67**, no. 4, pp. 1323-1332.

57.   Cole, R.A., Rudnicky, A., Zue, V. and Reddy, D.R., (1980),
      Speech as a pattern on paper in: *Perception and Production
      of Fluent Speech,* ed. by R. A. Cole, Arlbaum Associates.

58.   Cooper, F.S., Delattre, P.C. and Libermann, A.M., (1955)
      Acoustic loci and transitional cues for consonants. *J.A.S.A.,*
      vol. **27**, pp. 769-773.

59.   Cook, C.C., (1976), Word verification in a speech understanding
      system. *Proc. ICASSP,* Philadelphia, PA, pp. 553-556.

60.   Cook, C.C. and Schwarz, R.M., (1977), Advanced acoustic
      techniques in automatic speech understanding. *Proc. ICASSP,*
      Hartford, Conn., pp. 663-666.

61.   Coppo, M. and Saitta, L., (1976), Semantic support for a
      speech understanding system based on fuzzy relations. *Proc.
      IEEE Conf. on Cybernetics and Society,* Washington, D.C., pp.
      520-525.

62.   Cutler, A.C. and Foss, D.J., (1977), On the role of sentence
      processing. *Language and Speech,* vol. **20**, pp. 1-10.

63.   Damman, J.A., (1965), Application of adaptive threshold ele-
      ments to the recognition of acoustic-phonetic state. *J.A.S.A.,*
      vol. **38**, pp. 213-223.

64.   Datta, A.K., Ganguli, N.R. and Ray, S., (1980), Recognition
      of unaspirated plosives. A statistical approach. *IEEE
      Transaction,* vol. **ASSP-28**, no. 1, pp. 85-91.

65.   David, E.E. and **Denes**, P.B., (1972), *Human Communication. A Unified View*. McGraw Hill, New York.

66.   **Davis**, H.K., **Biddulph**, R. and **Balashek**, S., (1952), Automatic recognition of spoken digits. *J.A.S.A.*, vol. **24**, pp. 637-642.

67.   **Davis**, S.B. and **Mermelstein**, P., (1980), Comparison of parametric representations for monosyllabic word recognition in continuously spoken sentences. *IEEE Transactions*, vol. **ASSP-28**, no. 4, pp. 357-366.

68.   **Delgutte**, B., (1980), Representation of speech-like sounds in the discharge patterns of auditory-nerve fibers. *J.A.S.A.*, vol. **68**, pp. 843.

69.   **De Mori**, R., (1973), A descriptive technique for automatic speech recognition. *IEEE Transactions*, vol. **AU-21**, pp. 89-100.

70.   **De Mori**, R., (1974), Design for a syntax-controlled acoustic classifier. in: *Information Processing 74*, ed. by J.L. Rosenfield, North-Holland Publ. Co., pp. 753-757.

71.   **De Mori**, R., **Rivoira**, S. and **Serra**, A., (1975a), A speech understanding system with learning capabilities. *Proc. 4th IJCAI*, Tbilisi, USSR, pp. 468-475.

72.   **De Mori**, R., **Rivoira**, S. and **Serra**, A., (1975b), A special purpose computer for digital signal processing. *IEEE Transactions*, vol. **C-24**, pp. 1202-1211.

73.   **De Mori**, R. and **Laface**, P., (1976), Automatic detection of distinctive features in continuous speech. *Proc. IJCPR*, Coronado, CA, pp. 609-617.

74.   **De Mori**, R., **Laface**, P. and **Piccolo**, E., (1976), Automatic detection and description of syllabic features in continuous speech. *IEEE Transactions*, vol. **ASSP-24**, no. 5, pp. 365-379.

75.   **De Mori**, R. and **Torasso** P., (1976), Lexical classification in a speech understanding system using fuzzy relations. *Proc. ICASSP*, Philadelphia, PA, pp. 565-568.

76.   **De Mori**, R., **Laface**, P., **Makhonine**, A. and **Mezzalama**, M., (1977) A syntactic procedure for the recognition of glottal pulses in continuous speech. *Pattern Recognition*, vol. **9**, no. 4, pp. 181-189.

77.   **De Mori, R.** and **Laface, P.**, (1978), Representation of phonetic
        and phonemic knowledge in a speech understanding system.
        Proc. AISB/GI Conference on Artificial Intelligence, Hamburg,
        pp. 201-203.

78.   **De Mori, R.**, (1979), Recent advances in automatic speech
        recognition. *Signal Processing,* vol. **1**, no. 2, pp. 95-123.

79.   **De Mori, R.**, **Gubrinowicz, R.** and **Laface, P.**, (1979), Inference
        of a knowledge source for the recognition of nasals in
        continuous speech. *IEEE Transactions,* vol. **ASSP-27**, no. 5,
        pp. 538-549.

80.   **De Mori, R.** and **Saitta, L.**, (1979), Scheduling of processes
        in a speech understanding system based on approximate reason-
        ing. *Proc. 6th IJCAI,* Tokyo, Japan, pp. 204-207.

81.   **De Mori, R.** and **Giordano, G.**, (1980a), A parser for segmenting
        continuous speech into pseudo syllabic nuclei. *Proc. IEEE-
        ICASSP,* Denver, Colo., pp. 876-879.

82.   **De Mori, R.** and **Giordano, G.**, (1980b), Structural knowledge
        of the recognition of syllables in continuous speech. *Proc.
        World IFIP Congress, Information Processing-80,* North Holland
        Publ. Co., pp. 747-751.

83.   **De Mori, R.** and **Laface, P.**, (1980), Use of fuzzy algorithms
        for phonetic and phonemic labelling of continuous speech.
        IEEE Transactions, vol. **PAMI-2**, pp. 136-148.

84.   **De Mori, R.** and **Saitta, L.**, (1980), Automatic learning of
        fuzzy naming relations over finite languages. *Information
        Sciences,* **20**, pp. 93-139.

85.   **De Mori, R.**, **Laface, P.** and **Saitta, L.**, (1981), Inference of
        relations between phonetic and acoustic features in a speech
        understanding system. in: *Natural Communication with
        Computers,* ed. by L. Bolc [in Press].

86.   **Denes, P.** and **Mathews, M.V.**, (1960), Spoken digits recogni-
        tion using time-frequency pattern matching. *J.A.S.A.,* vol.
        32, pp. 1450-1455.

87.   **Denes, P.B.** and **Pinson, E.N.**, (1973), *The Speech Chain.*
        Doubleday, Garden City, New York.

88.   **Derkach, M.**, **Gumetsky, R.**, **Gura, B.** and **Mushin, L.**, (1975),
        Automatic recognition of simplified sentences constructed
        of the limited lexicon. in: *Speech Communication,* ed. by G.
        Fant, Almqvist and Wiksell, Stockholm.

89.   **Digital Signal Processing Committee IEEE-ASSP Society** (1979),
        Programs for Digital Signal Processing. IEEE Press, New York.

90.   **Dijkstra, E.**, (1959), A note on two problems in connection
        with graphs. *Numerische Mathematik* **1**, pp. 269-271.

91.   **Dixon, N.R.** and **Silverman, H.F.**, (1975), A description of
        the parametrically controlled modular structure for speech
        processing. *IEEE Transactions,* vol. **ASSP-23**, pp. 87-91.

92.   **Dixon, N.R.** and **Silverman, H.F.**, (1976a), A comparison of
        several speech spectra classification methods. *IEEE Transac-
        tions,* vol. **ASSP-25**, no. 4, pp. 289-298.

93.   **Dixon, N.R.** and **Silverman, H.F.**, (1976b), A general language-
        operated decision implementation system (GLODIS): its
        application to continuous speech segmentation. *IEEE Trans-
        actions,* vol. **ASSP-24**, no. 2, pp. 137-162.

94.   **Dixon, N.R.** and **Silverman, H.F.**, (1977), The 1976 modular
        acoustic processor (MAP). *IEEE Transactions,* vol. **ASSP-25**,
        no. 5, pp. 367-379.

95.   **Dixon, N.R.** and **Martin, T.B.**, (1979), *Automatic Speech and
        Speaker Recognition*. IEEE Press, New York.

96.   **Dorman, M.F.** and **Raphael, L.J.**, (1980), Distribution of
        acoustic cues for stop consonant place in VCV syllables.
        J.A.S.A., vol. **67**, no. 4, pp. 1333-1335.

97.   **Dubois, D.** and **Prade, H.**, (1978), *Fuzzy Algebra, Analysis,
        Logics*. School of Electrical Engineering, Purdue University,
        Lafayette, Ind., TR-EE 78-13.

98.   **Dubois, D.** and **Prade, H.**, (1979), Operations in a fuzzy-valued
        logic. *Information and Control,* vol. **22**, pp. 224-240.

99.   **Dubois, D.** and **Prade, H.**, (1980), *Fuzzy Sets and Systems*:
        Theory and Applications. Academic Press, New York.

100.   **Duda, R.** and **Hart, P.**, (1973), *Pattern Classification and
        Scene Analysis*. Wiley, New York.

101.   **Dudley, H.** and **Balashek, S.**, (1958), Automatic recognition
        of phonetic patterns in speech. *J.A.S.A.,* vol. **30**, pp.
        721-739.

102.   **Dukiewicz, L.**, (1967), *Polish Nasal Sounds* (in Polish). PWN,
        Warsaw.

103.   **Dutta Majumder, D.** and **Pal, S.K.**, (1978), On automatic plosive
       identification using fuzzyness in property sets. *IEEE
       Transactions,* vol. **SMC-8**, no. 4, pp. 302-308.

104.   **Erman, L.D.**, (1974a), *An environment and system for machine
       understanding of connected speech.* Ph. D. Thesis, Dept.of
       Computer Science, Carnegie-Mellon University, Pittsburg, PA.

105.   **Erman, L.D.**, (1974b), *Contributed Papers, IEEE Symposium on
       Speech Recognition.* IEEE Press, New York.

106.   **Erman, L.D.**, (1977), A functional description of the HEARSAY-
       II system. *Proc. IEEE-ICASSP,* Hartford, Conn., pp. 799-802.

107.   **Erman, L.D.**, **Hayes-Roth, F.**, **Lesser, V.R.** and **Raj Reddy, D.**,
       (1980), The HEARSAY-II speech understanding system. Inte-
       grating knowledge to resolve uncertainty. *Computing Surveys,*
       vol. **12**, no. 2, pp. 213-253.

108.   **Erman, L.D.** and **Smith, A.R.**, (1981), NOAH -- A bottom-up
       word hypothesizer for large vocabulary speech understanding
       systems. *IEEE Transactions,* vol. **PAMI-3**, no. 1, pp. 41-51.

109.   **Ernst. G.W.** and **Newell, A.**, (1969), *GPS: a Case Study in
       Generality and Problem Solving.* Academic Press, New York.

110.   **Evans, T.G.**, (1963), *A heuristic program to solve geometric-
       analogy problems.* Ph. D. Thesis, M.I.T., Cambridge, MA.

111.   **Fain, V.S.**, ed (1977), *Raspoznavanie Obrazov* (Pattern Recog-
       nition). Nauka, Moscow (in Russian).

112.   **Fant, G.**, (1960), *Acoustic Theory of Speech Production.*
       Mouton Co., The Hague.

113.   **Fant, G.**, (1973), *Speech Sounds and Features.* The MIT Press,
       Cambridge, MA.

114.   **Fant, G.**, ed. (1975), *Speech Communication.* Almqvist and
       Wiksell, Stockholm.

115.   **Fant, G.** and **Tatham, M.A.**, eds. (1975), *Auditory Analysis
       and the Perception of Speech.* Academic Press, New York.

116.   **Feigenbaum, E.**, ed. (1963), *Computers and Thought.* McGraw-
       Hill, New York.

117.   **Fennel, R.D.** and **Lesser, V.R.**, (1977), Parallelism in AI
       problem solving: a case study of HEARSAY-II. *IEEE Transac-
       tions,* vol. **C-26**, pp. 98-111.

118.   **Fikes, R.E.** and **Nilsson, N.J.**, (1971),STRIPS a new approach
       to the application of theorem proving to problem solving.
       Artificial Intelligence, vol. **2**, no. 3-4, pp. 189-208.

119.   **Fikes, D.**, **Hendrix, E.R.** and **Gary, G.**, (1977), A network-
       based knowledge representation and its natural deduction
       system. *Proc. 5th IJCAI,* Cambridge, MA, pp. 235-246.

120.   **Flanagan, J.L.**, (1972), *Speech Analysis, Synthesis and
       Perception* (2nd edition). Springer-Verlag.

121.   **Flanagan, J.L.** and **Rabiner, L.R.**, eds. (1973), *Speech
       Synthesis.* Dowden, Hutchinson and Ross, Stroudsberg, PA.

122.   **Forgie, J.W.** and **Forgie, C.D.**, (1959), Results obtained from
       a vowel recognition computer program. *J.A.S.A.,* Vol. **31**,
       pp. 1480-1489.

123.   **Forney, G.D.**, (1973), The Viterbi algorithm. *Proc. IEEE,*
       vol. **61**, pp. 268-278.

124.   **Fox, M.S.** and **Mostow, D.J.**, (1977), Maximal consistent inter-
       pretation of errorful data in hierarchically modeled domains.
       Proc. 5th IJCAI, Cambridge, MA, pp. 165-171.

125.   **Fray, D.B.**, (1959), Theoretical aspects of mechanical speech
       recognition. *Journal British Institute Radio Engineering,*
       vol. **19**, pp. 211-219.

126.   **Freeman, H.**, (1961), On the encoding of arbitrary geometric
       configurations. *IEEE Transactions,* vol. **EC-10**, pp. 260-268.

127.   **Fu, K.S.**, (1968), *Sequential Methods in Pattern Recognition
       and Machine Learning.* Academic Press, New York.

128.   **Fu, K.S.**, (1974), *Syntactic Methods in Pattern Recognition.*
       Academic Press, New York.

129.   **Fu, K.S.**, ed (1975), *Digital Pattern Recognition.* Springer-
       Verlag.

130.   **Fu, K.S.** and **Booth, T.L.**, (1975), Grammatical inference,
       Introduction and survey. *IEEE Transactions,* vol. **SMC-5**,
       Part I, no. 1, Part II, no. 4.

131.   **Fu, K.S.** and **Fung, L.W.**, (1975), Syntactic decoding for
       computer communication and pattern recognition. *IEEE
       Transactions,* vol. **C-24**, pp. 662-667.

132.   Fu, K.S., ed. (1977), *Syntactic Pattern Recognition. Applications.* Springer-Verlag.

133.   Fu, K.S. and Lu, S.Y., (1977), Stochastic error-correcting syntax analysis for recognition of noisy patterns. *IEEE Transactions,* vol. **C-26**, no. 12, pp. 1268-1276.

134.   Fu, K.S., ed. (1982), *Applications of Pattern Recognition.* CRC Press, Palm Beach, Fla.

135.   Fuchi, K. and Itahashi, S., (1976), Direct linear prediction for fundamental frequency analysis. *Proc. ICASSP,* Philadelphia, PA, pp. 318-322.

136.   Fujimura, D., (1962), Analysis of nasal consonants. *J.A.S.A.,* vol. **34**, no. 12, pp. 1865-1875.

137.   Fujimura, D., (1974), Syllables as units of speech recognition. *IEEE Symposium on Speech Recognition,* Carnegie-Mellon University, Pittsburg, PA, pp. 148-153.

138.   Fujisaki, H., Nakamura, N. and Yoshimune, K., (1970), Normalization and recognition of sustained Japanese vowels. *ASJ Transactions,* vol. **26**, no. 3, pp. 152-153.

139.   Fujisaki, H. and Kunisaki, O., (1978), Analysis, recognition and perception of voiceless fricative consonants in Japanese. *IEEE Transactions,* vol. **ASSP-26**, no. 1, pp. 21-37.

140.   Fujisaki, H., Higuchi, N. and Hosoya, K., (1980), *Analysis and recognition of voiced stop consonants.* Research on Human Information Processing. Annual report, Faculty of Engineering University of Tokyo, no. 1, pp. 99-107.

141.   Furui, S., (1980), A training procedure for isolated word recognition system. *IEEE Transactions,* vol. **ASSP-28**, no. 2, pp. 129-136.

142.   Gaines, B.R., (1975), Stochastic and fuzzy logic. *Electronic Letters,* vol. **21**, no. 9, pp. 188-189.

143.   Gaines, B.R., (1976), Foundations of fuzzy reasoning. *International Journal of Man-Machine Studies,* vol. **8**, pp. 623-668.

144.   Gardini, B. and Serra, A., (1974), Identification of speech parameters using a recursive method. *Proc. Speech Communication Seminar,* Almqvist and Wiksell, Stockholm, pp. 119-128.

145. Gerstman, L.J., (1968), Classification of self-normalized vowels. *IEEE Transactions,* vol. **AU-16,** no. 1, pp. 78-80.

146. Giles, R., (1979), A formal system for fuzzy reasoning. *Fuzzy Sets and Systems,* vol. **2,** pp. 233-257.

147. Gillmann, R.A., (1974), Automatic recognition of nasal phonemes. *IEEE Symposium on Speech Recognition,* Pittsburg, PA, pp. 74-79.

148. Giordana, A., Laface, P. and Saitta, L., (1980), Modelling control strategies for artificial intelligence applications. *Proc. IEEE International Conference on Parallel Processing,* Boyne Highlands, Mich., pp. 347-349.

149. Gold, B., (1966), *Word recognition computer program.* Res. Lab. Elec. MIT, Report no. 452.

150. Gold, B. and Rader, C.M., (1969), *Digital Processing of Signals.* McGraw-Hill, New York.

151. Goldberg, H.G., (1975), *Segmentation and labelling of speech: a comparative performance evaluation.* Techn. Rep. Computer Science, Carnegie-Mellon University, Pittsburg, PA.

152. Gonzales R.C. and Thomason, M.G., (1978), *Syntactic Pattern Recognition: an Introduction.* Addison-Wesley, Reading, MA.

153. Goodman, R., (1976), *Analysis of languages for man-machine voice communication.* Ph. D. Thesis, Stanford University.

154. Gray, A.H. and Markel, J.D., (1976), Distance measures for speech processing. *IEEE Transactions,* vol. **ASSP-24,** no. 5, pp. 380-391.

155. Gray, T., (1977), Articulatory movements in VCV sequences. *J.A.S.A.,* vol. **62,** no. 1, pp. 183-193.

156. Green, D.M. and Sweets, J.A., (1966), *Signal Detection: Theory and Psychophysics.* Wiley, New York.

157. Gresser, J.Y. and Mercier, G., (1975), Automatic segmentation of speech into syllabic and phonemic units: application to French words and utterances. in: *Auditory Analysis and Perception of Speech,* ed. by G. Fant and M.A.A. Thatam, Academic Press, New York, London, pp. 359-382.

158. Groc, B. and Tuffelli, D., (1980), A continuous speech recognition system for data base consultation. *Proc. ICASSP-80,* Denver, COLO, pp. 896-899.

159. Grosz, J.B., (1977), The representation and use of focus in
      a system for understanding dialogs. *Proc. 5th IJCAI,*
      Cambridge, MA, pp. 67-76.

160. Gueguen, C. and Le Roux, J., (1977), A fixed-point computation
      of partial correlation. *IEEE Transactions,* vol. **ASSP-25,**
      no. 3, pp. 257-258.

161. Gumetskii, R.J., (1973), Mashinoe raspoznavnie fraz s
      primeneniem nekotorykh lingvisticheskikh pravil (machine
      recognition of speech using certain linguistic rules).
      Trudy ARSO-UP, pp. 8-11, Alma-Ata, USSR (in Russian).

162. Gupta, V.N., Bryan, J.K. and Gowdy, J.N., (1978), A speaker
      independent system based on linear prediction. *IEEE Transac-
      tions,* vol. **ASSP-26,** no. 1, pp. 27-33.

163. Hall, P.V.A., (1971), Branch and bound and beyond. *Proc.
      2nd IJCAI,* Londond, pp. 641-650.

164. Hall, P.V.A., (1973), Equivalence between AND/OR graphs and
      context-free grammars. *Communications of the ACM,* vol. **16,**
      no. 7, pp. 445.

165. Halle, M., Hughes, G.W. and Redley (1957), Acoustic properties
      of stop consonants. *J.A.S.A.,* vol. **29,** pp. 107-116.

166. Halle, M. and Stevens, K.N., (1962), Speech recognition: a
      model and a program for research. *IEEE Transactions,* vol.
      IT-8, pp. 155-159.

167. Haltsonen, S., Bray, K.T. and Kohonen, T., (1978), Applica-
      tions of orthogonal projections principles to simultaneous
      phonemic segmentation and labelling of continuous speech.
      Proc. 4th IJCPR, Kyoto, Japan, pp. 1006-1008.

168. Harris, F.J., (1978), On the use of windows for harmonic
      analysis with the discrete fourier transform. *Proc. IEEE,*
      vol. **66,** no. 1, pp. 51-83.

169. Haton, J.P., (1976), Reconnaisance analytique de la parole
      aux niveaux accoustique, morphologique, lexical et syntaxique.
      RAIRO Informatique, vol. **10,** no. 9, pp. 57-75.

170. Haton, J.P. and Pierrel, J.M., (1976), Organization and
      operation of a connected speech understanding system at
      lexical, syntactic and semantic levels. *Proc. ICASSP,*
      Philadelphia, PA, pp. 430-433.

170. **Haton, J.P.**, (1980), The representation and use of the lexicon
      in automatic speech recongition. in: *Spoken Language
      Generation and Understanding,* ed. by J.C. Simon, Reidel,
      Leyden, The Netherlands.

171. **Hayes-Roth, F.** and **Mostow, D.J.**, (1976), Syntax and semantics
      in a distributed speech understanding system. *Proc. ICASSP,*
      Philadelphia, PA, pp. 421-424.

172. **Hayes-Roth, F.** and **Lesser, V.**, (1977), Focus of attention in
      a distributed logic speech understanding system. *Proc. 5th
      IJCAI,* Cambridge, MA, pp. 27-35.

173. **Hendrix, G.**, (1975), Expanding the utility of semantic
      networks through partitioning. *Proc. 4th IJCAI,* Tbilisi,
      USSR, pp. 115-121.

174. **Hersher, M.B.** and **Cox, R.B.**, (1976), Source data entry using
      voice input. *Proc, ICASSP,* Philadelphia, PA, pp. 190-193.

175. **Hess, W.J.**, (1976), A pitch synchronous digital feature
      extraction system for phonemic recognition of speech. *IEEE
      Transactions,* vol. **ASSP-24**, pp. 14-25.

176. **Hess, W.J.**, (1980), Pitch determination. An example for
      the application of signal processing methods in the speech
      domain. *Proc. EUSIPCO-80,* Lausanne, Switzerland.

177. **Hewitt, C.**, (1972), *Description and theoretical analysis
      (using schemata) of PLANNER: a language for proving theorems
      and manipulating models in a robot.* Ph. D. Thesis, Dept. of
      Math., MIT, Cambridge, MA.

178. **Hill, D.R.**, (1971), Man-machine interaction using speech in:
      Advances in Computers, ed. by Alt, Rubinoff and Yovits, vol.
      II, Academic Press, New York, pp. 165-230.

179. **Holden, A.D.C.** and **Strasbourger, E.**, (1976), A computer
      programming system using continuous speech input. *IEEE
      Transactions,* vol. **ASSP-24**, pp. 579-581.

180. **House, A.S.** and **Fairbanks, G.**, (1953), The influence of
      consonant environment upon the secondary acoustical charac-
      teristics of vowels. *J.A.S.A.,* vol. **25**, pp. 105-113.

181. **Hughes, G.W.** and **Hemdal, J.F.**, (1965), *Speech analysis*. Rep.
      AFCRL-65-681 (P13552), Purdue University.

182. **Hyde, S.R.**, (1972), Automatic speech recognition literature,
      survey and discussion. in: *Human Communication: a Unified
      View,* ed. by E.E. David and P.B. Denes, McGraw-Hill, New York.

183.   **Interstate Electronics**, (1978), Voice data entry computer
        terminal. *Computer Design,* June 1978, pp. 34-35.

184.   **Ishinazi, S.**, (1977), Interspeaker normalization using vocal
        tract length and vowel feature extraction. *Progress report
        on speech research 1977, Speech Processing Section, Informa-
        tion Sciences Division, Electrotechnical Laboratory, Tokyo,
        PIPS-R,* no. 18, pp. 66-70.

185.   **Ishinazi, S., Nakajima, T.** and **Ohmura, H.**, (1977), A method
        of continuous vowel segmentation and feature extraction.
        *Progress report on speech research 1977, Speeech Processing
        Section Laboratory, Information Sciences Division, Electro-
        technical Laboratory, Tokyo, PIPS-R,* no. 18, pp. 32-39.

186.   **Ishinazi, S.**, (1978), Vowel discrimination by use of articula-
        tory model. *Proc. 4th IJCPR,* Kyoto, Japan, pp. 1053-1055.

187.   **Itahashi, S., Makino, S.** and **Kido, K.**, (1973), Discrete word
        recognition utilizing a word dictionary and phonological
        rules.

188.   **Itahashi, S.** and **Yokoyama, S.**, (1979), *A formant extraction
        method utilizing mel scale and equal loudness contour.* STL-
        QPSR 4/78, Royal Institute of Technology, Stockholm.

189.   **Itakura, F.** and **Saito, S.**, (1968), Analysis synthesis tele-
        phony based on the maximum likelihood method. *Proc. 6th
        Congress on Acoustics,* Tokyo, paper C-5-5.

190.   **Itakura, F.**, (1975), Minimum prediction residual principle
        applied to speech recognition. *IEEE Transactions,* vol. **ASSP-
        23**, pp. 67-72.

191.   **Jakobson, R., Fant, G.M.** and **Halle, M.**, (1963), *Preliminary
        to Speech Analysis.* MIT Press, Cambridge, Ma.

192.   **Jelinek, F., Bahl, L.R.** and **Mercer, R.L.**, (1975), Design of
        a linguistic statistical decoder for the recognition of
        continuous speech. *IEEE Transactions,* vol. **IT-21**, pp. 250-
        256.

193.   **Jelinek, F.**, (1976), Continuous speech recognition by
        statistical method. *Proceedings IEEE,* vol. **64**, no. 4, pp.
        532-556.

194.   **Jelinek, F., Mercer, R.L., Bahl, L.R.** and **Baker, J.K.**, (1977),
        Perplexity. A measure of difficulty of speech recognition
        task. 9th Meeting of the Acoustical Society of America,
        Miami Beach, Dec. 1977.

195.   Johnson, D.H. and Weinstein, C.J., (1978), A phrase recognizer
       using syllable-based acoustic measurements. *IEEE Transactions*,
       vol. ASSP-26, no. 5, pp. 409-418.

196.   Kacprowski, J. and Mikiel, W., (1968), Simplified rules for
       parametric sythesis of nasals and stop consonants in CV
       syllables by means of the terminal analog speech synthesizer.
       Acoustica, vol. 16, no. 6, pp. 256-268.

197.   Kameny, I., (1975), Comparison of formant spaces of retro-
       flexed and non-retroflexed vowels. *IEEE Transactions*, vol.
       ASSP-23, pp. 38-49.

198.   Kanamori, Y. and Kido, K., (1975), Context effect in percep-
       tion of nasals in VCV syllables. *Journal of Acoustic Society
       of Japan*, pp. 269-270.

199.   Kaplan, G., (1980), Words into action. *IEEE Spectrum*, pp.
       22-29, June.

200.   Kashyap, R.L., (1979), Syntactic decision rules for recogni-
       tion of spoken words and phrases using a stochastic automaton.
       IEEE Transactions, vol. PAMI-1, no. 2, pp. 154-163.

201.   Kasuya, H. and Kido, K., (1968), *On properties of formant
       frequencies of vowels in meaningless words composed of three
       mores*. Technical report AE68-13 IECEH (in Japanese).

202.   Kasuya, H., Suzuki, H. and Kido, K., (1968), Changes in pitch
       and first three formant frequencies of five Japanese vowels
       with age and sex speakers. *Journal of the Acoustic Society
       of Japan*, vol. 24, no. 6, pp. 355-364.

203.   Kasuya. H. and Wakita, H., (1976), Speech segmentation and
       feature normalization based on area function. *Proc. IEEE-
       ICASSP*, Philadelphia, PA, pp. 29-32.

204.   Kasûya, H. and Wakita, H., (1978), *On segmentation of
       continuous speech*. Technical Report Acoustical Soceity of
       Japan, S68-10.

205.   Kaufmann, A., (1973), *Introduction à la théorie des sous-
       ensemble flous*. Masson, Paris. English translation published
       by Academic Press, New York, 1975.

206.   Kawaguchi, E., (1971), On a method of speech recognition with
       word dictionary. *Electronics and Communication in Japan*,
       vol. 54-C, no. 1, pp. 98-106.

207.   **Klatt, D.H.**, (1972), Acoustic theory of terminal analog speech
       synthesis. *Conference on Speech Communication and Processing*,
       Newton, Ma, Mem. D2, pp. 131-135.

208.   **Klatt, D.H.** and **Stevens, K.N.**, (1973), On the automatic
       recognition of continuous speech: implications of a spectro-
       gram-reading experiment. *IEEE Transactions*, vol. **AU-21**, pp.
       210-216.

209.   **Klatt, D.H.**, (1974a), On the design of speech understanding
       systems. *Proceedings of SCS-74*, Addison-Wesley, pp. 277-289.

210.   **Klatt, D.H.**, (1974b), The duration of /s/ in English words.
       Journal of Speech and Hearing Research, vol. **17**, pp. 51-63.

211.   **Klatt, D.H.**, (1975a), Word verification in a speech under-
       standing system. **See:** D.R. Reddy ed., 1975.

212.   **Klatt, D.H.**, (1975b), Voice onset time, frication and
       aspiration in word-initial consonant clusters. *Journal of
       Speech and Hearing Research*, vol. **18**, no. 4, pp. 686-706.

213.   **Klatt, D.H.**, (1976), A digital filter bank for spectral
       matching. *Proc. IEEE-ICASSP*, Philadelphia, PA, pp. 537-540.

214.   **Klatt, D.H.**, (1977), Review of the ARPA speech understanding
       project. *J.A.S.A.*, vol. **62**, pp. 1345-1366.

215.   **Klatt, D.H.**, (1979), Speech perception: a model of acoustic
       phonetic analysis and lexical access. *Journal of Phonetics*,
       vol. **7**, pp. 279-312.

216.   **Klein, W.**, **Plomp, R.H.** and **Pols, L.C.W.**, (1970), Vowel
       spectra, vowel spaces and vowel identification. *J.A.S.A.*,
       vol. **48**, pp. 999-1009.

217.   **Knipper, A.V.**, (1980), Formantnij svoistva sonorikh SG
       fragmentov. (Formant properties of sonorant CV fragments).
       Proc. ARCO-11, Yerevan, USSR, pp. 21-24 (in Russian).

218.   **Kobayashi, Y.** and **Niimi, Y.**, (1979), A procedural represen-
       tation of lexical entries in augmented transition network
       grammar. *Proc. 6th IJCAI*, Tokyo, Japan.

219.   **Kohda, M.**, **Hashimoto, S.** and **Saito, S.**, (1972), Spoken digit
       mechanical recognition system. *IECEJ Transactions*, vol.
       55-D, no. 3, pp. 186-193 (in Japanese).

220.   **Kohda, M.** and **Saito, S.**, (1973), Influence of long-term
        variations of learning and unknown samples on recognition
        rate of spoken digits. *Record of Joint Meeting Acoustic
        Society of Japan,* 1-3-23 (in Japanese).

221.   **Kohda, M., Nakatsu R.** and **Shikano, K.**, (1976), Speech
        recognition in the question-answering system operated by
        conversational speech. **Proc. IEEE-ICASSP**, Philadelphia, PA,
        pp. 442-445.

222.   **Kohda, M.** and **Nakatsu, R.**, (1978), An acoustic processor in
        a conversational speech system. *Review of the Electrical
        Communication Laboratories (NTT),* vol. **26**, no. 11-12, pp.
        1486-1504.

223.   **Kohonen, T.**, (1978), *Content Addressable Memories.* Springer-
        Verlag, Berlin.

224.   **Kohonen, T.** and **Reuhkala, E.**, (1978), A very fast associative
        method for the recognition and correction of misspelt words
        based on redundant hash addressing. *Proc. 4th IJCPR,* Kyoto,
        Japan, pp. 807-809.

225.   **Kohonen, T., Nemeth, G., Bry, K.J., Jalanko, M.** and **Rittinen,
        H.**, (1979), Spectral classification of phonemes by learning
        subspaces. *Proc. IEEE-ICASSP,* Washington DC, pp. 97-100.

226.   **Kohonen, T., Rittinen, H., Jalanko, M., Reuhkala, E.** and
        Haltsonen, S., (1980), A thousand word recognition system
        based on the learning subspace method and redundant hash
        addressing. *Proc. 5th IJCPR,* Miami, Fla.

227.   **Kopec, G.E., Oppenheim, A.V.** and **Tribolet, J.M.**, (1977),
        Speech analysis by homomorphic prediction. *IEEE Transactions,*
        vol. **ASSP-25**, pp. 40-49.

228.   **Kornfield, W.A.** and **Hewett, C.**, (1981), The scientific
        community metaphor. *IEEE Transactions,* vol. **SMC-11**, pp.
        81-96.

229.   **Kozhevnikov, V.A.** and **Chistovich, L.A.**, (1965), Speech:
        articulation and perception. JPRS 30, p. 543 (translated
        from Russian).

230.   **Krinov, S.N., Vasiljev, A.V., Saveljev, V.P.** and **Tsiemel, G.I.**,
        (1975), *A speech recognition system. Rechevoe Obshchenie,*
        Moscow, Nauka (in Russian).

231.   **Kunt, M.**, (1980), *Traitement numérique des signaux.* Giorgi
        publ., Lausanne, Switzerland.

232.   **Kuwahara, H.** and **Sakai, T.**, (1973), Normalization of
       coarticulation effect for a sequence of vowels in continuous
       speech. *Transaction ASJ,* vol. **29**, no. 2, pp. 91-99 (in
       Japanese).

233.   **Ladefoget, P.**, (1975), *A Course in Phonetics.* Harcourt Brace
       Jovanovich, New York.

234.   **Laface, P.**, (1980), A formant tracking system toward automatic
       recognition of speech. *Signal Processing,* vol. **2**, no. 2,
       pp. 113-130.

235.   **Lea, W.A.**, **Medress, M.F.** and **Skinner, T.E.**, (1975), A
       prosodically guided speech understanding system. *IEEE
       Transactions,* vol. **ASSP-23**, pp. 30-38.

236.   **Lea, W.A.**, (1976), *Prosodic aids to speech recognition -
       VIII. Listeners' perceptions of selected English stress
       patterns.* Univac-Report no. PX-11711, Univac Park, St. Paul,
       Minnesota.

237.   **Lea, W.A.** and **Soup, J.E.**, (1979), *Review of the ARPA SUR
       project and survey of current technology in speech understand-
       ing.* Speech Communication Research Laboratory, Los Angeles,
       CA.

238.   **Lea, W.A.**, Ed. (1980), *Trends in Automatic Speech Recognition.*
       Englewood Cliffs, Prentice-Hall, New York.

239.   **Lee, R.C.T.**, (1972), Fuzzy logic and the resolution principle.
       Journal of the ACM, vol. **19**, pp. 109-119.

240.   **Lennig, M.** and **Mermelstein, P.**, (1980), Entrainement lexical
       semi-automatique d'un système de reconnaissance à base
       syllabique. *Proc. XIèmes Journées d'Etude sur la Parole,*
       GALF, Strasbourg, France.

241.   **Lesser, V.R.**, **Fennel, R.D.**, **Erman, L.D.** and **Reddy, D.R.**,
       (1975), Organization of the HEARSAY-II speech understanding
       system. *IEEE Transactions,* vol. **ASSP-23**, pp. 11-23.

242.   **Lesser, V.R.** and **Corkill, D.D.**, (1981), Functionally-accurate,
       cooperative distributed systems. *IEEE Transactions,* vol.
       SMC-11, pp. 81-96.

243.   **Levenshtein, V.I.**, (1966), Binary codes capable of corrections,
       deletions, insertions and reversals. *Soviet Physics-Doklady,*
       vol. **10**, pp. 707-710.

244.   **Levinson, S.E.**, (1977), The effect of syntax on word
       recognition accuracy. *Proc. IEEE-ICASSP,* Hartford, Connecti-
       cut.

245.   **Levinson, S.E.** and **Shipley, K.L.**, (1980), A conversational
       mode airline information and reservation system using speech
       input and output. *Bell System Technical Journal,* vol. **59,**
       pp. 119-137.

246.   **Liberman, A.M., Cooper, F.S., Harris, K.S., MacNeilage, P.F.**
       and **Studdert-Kennedy, M.**, (1967), **in:** *Some observations on
       a model for speech perception. Model for the perception of
       Speech and Visual Form,* ed. by W.W. Dunn, MIT Press, pp. 68-77.

247.   **Liberman, A.M.**, (1970), The grammars of speech and language.
       Cognitive Psychology, vol. **1,** no. 4, pp. 301-323.

248.   **Liberman, A.M., Cooper, F.S., Shankweiller, D.P.** and **Studdert-
       Kennedy, M.**, (1971), Perception of the speech code. in:
       Human Communication: A Unified View, ed. by E. David Jr. and
       P. Denes, McGraw-Hill, New York.

249.   **Lienard, J.S., Mlouka, M.** and **Mariani, J.J.**, (1974), Real-
       time segmentation of speech. *Proc. Speech Communication
       Seminar, SCS-74,* Stockholm, Sweden, pp. 183-187.

250.   **Lindblom, B.** and **Ohman, S.**, (1979), *Frontiers of Speech
       Communication Research.* Academic Press, New York.

251.   **Lisker, L.** and **Abramson, A.S.**, (1964), A cross-language study
       of voicing in initial stops: acoustic measurements. *Word,*
       vol. **20,** no. 3, pp. 384-422.

252.   **Lowerre, B.T.**, (1976), *The HARPY speech recognition system.*
       Ph. D. Dissertation, Tech. Rept. Comp. Sci. Dept., Carnegie-
       Mellon University, Pittsburg, Pa.

253.   **Maeda, S.**, (1979), An articulatory model of the tongue based
       on statistical analysis. *Speech Communication papers,*
       Acoustical Society of America, pp. 67-70.

254.   **Makhoul, J.**, (1970), *Speaker-machine interaction in automatic
       speech recognition.* MIT RLE technical report no. 480.

255.   **Makhoul, J.** and **Wolf. J.**, (1972), *Linear prediction and the
       spectral analysis of speech.* Technical report no. 2304. Bolt
       Baranek and Newman, Cambridge, Ma.

256.   **Makhoul, J.** and **Wolf, J.**, (1973), *The use of a two-pole linear prediction model in speech recognition.* Tecnical report no. 2357, Bolt Baranek and Newman, Cambridge, Ma.

257.   **Makhoul, J.**, (1975), Linear prediction: a tutorial review. *Proceedings IEEE,* vol. **63**, pp. 561-580.

258.   **Makhoul, J.**, (1977), Stable and efficient lattice methods for linear prediction. *IEEE Transactions,* vol. **ASSP-25**, no. 5, pp. 423-427.

259.   **Makhoul, J.**, **Viswanathan, R.**, **Schwartz, R.** and **Huggins, A.W.F.**, (1978), A mixed-source model for speech comprehension and synthesis. *Proc. IEEE-ICASSP,* Tulsa, Ok, pp. 83-86.

260.   **Malécot, A.**, (1956), Acoustic cues for nasal consonants. *Language,* vol. *32,* no. 2, pp. 274-284.

261.   **Mariani, J.**, (1980), Some points concerning speech communication with computers. **see**: J.C. Simon ed., 1980.

262.   **Markel, J.D.**, (1972), Digital inverse filtering - A new tool for formant trajectory estimation. *IEEE Transactions,* vol. **AU-20**, pp. 129-137.

263.   **Markel, J.D.** and **Gray, Jr., A.H.**, (1975), Cepstral distance and the frequency domain. *90th Meeting of the Acoustical Society of America,* vol. **58**, p. 597.

264.   **Markel, J.D.** and **Gray, Jr.,A.H.**, (1976), *Linear Prediction of Speech.* Springer-Verlag, Berlin.

265.   **Marslen-Wilson, W.D.**, (1975), Sentence perception as an interactive parallel processing. *Science,* vol. **189**, pp. 487-501.

266.   **Martin, T.B.**, (1976), Practical applications of voice input to machine. *Proceedings IEEE,* vol. **64**, no. 4, pp. 487-501.

267.   **Massaro, D.W.** and **Oden. G.C.**, (1978), Integration of featural information in speech perception. *Psychological Review,* vol. **85**, no. 3, pp. 172-191.

268.   **Massaro, D.W.** and **Oden. G.C.**, (1980), Evaluation and integration of acoustic features in speech perception. *J.A.S.A.,* vol. **67**, no. 3, pp. 996-1013.

269.   **McCandless, S.S.**, (1974), An algorithm for automatic formant extraction using linear prediction spectra. *IEEE Transactions,* vol. **ASSP-22**, pp. 135-140.

270.   McDermott, D.V. and Sussman, G.J., (1972), *The CONNIVER reference manual*. MIT Artificial Intelligence Laboratory, memo no. 259, Cambridge, Ma.

271.   McDermott, D.V., (1978), The last survey of representation of knowledge. *Proc. AISB Conference on Artificial Intelligence,* Hamburg, pp. 147-174.

272.   McKeown, D.M., (1977), Word verification in the HEARSAY-II speech understanding system. *Proc. IEEE-ICASSP,* Hartford, Connecticut, pp. 795-798.

273.   Medress, M.F., Cooper, F.S., Forgie, J.W., Green, C.C., Klatt, D.H., O'Malley, M.H., Neuburg, E.P., Newell, A., Reddy, D.R., Ritea, B., Shoup-Hummel, J.E., Walker, D.E. and Woods, W.A., (1977), Speech understanding systems: report of steering-committee. *Artificial Intelligence,* vol. **9**, pp. 158-166.

274.   Meloni, H., (1979), Formalisation d'un lexique du premier ordre pour la reconnaissance automatique de la parole. *Proc. X Journées d'Etude sur la Parole, GALF,* Grenoble, France, pp. 337-340.

275.   Mercier, G., (1978), Evaluation des indices acoustiques utilisés dans l'analyse phonétique du systeme KEAL. *Proc. IX Journées d'Etude sur la Parole, GALF,* Lannion, France, pp. 321-342.

276.   Mercier, G. and Quinton, P., (1980), The KEAL speech understanding system. **see:** J.C. Simon ed., 1980.

277.   Mermelstein, P., (1975), Automatic segmentation of speech into syllabic units. *J.A.S.A.,* vol. **58**, pp. 880-883.

278.   Mermelstein, P., (1977), On detecting nasals in continuous speech. *J.A.S.A.,* vol. **61**, pp. 581-587.

279.   Miller, P.L., (1973), *A locally organized parser for spoken input*. Technical Report 503, Lincoln Laboratory, MIT, Lexington, Ma.

280.   Minsky, M., (1968), *Semantic Information Processing*. MIT press, Cambridge, Mass.

281.   Minsky, M., (1974), *A framework for representing knowledge*. AI Memo 306, Artificial Intelligence Laboratory, MIT, Cambridge, Ma.

282.   **Mohlo, L.M.**, (1976), Automatic acoustic-phonetic analysis of
        fricatives and plosives. *Conference Record, IEEE-ICASSP*,
        pp. 182-185.

283.   **Moll, K.L.**, (1962), Velopharyngeal closure on vowels. *Journal
        of Speech and Hearing Research,* vol. **5**, pp. 30-37.

284.   **Moore, R.C.**, (1975), *Reasoning from incomplete knowledge in
        a procedural deduction system.* AI-TR-347, Artificial
        Intelligence Laboratory, MIT, Cambridge, Ma.

285.   **Morton, J.**, (1970), A functional model of memory. in: *Models
        of Human Memory,* ed. by D.A. Norman, Academic Press, New York.

286.   **Myers, C.S.**, **Rabiner, L.R.** and **Rosenberg, A.E.**, (1980), An
        investigation of the use of dynamic time-warping for word
        spotting and connected speech recognition. *IEEE Transactions,*
        vol. **ASSP-28**, pp. 622-635.

287.   **Myers, C.S.** and **Rabiner, L.R.**, (1981), A dynamic time-
        warping algorithm for connected word recognition. *IEEE
        Transactions,* vol. **ASSP-29**.

288.   **Nakagawa, S.**, (1976), *A machine understanding system for
        spoken Japanese sentences.* Ph. D. Thesis, Dept. Information
        Sciences, Kyoto University.

289.   **Nakajima, T., ed.** (1976), *Progress report on speech research
        76.* Information Sciences Division, Electrotechnical Labora-
        tory, Tokyo.

290.   **Nakajima, T.** and **Suzuki, T.**, (1978), *Application of the
        articulatory feature vowel system to continuous speech.*
        PIPS-R, no. 20, pp. 67-70.

291.   **Nakata, K.**, (1959), Synthesis and perception of nasals
        consonants. *J.A.S.A.,* vol. **31**, no. 6, pp. 661-666.

292.   **Nakatsu, R.** and **Kohda, M.**, (1978), Speech recognition of
        connected words. *Proc. 4th IJCPR,* Kyoto, Japan.

293.   **Nakatsu, R.**, (1980), A speech recognition machine for connec-
        ted words. *Proc. IEEE-ICASSP,* Denver, Colo., pp. 199-202.

294.   **Nakatsui, M.** and **Suzuki, J.**, (1969), Methods of observation
        of glottal source wave using digital inverse filtering in
        time domain. *J. Radio Research Labs.,* vol. **16**, pp. 95-98.

295.   Nash-Webber, B., (1975), Semantic support for a speech
        understanding system. *IEEE Transactions,* vol. **ASSP-23**, pp.
        124-128.

296.   Neely, R.B. and White, G.M., (1974), A real time low cost
        speech understanding system. in: *Information Processing,*
        ed. by A. Rosenfeld, North-Holland, Amsterdam, New York, pp.
        748-752.

297.   Neuburg, E.P., (1975), Philosophies of speech recognition.
        in: *Speech Recognition,* ed. by D.R. Reddy, Academic Press,
        New York.

298.   Newell, A., (1973), Production systems: models of control
        structures. in: *Visual Information Processing,* ed. by W.C.
        Chase, Academic Press, New York, pp. 463-526.

299.   Newell, A., Barnett, J., Forgie, J., Green, C., Klatt, D.H.,
        Licklider, J.C.R., Munson, J., Reddy, D.R. and Woods, W.A.,
        (1973), *Speech understanding systems - Final report of a
        study group.* North-Holland, Amsterdam, New York.

300.   Newell, A., (1975), A tutorial on speech understanding
        systems. *Speech Recognition, Invited Papers of the 1974
        IEEE Symposium,* ed. by D.R. Reddy, Academic Press, New York,
        pp. 3-54.

301.   Newell, A., (1980), HARPY, production systems and human
        cognition. in: *Perception and Production of Fluent Speech,*
        ed. by R.A. Cole, Erlbaum Associates.

302.   Nii, H.P. and Feigenbaum,E.A., (1978), Rule-based understand-
        ing of signals. in: *Pattern Directed Inference Systems,*
        ed. by F. Hayes Roth and D.A. Watermann, Academic Press, New
        York.

303.   Niimi, Y., (1978), A method for forming universal reference
        patterns in an isolated word recognition system. *Proc. 4th
        IJCPR,* Kyoto, Japan, pp. 1022-1032.

304.   Niimi, Y., (1979), *On sei nin shiki (speech recognition).*
        Kyo-ritsu publishing co., Tokyo (in Japanese).

305.   Nilsson, N.J., (1971), *Problem Solving Methods in Artificial
        Intelligence.* McGraw-Hill, New York.

306.   Nilsson, N.J., (1977), *A production system for automatic
        deduction.* SRI International, Technical Note 148, Menlo
        Park, Ca.

307. **Oden, G.C.**, (1978), Integration of place and voicing infor-
      mation in the identification of synthetic stop consonants.
      Journal of Phonetics, vol. **6**, pp. 83-93.

308. **Ohman, S.E.G.**, (1966), Coarticulation in VCV utterances:
      spectrographic measurements. *J.A.S.A.,* vol. **39**, no. 1, pp.
      151-168.

309. **Oka, R.**, (1978), *A study of pattern matching algorithms using
      dynamic programming for the purpose of continuous word
      recognition.* Progress report on Speech Research, Speech
      Processing Section, Information Sciences Division, Electro-
      technical Laboratory, Tokyo, PIPS-R, no. 20, pp. 32-36.

310. **O'Malley, M.**, (1976), *Lecture Notes on Automatic Speech
      Recognition.* CISM, Udine, Italy.

311. **Oppenheim, A.V.** and **Shafer, R.W.**, (1975), *Digital Signal
      Processing.* Prentice Hall, Englewood Cliffs, NJ.

312. **Oppenheim, A.V., Kopec, G.E.** and **Tribolet, J.M.**, (1976),
      Signal analysis of homomorphic prediction. *IEEE Transactions,*
      vol. **ASSP-24**, no. 4, pp. 327-332.

313. **Oshika, B.T., Zue, V.W., Weeks, R.V., Nue, H.** and **Auerbach, J.**,
      (1975), The role of phonological rules in speech understanding
      research. *IEEE Transactions,* vol. **ASSP-23**, pp. 104-112.

314. **Pavlidis, T.**, (1977), *Structural Pattern Recognition.*
      Springer-Verlag, Berlin, New York.

315. **Paxton, W.H.** and **Robinson, A.E.**, (1973), *A parser for speech
      understanding systems.* Advance Papers, IJCAI, Stanford, CA,
      pp. 216-222.

316. **Paxton, H.W.**, (1975), A best-first parser. *IEEE Transactions,*
      vol. **ASSP-23**, pp. 426-432.

317. **Paxton, H.W.**, (1977), *A framework for speech understanding.*
      Ph. D. Dissertation, Stanford University, Stanford, CA.

318. **Perrenou, G.** and **Tep, G.**, (1979), Grandes lexiques et traite-
      ments phonologiques: une structure de composante phonologique
      adaptee au traitement automatique. *Proc. X Journees d'Etude
      sur la Parole, GALF,* Grenoble, France, pp. 344-352.

319. **Pierce, J.R.**, (1969), Whither speech recognition? *J.A.S.A.,*
      vol. **46**, pp. 1049-1051.

320.   Pols, L.C.W., (1971), Real-time recognition of spoken words.
       IEEE Transactions, vol. **C-20**, pp. 1972-1978.

321.   Quinton, P., (1980), *Contribution à la reconnaissance de la
       parole. Utilisation de méthodes pour la reconnaisance de
       phrases.* Ph. D. Thesis, University of Rennes, France.

322.   Rabiner, L.R. and Sambur, M.R., (1975), An algorithm for
       determining the endpoints of isolated utterances. *Bell
       System Technical Journal,* vol. **54**, no. 2, pp. 297-315.

323.   Rabiner, L.R., Atal, B.S. and Sambur, M.R., (1977), LPC
       prediction error - Analysis of its variation with the position
       of the analysis frame. *IEEE Transactions,* vol. **ASSP-25**, no.
       5, pp. 434-440.

324.   Rabiner, L.R. and Sambur, M.R., (1977), Application of the
       LPC distance measure to the voiced-unvoiced silence detection
       problem. *IEEE Transactions,* vol. **ASSP-25**, no. 4, pp. 338-
       345.

325.   Rabiner, L.R., (1978), On creating reference templates for
       speaker independent recognition of isolated words. *IEEE
       Transactions,* vol. **ASSP-26**, no. 1, pp. 34-42.

326.   Rabiner, L.R., Rosenberg, A.E. and Levinson, S.E., (1978),
       Considerations in dynamic time warping for discrete word
       recognition. *IEEE Transactions,* vol. **ASSP-26**, pp. 575-582.

327.   Rabiner, L.R. and Shafer, R.W., (1978), *Digital Processing
       of Speech Signals.* Prentice Hall, Englewood Cliffs, NJ.

328.   Rabiner, L.R., Levinson, S.E., Rosenberg, A.E. and Wilpon,
       J.G., (1979), Speaker-independent recognition of isolated
       words using clustering techniques. *IEEE Transactions,* vol.
       ASSP-27, pp. 336-349.

329.   Rabiner, L.R. and Wilpon, J.G., (1979), Speaker independent
       isolated word recognition for a moderate size (54 word)
       vocabulary. *IEEE Transactions,* vol. **ASSP-27**, pp. 583-587.

330.   Rabiner, L.R. and Schmidt, C.E., (1980), Application of
       dynamic time warping to connected digit recognition. *IEEE
       Transactions,* vol. **ASSP-28**, pp. 337-388.

331.   Rabiner, L.R. and Wilpon, J.G., (1981), A two pass system
       for isolated word recognition. *Bell System Technical Journal,*
       vol. **60**, pp. 739-766.

332. Reddy, D.R., (1967), Computer recognition of connected speech. *J.A.S.A.*, vol. **42**, pp. 329-437.

333. Reddy, D.R. ed., (1975), *Speech Recognition - Invited Papers of the 1974 IEEE Symposium*. Academic Press, New York.

334. Reddy, D.R., (1976), Speech recognition by machine: a review. *Proceedings IEEE*, vol. **64**, pp. 501-531.

335. Reddy, D.R. ed., (1977), *HEARSAY speech understanding systems: summary of results of the five-years research effort at Carnegie-Mellon University*. Dept. Computer Sciences, Carnegie-Mellon University, Pittsburg, Pa.

336. Rescher, N., (1969), *Many-Valued Logic*. McGraw Hill, New York.

337. Ritea, H.B., (1975), Automatic speech understanding systems. *Proc. 11th Annual IEEE Computer Society Conference*, Washington, D.C., pp. 319-322.

338. Rivoira, S. and Torasso, P., (1978), An isolated word recognizer based on grammar classification processes. *Pattern Recognition*, vol. **10**, no. 2, pp. 73-84.

339. Robinson, J.J., (1975a), Performance grammars. in: *Speech Recognition*, ed. by D.R. Reddy, New York, Academic Press, pp. 401-427.

340. Robinson, J.J., (1975b), A tuneable performance grammar. *American Journal of Computational Linguistics*, microfiche 34.

341. Rosenberg, A.E., Rabiner, L.R., Levinson, L.R. and Wilpon, J.G., (1981), A preliminary study on the use of demisyllables in automatic speech recognition. *Proc. ICASSP-81*, Atlanta, pp. 967-970.

342. Rosenfeld, A., (1975), Fuzzy graphs. in: *Fuzzy Sets and Their Applications to Cognitive and Decision Process*, ed. by L.A. Zadeh, K.S. Fu, K. Tanaka and K. Shimura, Academic Press, New York.

343. Ruske, G. and Shotola, T., (1977), An automatic speech recognition system using syllable nuclei and consonant clusters. *9th International Congress on Acoustics*, Madrid, Spain, p. 489.

344. Ruske, G., (1982), Auditory perception and its application to computer analysis of speech. in: *Computer Models for Perception of Vision and Auditory Signals*, ed. by C.Y. Suen and R. De Mori, CRC Press.

345.   Sacerdoti, E.D., (1977), *A structure for Plans and Behaviour.*
        Elsevier North-Holland, New York.

346.   Saito, S. and **Kohda, M.**, (1976), Spoken word recognition
        using the restricted number of learning samples. *Proc. IEEE-
        ICASSP,* Philadelphia, PA, pp. 229-233.

347.   Sakai, T. and **Nakagawa, S.**, (1977), A speech understanding
        system of simple Japanese sentences in a task domain. *IECEJ
        Transactions,* vol. **E-60**, no. 1, pp. 13-16.

348.   Sakai, T., (1980), Automatic mapping of acoustic features
        into phonemic labels. in: *Spoken Language Generation and
        Understanding,* ed. by J.C. Simon, Nordoff, Leyden.

349.   Sakoe, H. and **Chiba, S.**, (1978), Dynamic programming algorithm
        optimization for spoken word recognition. *IEEE Transactions,*
        vol. **ASSP-26**, no. 1, pp. 43-49.

350.   Sakoe, H., (1979), Two-level dp-matching - A dynamic program-
        ming based pattern matching algorithm for connected word
        recognition. *IEEE Transactions,* vol. **ASSP-27**, no. 6, pp.
        588-595.

351.   Sambur, M.R. and **Rabiner, L.R.**, (1975), A speaker independent
        digit recognition system. *Bell System Technical Journal,*
        vol. **54**, pp. 81-102.

352.   Sambur, M.R. and **Rabiner, L.R.**, (1976), A statistical
        decision approach to the recognition of connected digits.
        IEEE Transactions, vol. **ASSP-24**, no. 6, pp. 550-558.

353.   Sato, Y. and **Fujisaki, H.**, (1978), Formulation of the process
        of coarticulation in terms of formant frequencies and its
        application in automatic speech recognition. *Journal of the
        Acoustical Society of Japan,* vol. **34**, no. 3, p. 1777.

354.   Schroeder, M.R., (1975), Models of hearing. *Proceedings IEEE,*
        vol. **63**, no. 9, pp. 1322-1352.

355.   Schwartz, R.M., (1971), *Automatic normalization for recogni-
        tion of vowels of all speakers.* S.B. thesis, MIT, Cambridge,
        MA.

356.   Schwartz, R.M., (1976), Acoustic-phonetic experiment facility
        for the study of continuous speech. *Conference Record, IEEE-
        ICASSP* pp. 1-4.

357.   Schwartz, R.M. and Zue, V.W., (1976), Acoustic-phonetic
        recognition in BBN-SPEECHLIS. *Conference Record, IEEE- ICASSP*
        pp. 21-24.

358.   Schwartz, R.M., Klovstad, J., Makhoul, J. and Sorensen, J.,
        (1980), A preliminary design of a phonetic vocoder based on
        a diphone model. *Proc. IEEE-ICASSP*, Denver, Colo., pp. 32-
        35.

359.   Searle, J.R., (1969), *Speech Acts: An Essay In The Philosophy
        Of Language*. Cambridge University Press, Cambridge.

360.   Searle, C.L., Jacobson, J.Z. and Rayment, S.G., (1979),
        Stop consonant discrimination based on human audition.
        J.A.S.A., vol. **65**, no. 3, pp. 799-809.

361.   Sekiguchi, Y., Oowa, H., Aoki, K. and Shigenaga, M., (1977),
        Speech recognition system for FORTRAN programs. *Information
        Processing*, vol. **18**, no. 5, pp. 445-450.

362.   Shafer, R.W. and Rabiner, L.R., (1970), System for Automatic
        Formant Analysis of Voiced Speech. *J.A.S.A.*, vol. **47**, pp.
        634-648.

363.   Shafer, R.W., (1979), *Speech Analysis*, IEEE Press, New York.

364.   Shigenaga, M. and Sekiguchi, Y., (1979), Speech recognition
        of connectedly spoken FORTRAN programs. *IECEJ Transactions*,
        vol. **E-62**, no. 7, pp. 466-473.

365.   Shikano, K. and Kohda, M., (1978), *On The LPC Distance
        Measures For Vowel Recognition In Continuous Utterances*.
        Technical Report of the Acoustical Society of Japan, pp. 578-
        619.

366.   Shirai, K. and Honda, M., (1978), Feature extraction for
        speech recognition based on articulatory model. *Proc. 4th
        IJCPR*, Kyoto, Japan, pp. 1064-1068.

367.   Shokey, L. and Reddy, D.R., (1975), Quantitative analysis
        of speech perception: results from transcription of connected
        speech from unfamiliar languages. in: G. Fant Ed.: *Speech
        Communication*, Almqvist and Wiksell, Stockholm.

368.   Siegel, L.J., (1979), A Procedure for Using Pattern Classifi-
        cation techniques to Obtain a Voiced/Unvoiced Classifier.
        IEEE Transactions, vol. **ASSP-27**, no. 1, pp. 83-88.

369.   Silverman, H.F. and Dixon, N.R., (1980), State constrained
        dynamic programming (SCDP) for discrete utterance recognition.
        Proc. IEEE-ICASSP, Denver, Colo., pp. 169-172.

370.   Simon, J.C. ed., (1980), *Spoken Language Generation and
        Understanding.* Reidel, Leyden, The Netherlands.

371.   Simon, J.C., Backer, E. and Sallantin, J., (1980), A
        structural approach of pattern recognition. *Signal Proces-
        sing,* vol. 2, no. 1, pp. 5-22.

372.   Skinner, T.E., (1977), Speaker-invariant characteristics of
        vowels, liquids and glides using relative formant frequencies.
        J.A.S.A., vol. 62, suppl. 1, p. 55.

373.   Smith, D.R., (1976), Word hypothesization in the HEARSAY-II
        speech system. *Proc. IEEE-ICASSP,* Philadelphia, PA, pp.
        549-552.

374.   Sondhi, M.H. and Levinson, S.E., (1978), Computing relative
        redundancy to measure grammatical constraint in speech
        recognition task. *Proc. IEEE-ICASSP,* Tulsa, OK, pp. 409-
        412.

375.   Steele, G.L. and Sussman, G.J., (1980), Design for a LISP-
        based microprocessor. *Communications of the ACM,* vol. 23,
        no. 11, pp. 628-645.

376.   Steiglitz, K., (1977), On the simultaneous estimation of
        poles and zeros in speech analysis. *IEEE Transactions,* vol.
        ASSP-25, no. 3, pp. 229-235.

377.   Stevens, K.N. and House, A.S., (1955), Development of a
        quantitative description of vowel articulation. *J.A.S.A.,*
        vol. 27, no. 3.

378.   Stevens, K.N., (1973), *Potential Role of Property Detectors
        in the Perception of Consonants.* MIT-RLE Progress Rep. 110,
        pp. 155-168.

379.   Stevens, K.N. and Blumstein, S.E., (1978), Invariant cues
        for place of articulation in stop consonants. *J.A.S.A.,*
        vol. 64, no. 5, pp. 1358-1368.

380.   Stockman, G.C., (1977), *A problem reduction approach to the
        linguistic analysis of waveforms.* Ph. D. thesis, University
        of Maryland, College Park, MD.

381.   Suen, C.Y. and De Mori, R. Eds., (1981), *Computer Analysis and Perception of Visual and Auditory Signals*. CRC press, Raton Boca, Fla., U.S.A.

382.   Tabata, K. and Sakai, T., (1977), Evaluation of the speech factors in Japanese VCV utterances. *IECEJ Transactions,* vol. E-60, no. 6, pp. 284-289.

383.   Takeuchi, S., Kasuya, H. and Kido, K., (1975), A method for extraction of the spectral cues of nasal consonants. *Journal of the Acoustical Society of Japan,* vol. 31, no. 12, pp. 739-740.

384.   Takeya, S. and Kawajuchi, E., (1976), Simulation of a recognition system for connected speech sounds using linguistic information. *Electronics and Communication in Japan,* vol. A-56, no. 9, pp. 38-41.

385.   Tanaka, K., (1976), A dynamic programming approach to extraction and categorization of phonetic information. *Proc. IEEE-ICASSP,* Philadelphia, PA, pp. 5-8.

386.   Tanaka, K., (1978), A standard category pattern matching method with application to phoneme recognition. *Proc. 4th IJCPR,* Kyoto, pp. 1030-1032.

387.   Tappert, C.C., Dixon, N.R. and Rabinowitz, A.S., (1973), Application of sequential decoding for converting phonemic to graphemic representation in automatic recognition of continuous speech. *IEEE Transactions,* vol. AU-21, pp. 225-228.

388.   Tappert, C.C., (1977), A Markov model acoustic phonetic component for automatic speech recognition. *International Journal of Man-Machine Studies,* pp. 363-373.

389.   Tou, J.T. and Gonzales, R.C., (1974), *Pattern Recognition Principles*. Addison-Wesley, Reading, Ma.

390.   Trunin-Donskoi, V.N., (1975), *Analyz i raspoznavanie rechevich signalov na EVM. (Computer Analysis and Recognition of Speech Signals)*. Academy of Sciences, Moscow, USSR (in Russian).

391.   Tsiemel', G.I., (1971), *Opoznavanie Rechevich Signalov. (Recognition of speech signals)*. Academy of Sciences, Moscow, USSR (in Russian).

392.   Tsuruta, S., Sakoe, H. and Chiba, S., (1977), *DP voice recognition system*. Denshi, Tokyo, IEEE Tokyo section, vol. 16.

393. **Tsuruta, S.**, (1978), DP-100 voice recognition system achieves
      high efficiency. *JEE,* pp. 50-54.

394. **Umeda, N.**, (1977), Consonant duration in American English.
      J.A.S.A., vol. **61**, pp. 846-853.

395. **Vaissière, J.**, Une procedure de segmentation automatique de
      la parole en mots prosodiques. *C.R. des 7ème Journées
      d'Etudes sur la Parole, GALF,* Nancy, France, pp. 103-114
      (in French).

396. **Vaissière, J.**, (1980), La structure acoustique de la phrase
      française. (The acoustical structure of the French sentence).
      Annali della Scuola Normale di Pisa, Pisa, Italy, vol. x,2,
      pp. 529-560, (in French).

397. **Vaissière, J.**, (1981), Speech recognition programs as models
      of speech preception. in: *The cognitive Representation of
      Speech,* ed. by T. Myers, J. Lavev and J. Anderson (in press).

398. **Velichiko, V.M.** and **Zagoruiko, N.G.**, (1970), Automatic
      recognition of 200 words. *International Journal of Man-
      Machine Studies,* vol. **2**, pp. 223-234.

399. **Velichiko, V.M.** and **Zagoruiko, N.G.**, (1974), Raspoznavanie
      bol'shikh slovarej. (Recognition of large vocabularies).
      Tezisy dokladov ARSO-USH, chast'3, L'vov, pp. 9-13 (in
      Russian).

400. **Vicens, P.J.**, (1969), *Aspects of speech recognition by
      computer.* Ph. D. thesis, Stanford University, Stanford, CA.

401. **Vintsjuk, T.K., Gavriljuk, O.N.** and **Tuchkov, N.G.**, (1974),
      Algoritmy raspoznavanie slov i slitnykh fraz i rezul'taty
      ikh modelirovania. (Algorithms for recognition of words
      and continuous sentences, and their simulation results).
      Tezisy dokladov ARSO-USH, chast'3 L'vov, pp. 33-37 (in
      Russian).

402. **Vintsjuk, T.K.** and **Shinkazh, A.G.**, (1974), Pofonemnoe
      raspoznavanie slov ustnoj rechi. (Phoneme-by-phoneme
      recognition of spoken words). *Tezisy dokladov, ARSO-USh,
      Chast'3,* L'vov, pp. 19-24 (in Russian).

403. **Vintsjuk, T.K.**, (1976), Generative grammars and dynamic
      programming in speech recognition with learning. *Proc.
      IEEE-ICASSP,* Philadelphia, PA, pp. 446-449.

404. **Viterbi, A.J.**, (1967), Error bounds for convolutional codes and an asymptotically optimum decoding algorithm. *IEEE Transactions,* vol. **IT-13**, pp. 260-269.

405. **Vives, R.** and **Gresser, J.Y.**, (1973), A similarity index between strings of symbols - Application to automatic word and language recognition. *Proc. 1st IJCPR,* Washinton, D.C.

406. **Vives, R.**, (1976), L'analyse lexicale dans le systeme KEAL pour la reconnaisance de la parole continue. *C.R. des 7èmes Journées d'Etude de la Parole, GALF,* Nancy, France, pp. 115-128 (in French).

407. **Vysotskij, Ja.G., Rudnij, B.** and **Trunin-Donskoj, V.**, (1970), Osobennosti tematicheskogo analiza tipovykh fraz slitnoj rechi. (Features of subject-matter analysis of typical sentences in continuous speech). *Trudy ARSO-UP,* Alma-Ata, pp. 4-7 (in Russian).

408. **Vysotski, Ja G., Rudnij, B., Trunin-Donskoj, V.** and **Tsiemel, G.I.**, (1970), Opyt rechevogo upravlenija vichislitel'noj mashinoi. (Experiments with speech control of a computer). *Izv. AN SSSR Technicheskaja Kibernetika,* no. 2, pp. 134-143 (in Russian).

409. **Wakita, H.**, (1972), *Estimation of vocal tract shape by optimal inverse filtering and acoustic articulatory conversation methods.* SCRL Monograph 9, Speech Communication Research Laboratory, Santa Barbara, CA.

410. **Wakita, H.**, (1973), Direct estimation of the vocal tract shape by inverse filtering of acoustic speech waveform. *IEEE Transactions,* vol. **AU-21**, pp. 417-427.

411. **Wakita, H.**, (1977), Normalization of vowels by vocal tract length and its application to vowel identification. *IEEE Transactions,* vol. **ASSP-25**, no. 2, pp. 183-192.

412. **Walker, D.E.**, (1975), The SRI speech understanding system. *IEEE Transactions,* vol. **ASSP-23**, pp. 397-416.

413. **Walker, D.E. ed.**, (1976), *Speech understanding research.* Final technical report project 4762, Artificial Intelligence Center, Stanford Research Institute, Menlo Park, CA.

414. **Walker, D.E., Paxton, W.H., Grosz, B.J., Hendrix, G.G., Robinson, A.E., Robinson, J.J.** and **Slocum, J.**, (1977), Procedure for integrating knowledge in a speech understanding system. *Proc. 5th IJCAI,* Cambridge, Ma, pp. 36-42.

415.   **Walker, D.E. ed.**, (1978), *Understanding Spoken Language*.
        Elsevier North-Holland, New York.

416.   **Webster, W.R.** and **Aitkin, L.M.**, (1975), Central auditory
        processing. in: *Handbook of Psychobiology*, Academic Press,
        New York.

417.   **Weinstein, C.J.**, **McCandless, S.S.**, **Mondshein, L.F.** and **Zue,
        V.W.**, (1974), A system for acoustic-phonetic analysis of
        continuous speech. *IEEE Symposium on Speech Recognition*,
        Carnegie-Mellon University, Pittsburg, Pa, pp. 89-100.

418.   **Welch, P.D.** and **Wimpress, R.S.**, (1971), Two multivariate
        statistical computer programs and their application to the
        vowel recognition problem. *J.A.S.A.*, vol. **33**, no. 4.

419.   **White, G.M.**, (1976), Speech recognition: a tutorial over-
        view. *Computer Magazine*, pp. 40-53.

420.   **White, G.M.** and **Neely, R.B.**, (1976), Speech recognition
        experiments with linear prediction, bandpass filtering and
        dynamic programming. *IEEE Transactions*, vol. **ASSP-24**, pp.
        183-188.

421.   **White, G.M.**, (1978), Dynamic programming, the Viterbi
        algorithm and low cost speech recognition. *Proc. IEEE-
        ICASSP*, Tulsa, OK, pp. 413-417.

422.   **Wickelgren, W.A.**, (1976), Phonetic coding and serial
        ordering. *Handbook of Perception, vol. VII*, pp. 227-264,
        Academic Press, New York.

423.   **Wilber, B.M.**, (1976), *A QLISP reference manual*. SRI
        Artificial Intelligence Center, Technical Note no. 118,
        Stanford Research Institute, Menlo Park, CA.

424.   **Winograd, T.**, (1971), *Procedures as a representation for
        data in a computer program for understanding natural language*.
        Report MAC-TR-84, Project MAC, MIT, Cambridge, MA (published
        as: *Understanding Natural Language*, Academic Press, New York.

425.   **Winston, P.H.**, (1977), *Artificial Intelligence*. McGraw Hill,
        New York.

426.   **Wiren, J.** and **Stubbs, H.L.**, (1956), Electronic binary
        selection for phoneme classification. *J.A.S.A.*, vol. **26**.

427.   **Wolf, J.J.**, (1976), Speech recognition and understanding.
        in: *Digital Pattern Recognition*, ed. by K.S. Fu, Springer-
        Verlag, New York.

428.   **Wolf. J.J.**, (1977), HWIM, a natural language speech under-
        stander. *Proc. IEEE Conference on Decision and Control,
        New Orleans, La, vol. 1,* pp. 560-565.

429.   **Woods, W.A.**, (1970), Transition network grammars for natural
        language analysis. *Communications of the ACM,* vol. **13**, no.
        10, pp. 591-606.

430.   **Woods, W.A.** and **Makhoul, J.**, (1974), Mechanical inference
        problems in continuous speech understanding. *Artificial
        Intelligence,* vol. **5**, pp. 73-91.

431.   **Woods, W.A.**, (1975), What's in a link: foundations for
        semantic networks. in: *Representation and Understanding,*
        ed. by D.G. Bobrow and A.M. Collins, Academic Press, New
        York, pp. 35-82.

432.   **Woods, W.A., Bates, M., Brown, G., Bruce, B., Cook, C.,
        Klovstad, J., Makhoul, J., Nash-Webber, B., Schwartz, R.,
        Wolf, J.** and **Zue, V.**, (1976), *Speech understanding systems.
        Final technical progress report. Volumes I-V.* Report no.
        3438, Bolt Baranek and Newman, Cambridge, Ma.

433.   **Woods, W.A.** and **Zue, V.W.**, (1976), Dictionary expansion via
        phonological rules for a speech understanding system. *Proc.
        IEEE-ICASSP,* Philadelphia, Pa, pp. 561-564.

434.   **Woods, W.A.**, (1977), Shortfall and density scoring strategies
        for speech understanding control. *Proc. 5th IJCAI,* MIT,
        Cambridge, Ma, pp. 18-26.

435.   **Yegnanarayana, B.**, (1981), Speech analysis by pole-zero
        decomposition of short time spectra. *Signal Processing,*
        vol. **3**, no. 1, pp. 5-18.

436.   **You, K.C.** and **Fu, K.S.**, (1979), A syntactic approach to
        shape recognition using attributed grammars. *IEEE Transac-
        tions,* vol. **SMC-9**, no. 6, pp. 334-345.

437.   **Zadeh, L.A.**, (1965), Fuzzy sets. *Information and Control,*
        vol. **8**, pp. 338-353.

438.   **Zadeh, L.A.**, (1973), Outline of a new approach to the
        analysis of complex systems and decision processes. *IEEE
        Transactions,* vol. **SMC-3**, pp. 28-44.

439.   **Zadeh, L.A.**, (1975), The concept of a linguistic variable
        and its application to approximate reasoning - II.
        Information Sciences, vol. **8**, pp. 301-357.

440.   **Zadeh, L.A.**, (1976), A fuzzy algorithmic approach to the
        definition of complex and unprecise concepts. *International
        Journal of Man-Machine Studies,* vol. **8**, pp. 249-291.

441.   **Zadeh, L.A.**, (1978), Fuzzy sets as a basis for a theory of
        possibility. *Fuzzy Sets and Systems,* vol. **1**, pp. 3-28.

442.   **Zadeh, L.A.**, (1979), A theory of approximate reasoning. in:
        Machine Intelligence, ed. by J.E. Hayes, D. Michie and
        L.I. Mikulich, vol. **19**, pp. 149-194.

443.   **Zagoruiko, N.G. ed.**, (1976), *Empiricheskoe predskazanie i
        raspoznavanie obrazov. (Emirical prediction and pattern
        recognition).* Akademija Nauk, USSR (in Russian).

444.   **Zagoruiko, N.G.**, (1977), Sostojanie problemy raspoznavanija
        rechi (The state of the speech recognition problem). *IX
        Vseosojuznaja Akusticheskaha Konferentsija,* Moscow, pp. 143-
        156 (in Russian).

445.   **Zigangirov, K.S.**, (1974), *Procedury posledovatel'nogo
        dekodirovanija. (Sequential decoding procedures).* 'Svjaz',
        Moscow, USSR (in Russian).

446.   **Zigangirov, K.S.** and **Sorokin, V.N.**, (1977), On application
        of sequential decoding for continuous speech recognition.
        Information Transmission Problems, vol. **13**, p. 4.

447.   **Zwicker, E., Terhardt, E.**and **Paulus, E.**, (1979), Automatic
        speech recognition using psychoacoustic models. *J.A.S.A.,*
        vol. 65, no. **2**, pp. 487-498.

Acoustic-phonetic non-invariance, 272
    preconditions, 418
    processor, 283
Active model, 7
Actor, 165
Actor creation, 165
Admissible, 349
Admissible degradations, 310
    search strategy, 323
Affricate, 151
Algebraic form, 174
Aliasing, 31
Allophonic variations, 303
    variations of a word, 327
Ambiguities, 13
Analysis-by-synthesis, 273
AND graph, 196
AND/OR graphs, 20
Antiresonances, 210
Antiresonance effects, 173
Approximate reasoning, 22, 67, 123, 137, 360
Articulatory parameters, 37
Artificial intelligence, 11
ATNG, 142, 174
Attributed grammars, 108, 282
Auditory expert society, 161
Augmented phrase-structure rules, 63, 388
    transition network, 63, 291
Axis-crossings, 28

Backward DP-matching, 283
Beam-search, 129, 279
Beam-search algorithm, 353
Best-first search, 59
Blackboard, 14, 179, 303
Boolean algebra, 96
Bottom-up, 16, 71
Bounded sum, 365
Branch-and bound, 347
Branching questionnaire, 108 ff
Breadth-first, 351

Burst spectra, 289
Buzz-bar, 71

Cartesian product, 102
Categorical classification, 71
Centisecond labels, 291
Cepstral analysis, 33
Cepstrum, 33, 37
Classification of liquids, 207
CLINKS, 290
Coarticulation, 70 ff, 138, 172
Coarticulation effects, 26
    instances, 79, 288
Coda, 290
Cognates, 228
Cognition, 7
Composite, 108
Composite fuzzy restrictions, 403
Composition of evidences, 307
Conflict set, 357
Connected speech, 6
    words, 275
Consonant-vowel segments, 173
Content-addressable memory, 310
Context-constraints, 198
    free grammar, 84
    independent-features, 70
Continuant, 151
Continuant sounds, 228
Continuous natural language, 8
    speech, 57
Control rules, 154
    strategy, 326
Cooperating processes, 416
COSH measure, 167
Cross-correlation, 51
Cylindrical extension, 103

Data-driven constraints, 292
    KS instantiation, 14
Default conditions, 423
Degree(s) of compatibility, 75
    worthiness, 20
Deletion or insertion errors, 327

Dependency grammar, 60
Depth-first, 351
Depth-first search, 351
Descriptions of acoustic
    patterns, 48
Descriptor, 104
Digits, 51
Dipthongs, 179
Distance measures, 167
Distributed knowledge systems,

            problem solving, 160
Domain of a fuzzy relation, 102
Dynamic branching factor, 384
        programming, 52, 56, 275

Edge-fixed DP matching, 169
Edge-free DP matching, 169
Edges, 350
Energoids, 142
Entropy, 26, 384
Equivalent vocabulary, 386
Error correcting parsers, 282
Evidence evaluations, 324
            judgment, 135
            measure, 341
Experts, 160, 416
Extension principle, 103

Fact-tree, 305, 324
False alarms, 293, 325
Fast fourier transform, 31, 33
Feature detectors, 248
            distance, 56
Filter bank, 29
First complete syllable, 304
Focused state, 181
Focus model of search, 62, 387
    of attention, 152, 303,

Formant(s), 30, 37, 46, 173
Formant graph simplification,

        loci, 196
        tracking, 183
        transitions, 210, 253
Forward DP-matching, 283
Frame, 423
Frame-structure grammar, 424

French language, 273, 291
Fricative consonants, 228
Function words, 369
Fuzzy algebra, 20, 95
    algorithms, 75, 123
    concepts, 107
    graph, 183
    language, 104
    linguistic variable, 104,
                                401
    logic, 22, 361
    naming, 403
    number, 122
    recognition algorithm, 107
    relations, 75, 101 ff
    restrictions, 75, 97
    rules, 252
    set, 94, 122
    singleton, 94
    structured language, 105
    subset, 94
    truth value, 362

Glottal waveform, 26
Goal-tree, 305, 324
Grade of membership, 94
Grammar, 9
Grammatical inference, 9, 64
Gross spectral features, 74

HARPY, 357
Hash functions, 56
Heterarchical structure, 14
Heuristic graph search, 350
Hierarchically organised plan-
    ning, 293
Hierarchical structure, 14
Homomorphic prediction, 36
Human factors, 6
HWIM, 389
HWIM system, 288
Hypothesis formation, 76
Hypothesize-and-test paradigm,
    12, 108, 285

Implied composite attribute, 403
Imprecise descriptions, 403
            knowledge, 20
Imprecision, 20

Inference triggering, 330
Instantiation, 160, 356
Integrated knowledge models, 331
Intelligent program, 329
Interrupted-lax, 71
          sounds, 228
Interval functions, 124
Intervocalic sonorant, 151
Isolated words, 4, 51
Itakura distance, 58
Italian language, 308, 310
Iterative computational struc-
    tures, 416

Knowledge sources, 7
KS controller, 74
    instantiation, 14, 74

Language, 43
Language complexity, 382
          model, 284, 323, 331
Laryngeal actions, 229
Lattice, 336
Lattice of phoneme, 286
          phoneme hypotheses,
                              169
Lax, 71, 151, 228
Learning, 9, 314, 327, 332, 408
Learning the membership, 403
Left-hybrid strategy, 393
Level of problem solution, 318
Levels of solutions, 314
Lexical access, 273, 291, 417
          knowledge, 294
          knowledge source, 271
          retrieval, 291
          tree, 286, 289
Lexicon, 418
Lexicon access, 308
Likelihood ratio measure, 167
Limited-vocabulary continous
    speech, 8
Linear grammar, 357
          predictive coding, 33
Linguistic decoder, 61
          modifiers, 98
          probability, 123,
                              130 ff
          probability values,
                              124, 360

Liquid-nasal classification, 202
LISP, 142, 157
Locus theory, 248
Log-area ratios, 57
Long-term memory, 356
Lower bounding function, 349
LPC cepstrum distance, 167

Manner of articulation, 41, 85,
    195, 230
Markov source, 291, 341, 344
Mathematical theory of lan-
    guages, 75
Meaning, 97
Meaning of a naming, 403
          the meaning, 123
Mel-scale cepstrum, 30, 41
Membership, 94
Model-driven constraints, 292
          hypotheses, 173
          KS instantiation,
                              14
Multiple-task capability, 2

Naming problem, 403
          relation, 401
Nasal consonants, 209
          intervocalic consonants,
                              201
          resonances, 209
          sounds, 172 ff
Network of acoustic states, 288
          plans, 424
Neurogram, 42
NOAH system, 290
Non-admissible search algorithm,
                              351
Non-directional parser, 225
          strategy, 84
Nonsonorant, 71, 111, 115, 135
Nonsonorant in a cluster, 151
Normalization of speaker differ-
    ences, 170
Normalized residual power meas-
    ure, 167

Onset, 290
Operations on fuzzy sets, 95
Opportunistic strategies, 331

Parallel computation model, 160
Parallelism degree, 166
Parametric graphs, 190
        representation, 9
Parser, 45
Parse tree, 45
Parsimonious strategy, 424
Parsing, 45
Partial interpretations, 400
Partition of preconditions, 304,
    310
Passive model, 7
Path, 350
Pattern directed invocation, 332
        matching, 276
Perceptual model, 274
Perplexity, 384
Phrase-structure grammar, 42
Phone, 61, 169, 230, 283 ff
Phoneme, 11, 71, 284
Phone network, 288
Phonetic arc labels, 144
        features, 52
Phonological model, 283 ff
        recoding, 272
        rules, 60, 272, 303
Phonology, 7, 11, 230
Pitch contour, 74
    extraction, 37
    period, 26
    synchronous analysis, 32
        FFT, 74
Place of articulation, 41, 195,
                        229
Plans, 292
Pole-zero modeling, 36
Possibility density, 99
        distribution, 98
        theory, 20
        values, 20
Postvocalic sonorant, 151, 179
Power of a fuzzy set, 98
Pragmatic(s), 7, 13, 329
Pragmatic grammar, 63
Precategorical classification,
    71, 76, 109
Precondition evidences, 316
Preconditions of a lexical item,
    308

Prediction error, 34
Prevocalic sonorant, 151, 173,
                        179
Primary problems, 358
        stimuli, 303 ff
Principal components, 45
Problem decomposition, 12
        reduction representa-
    tion, 12, 81, 154, 274, 294
        solver, 18, 292
        solving, 196
Procedural knowledge, 7, 18
        semantic, 381
Production rules, 356
Projection of a fuzzy subset,
    102
Pronunciation speed, 283
Property detectors, 251
Prosodic cues, 279, 303, 305
        information, 390
        pattern, 302
Prosody, 7, 32
PSS problem, 152
Pseudo-loci, 253
        -syllabic segments, 70,
            80, 125, 137, 302
Psychoacoustic models, 42

Quadratic discriminant function,
                        168
Quantizing, 72
Quefrencies, 33

Real-time, 161
Reasoning program, 423
Recency, 357
Redundancy, 231
Reference patterns, 168
Refraction, 357
Relational graph, 79, 152
Review of phoneme recognition,
                        167
Rewriting rules, 42
Rule-based systems, 4

Sampling, 72
Scheduler, 181

Scheduling, 318
Search algorithm, 349
Secondary preconditions, 309
Seed-theories, 367
Segmentation, 70, 80, 125, 280
Segmentation algorithm, 126
        control knowledge
  source, 125
        grammar, 80, 138,
            142
        knowledge source,
            125
        network, 150
        parser, 155
Semantic(s), 7, 58, 329
Semantic ATNG, 332, 373
      knowledge, 332
      rules, 13, 91
Semiotic information, 11
Sequential decoding method, 323
Short-fall densities of loga-
  rithms of probabilities, 323
Shortfall density score, 395
Short term memory, 161, 356
        spectra, 70
Signal preprocessing, 71
Simultaneous estimation of the
  poles and zeros, 36
Single intervocalic sonorant,
  151, 179
Slot filler, 423
Soft-match, 305
Sonorant, 71, 111, 115, 135, 172
Sonorant sounds, 47
Speaker difference, 170
Speaking rate, 4
Special case order, 357
Spectral templates, 288
Spectrogram, 37, 68
Spectrogram reader, 274
Speech chain, 5
     pattern, 9
     states, 280
Stable zones, 190
State-space representation, 15,
        83, 293
Static branching factor, 383
Stimuli, 71
Stochastic automation, 283, 344
      model, 283

Stop consonants, 228
Structural knowledge source, 18
Support, 102
Suprasegmental cues, 273, 302
        features, 25
Surface forms, 283
Syllabic ATNG, 143, 179
        blackboard, 305
        control KS, 182
        controller, 180
        directory, 179
        expert, 420
        expert society, 161
Syllables, 11
Syllable-type tree, 304
Sylparts, 290
Syntactic knowledge, 379
        pattern recognition,
          79
Syntax, 7, 13, 58, 329
Synthesis-by-rule, 289

Talker normalization, 10
Task-dependent knowledge, 57,
        329
        knowledge
  sources, 13
Tense, 228
Terminal alphabet, 42
Terminations, 302
Theorem proving, 326
Theory of possibility, 21, 75
Time normalization, 9, 52
T-level set, 95
Top-down, 16
Top-down systems, 274
Traffic controller, 181
Trainable classifier, 52
Tree of syllable types, 312
Two-level-dp-matching, 276
Two-level matching, 57
Type-2 fuzzy sets, 122

Unconnected natural language, 8

Vague data, 20
Vagueness of the knowledge
  source, 406
Verification process, 307
Vista, 307, 367

Vista interval, 307
Viterbi algorithm, 291, 344
Vocalic, 71, 87, 91, 99, 150
Vocal tract, 25
    tract shape, 36
Voiced consonants, 26
Voice-input terminals, 2
    onset, 289
Vowel(s), 51, 190
Vowel classification, 194
    -consonant-vowel, 173
    hypothesis, 195
    loci, 198
    reduction, 172
    transitions, 209

Warping function, 56
Weighted directed graphs, 349
Windows, 32
Word(s), 271
Word automata, 336
    boundaries, 271
    -boundary effects, 288
    hypotheses, 324
    -junction phonological rules,
                                  276
    recognition, 271
    stimuli, 303
    verification, 324

Z-transform, 34